METHODS IN MOLECULAR BIOLOGY™

Series Editor
John M. Walker
School of Life Sciences
University of Hertfordshire
Hatfield, Hertfordshire, AL10 9AB, UK

For further volumes:
http://www.springer.com/series/7651

Platelets and Megakaryocytes

Volume 3, Additional Protocols and Perspectives

Edited by

Jonathan M. Gibbins

*Institute for Cardiovascular and Metabolic Research, School of Biological Sciences,
The University of Reading, Reading, UK*

Martyn P. Mahaut-Smith

Department of Cell Physiology and Pharmacology, University of Leicester, Leicester, UK

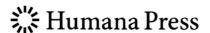 Humana Press

Editors
Jonathan M. Gibbins
Institute for Cardiovascular and Metabolic
Research, School of Biological Sciences
The University of Reading
Reading, UK
j.m.gibbins@reading.ac.uk

Martyn P. Mahaut-Smith
Department of Cell Physiology and Pharmacology
University of Leicester
Leicester, UK
mpms1@leicester.ac.uk

ISSN 1064-3745   e-ISSN 1940-6029
ISBN 978-1-61779-306-6   e-ISBN 978-1-61779-307-3
DOI 10.1007/978-1-61779-307-3
Springer New York Dordrecht Heidelberg London

Library of Congress Control Number: 2004002112

Cover illustration: Cross-section of bone marrow showing megakaryocytes in their native environment. For further
details see Chapter 13 by Anita Eckly and colleagues.

Printed on acid-free paper

Humana Press is part of Springer Science+Business Media (www.springer.com)

Preface

New techniques to study cell signaling and function can develop at a staggering pace; however, many approaches may be as valid as the day they were established. Thus, the main aim of this third volume of "*Platelets and Megakaryocytes*" within the *Methods in Molecular Biology* series is to complement the first two volumes published in 2004 by adding recently developed state-of-the-art techniques. Volume 3 has been divided into three parts: Part I focusing on techniques to study platelet function and Part II covering approaches to investigate megakaryocyte function, although inevitably, some chapters (e.g., Chapter 9 on the visualization and manipulation of the cytoskeleton) describe techniques that are applicable to both the megakaryocyte and its anuclear product. In Part III, a number of perspectives chapters discuss important overall concepts and new developments in the field of megakaryocyte and platelet biology.

One area that has undergone substantial development over the past decade is the in vitro culture of megakaryocytes from stem cells. This reflects the significant demand for an in vitro system that allows transfection/knock-down of specific cellular targets in the megakaryocyte and thus to produce genetically modified platelets without the need to use transgenic animal models. In addition, there is enormous potential for such cultures in the clinical treatment of inherited diseases of megakaryocyte or platelet function. Several megakaryocyte culture systems have been described, with starting material ranging from induced pluripotent stem cells to fetal liver cells. It is worth emphasizing that megakaryocytes and platelets derived in vitro from all sources remain to be fully characterized. Thus, at present, it is unclear which is the "best" or most "in vivo"-like culture system, and we have attempted to be as all-inclusive as possible with a total of five chapters (Chapters 9 and 14–17) covering this topic. A number of chapters in Part II also describe how to measure specific aspects of megakaryocyte function, ranging from its unique morphology (Chapter 13) to migration from explants or across gradients of chemotactic stimuli (Chapters 13 and 19). Furthermore, a perspectives chapter (Chapter 20, in Part III) summarizes our current understanding of cytokines and surface markers involved in megakaryocyte development.

While many groups are working hard to improve the fidelity of platelet-producing megakaryocyte culture systems, many scientific questions cannot currently be addressed without the use of animal models, particularly given the power of transgenic approaches. Over the last decade, a number of in vivo models of thrombosis have become established, ranging from studies of cerebral artery occlusion (Chapter 3) to the real-time imaging of thrombus development following ultrasonic disruption of atherosclerotic plaques (Chapter 1). However, the complexity of such in vivo models, together with technical constraints, can limit the type of information gained, and therefore "extracorporeal" or "ex vivo" systems (Chapter 4) represent further useful methodologies to investigate platelet activation in the intact vessel.

In the postgenomic era, scientists have the potential to rapidly screen for expression of all cellular proteins expressed in an individual cell type and thus to identify those involved in specific physiological or pathophysiological processes. Indeed, recent studies have revealed the ability of platelets to process RNA and to translate proteins in an activation-dependent manner. Methods to study protein production by platelets are described in Chapter 11. Many important questions arise when applying screening techniques, from the level of purification required of the starting material to whether a screen of mRNA transcripts is as indicative of protein expression as proteomics. In the platelet and megakaryocyte, further questions arise, such as the level of transcriptional modification in the anuclear platelet and how to purify the megakaryocyte from native tissue. A number of methods and perspectives chapters (e.g., Chapters 12, 18, 24) cover these complex issues.

We wish to express our enormous gratitude to the 75 authors who have contributed to the 24 chapters of Volume 3. This brings the total number of chapters in the three volumes to 80, contributed by a total of more than 130 different authors. Their willingness to share extensive expertise, including many tricks of the trade learned from countless hours of hard work in the laboratory, represents the foundation of these books.

Reading, UK   *Jonathan M. Gibbins*
Leicester, UK   *Martyn P. Mahaut-Smith*

Contents

Contributors

MOHAMMAD AL-TAMIMI • *Australian Centre for Blood Diseases, Monash University, Melbourne, VIC, Australia*

STEFAN AMISTEN • *Oxford Centre for Diabetes, Endocrinology, and Metabolism, Oxford University, The Churchill Hospital, Oxford, UK*

ROBERT K. ANDREWS • *Australian Centre for Blood Diseases, Monash University, Melbourne, VIC, Australia*

JOSEPH E. ASLAN • *Department of Biomedical Engineering, Department of Cell and Developmental Biology, Oregon Health & Science University, Portland, OR, USA*

LINA BADIMON • *Centro de Investigación Cardiovascular, CSIC-ICCC, Hospital de la Santa Creu i Sant Pau, Barcelona, Spain; CIBEROBN Instituto de Salud Carlos III, Cátedra de Investigación Cardiovascular, (UAB-HSCSP-Fundación Jesús Serra), Barcelona, Spain*

NATASHA E. BARRETT • *Institute for Cardiovascular and Metabolic Research, School of Biological Sciences, The University of Reading, Reading, UK*

MICHAEL C. BERNDT • *Biomedical Diagnostics Institute, National Centre for Sensor Research, Dublin City University, Dublin, Ireland*

STEFAN BRAEUNINGER • *Department of Neurology, University of Wuerzburg, Wuerzburg, Germany*

ROBERT A. CAMPBELL • *Program in Molecular Medicine, University of Utah, Salt Lake City, UT, USA*

ALAN B. CANTOR • *Division of Hematology-Oncology, Children's Hospital Boston and Dana Farber Cancer Institute, Harvard Medical School, Boston, MA, USA*

RICHARD N. CARTER • *Department of Physiology, Development and Neuroscience, University of Cambridge, Cambridge, UK*

JEAN-PIERRE CAZENAVE • *UMR_S949 Inserm-Université de Strasbourg, Strasbourg, France*

VALÉRIE CORTIN • *Département de Recherche et Développement, Héma-Québec, Québec City, QC, Canada*

FRANÇOISE DIGNAT-GEORGE • *UMR-S608 INSERM, Marseille, France; Faculté de Pharmacie, Université de la Méditerranée, Marseille, France; Service d'hématologie, CHU La Conception, AP-HM, Marseille, France*

MARK R. DOWLING • *Immunology Division, The Walter and Eliza Hall Institute of Medical Research, Parkville, VIC, Australia; Department of Medical Biology, The University of Melbourne Melbourne VIC, Australia*

ANITA ECKLY • *UMR_S949 Inserm-Université de Strasbourg, Strasbourg, France*

MICHAEL EMERSON • *National Heart and Lung Institute, Imperial College London, London, UK*

KOJI ETO • *Clinical Application Department, Center for iPS Cell Research and Application (CiRA), Kyoto University 53 Kawahara-cho, Shogoin, Sakyo-ku Kyoto, 606-8507, Kyoto, Japan*

BARBARA C. FURIE • *Division of Hemostasis and Thrombosis, Beth Israel Deaconess Medical Center, Harvard Medical School, Boston, MA, USA*

BRUCE FURIE • *Division of Hemostasis and Thrombosis, Beth Israel Deaconess Medical Center, Harvard Medical School, Boston, MA, USA*

CHRISTIAN GACHET • *UMR_S949 Inserm-Université de Strasbourg, Strasbourg, France*

ÁNGEL GARCÍA • *Departamento de Farmacoloxía, Facultade de Farmacia, Universidade de Santiago de Compostela, Santiago de Compostela, Spain*

ELIZABETH E. GARDINER • *Australian Centre for Blood Diseases, Monash University, Melbourne, VIC, Australia*

ALAIN GARNIER • *Department of Chemical Engineering, Université Laval, Québec City, QC, Canada*

JACQUELINE M. GERTZ • *Department of Biomedical Engineering, Oregon Health & Science University, Portland, OR, USA*

JONATHAN M. GIBBINS • *Institute for Cardiovascular and Metabolic Research, School of Biological Sciences, The University of Reading, Reading, UK*

IAN S. HARPER • *The Australian Centre for Blood Diseases, Monash University, Melbourne, VIC, Australia*

JOHAN W.M. HEEMSKERK • *Department of Biochemistry, CARIM, Maastricht University, Maastricht, The Netherlands*

YASUO IKEDA • *Faculty of Science and Engineering, Life Science and Medical Bioscience, Waseda University, Tokyo, Japan*

ASAKO ITAKURA • *Department of Cell and Developmental Biology, Oregon Health & Science University, Portland, OR, USA*

JOSEPH E. ITALIANO • *Hematology Division, Brigham and Women's Hospital, Boston, MA, USA; Harvard Medical School, Boston, MA, USA; Vascular Biology Program, Department of Surgery, Children's Hospital, Boston, MA, USA*

DENISE E. JACKSON • *Health Innovations Research Institute, School of Medical Sciences, RMIT University, Bundoora, VIC, Australia*

SHAUN P. JACKSON • *The Australian Centre for Blood Diseases, Monash University, Melbourne, VIC, Australia*

CHRIS I. JONES • *Institute for Cardiovascular and Metabolic Research, School of Biological Sciences, The University of Reading, Reading, UK*

EMMA C. JOSEFSSON • *Cancer and Haematology Division, The Walter and Eliza Hall Institute of Medical Research, Parkville, VIC, Australia*

BENJAMIN T. KILE • *Cancer and Haematology Division, The Walter and Eliza Hall Institute of Medical Research, Parkville, VIC, Australia; Department of Medical Biology, The University of Melbourne, Melbourne, VIC, Australia*

CHRISTOPH KLEINSCHNITZ • *Department of Neurology, University of Wuerzburg, Wuerzburg, Germany*

MARIJKE J.E. KUIJPERS • *Department of Biochemistry, CARIM, Maastricht University, Maastricht, The Netherlands*

Romaric Lacroix • *UMR-S608 INSERM, Marseille, France; Faculté de Pharmacie, Université de la Méditerranée, Marseille, France; Service d'hématologie, CHU La Conception, AP-HM, Marseille, France*

François Lanza • *UMR_S949 Inserm-Université de Strasbourg, Strasbourg, France*

Catherine Léon • *UMR_S949 Inserm-Université de Strasbourg, Strasbourg, France*

Martyn P. Mahaut-Smith • *Department of Cell Physiology and Pharmacology, University of Leicester, Leicester, UK*

Yumiko Matsubara • *Department of Laboratory Medicine, School of Medicine, Keio University, Tokyo, Japan*

Alexandra Mazharian • *Centre for Cardiovascular Sciences, Institute of Biomedical Research, School of Clinical and Experimental Medicine, College of Medical and Dental Sciences, University of Birmingham, Birmingham, UK*

Owen J.T. McCarty • *Department of Biomedical Engineering, Department of Cell and Developmental Biology, Oregon Health & Science University, Portland, OR, USA*

Nigel Miller • *Department of Pathology, University of Cambridge, Cambridge, UK*

Christopher Moore • *National Heart and Lung Institute, Imperial College London, London, UK*

Leonardo A. Moraes • *Institute for Cardiovascular and Metabolic Research, School of Biological Sciences, The University of Reading, Reading, UK*

Mitsuru Murata • *Department of Laboratory Medicine, School of Medicine, Keio University, Tokyo, Japan*

Warwick S. Nesbitt • *The Australian Centre for Blood Diseases, Monash University, Melbourne, VIC, Australia*

Bernhard Nieswandt • *Rudolf Virchow Center, DFG Research Center for Experimental Biomedicine, University of Wuerzburg, Wuerzburg, Germany*

Teresa Padro • *Centro de Investigación Cardiovascular, CSIC-ICCC, Hospital de la Santa Creu i Sant Pau, Barcelona, Spain*

Nicolas Pineault • *Département de Recherche et Développement, Héma-Québec, Québec City, QC, Canada; Department of Biochemistry and Microbiology, Université Laval, Québec City, QC, Canada*

Alastair W. Poole • *Department of Physiology & Pharmacology, School of Medical Sciences, University of Bristol, Bristol, UK*

Amélie Robert • *Département de Recherche et Développement, Héma-Québec, Québec City, QC, Canada*

Jesse W. Rowley • *Program in Molecular Medicine, University of Utah, Salt Lake City, UT, USA*

Tanya Sage • *Institute for Cardiovascular and Metabolic Research, School of Biological Sciences, The University of Reading, Reading, UK*

Simone M. Schoenwaelder • *The Australian Centre for Blood Diseases, Monash University, Melbourne, VIC, Australia*

Harald Schulze • *Labor für Pädiatrische Molekularbiologie, Charité – Universitätsmedizin, Berlin, Germany*

Hansjörg Schwertz • *Program in Molecular Medicine, Department of Surgery, University of Utah, Salt Lake City, UT, USA*

YOTIS SENIS • *Centre for Cardiovascular Sciences, Institute of Biomedical Research, School of Clinical and Experimental Medicine, University of Birmingham, Birmingham, UK*

GUIDO STOLL • *Department of Neurology, University of Wuerzburg, Wuerzburg, Germany*

CATHERINE STRASSEL • *UMR_S949 Inserm-Université de Strasbourg, Strasbourg, France*

NAOYA TAKAYAMA • *Clinical Application Department, Center for iPS Cell Research and Application (CiRA), Kyoto University, Kyoto, Japan*

JONATHAN N. THON • *Hematology Division, Brigham and Women's Hospital, Boston, MA, USA; Harvard Medical School, Boston, MA, USA*

GWEN TOLHURST • *Department of Clinical Biochemistry, University of Cambridge, Cambridge, UK*

NEAL D. TOLLEY • *Program in Molecular Medicine, University of Utah, Salt Lake City, UT, USA*

KATHERINE L. TUCKER • *Institute for Cardiovascular and Metabolic Research, School of Biological Sciences, The University of Reading, Reading, UK*

GEMMA VILAHUR • *Centro de Investigación Cardiovascular, CSIC-ICCC, Hospital de la Santa Creu i Sant Pau, Barcelona, Spain*

ANDREW S. WEYRICH • *Program in Molecular Medicine, Department of Internal Medicine, University of Utah, Salt Lake City, UT, USA*

MICHAEL J. WHITE • *Cancer and Haematology Division, The Walter and Eliza Hall Institute of Medical Research, Parkville, VIC, Australia; Department of Medical Biology, The University of Melbourne, Melbourne, VIC, Australia*

CHRISTOPHER M. WILLIAMS • *Department of Physiology & Pharmacology, School of Medical Sciences, University of Bristol, Bristol, UK*

MING YU • *Laboratories of Biochemistry and Molecular Biology, The rockefeller University New york, NY, USA*

YUPING YUAN • *The Australian Centre for Blood Diseases, Monash University, Melbourne, VIC, Australia*

JEFFREY I. ZWICKER • *Division of Hemostasis and Thrombosis, Beth Israel Deaconess Medical Center, Harvard Medical School, Boston, MA, USA*

Part I

Additional Protocols for the Study of Platelet Function

Chapter 1

Intravital Imaging of Thrombus Formation in Small and Large Mouse Arteries: Experimentally Induced Vascular Damage and Plaque Rupture In Vivo

Marijke J.E. Kuijpers and Johan W.M. Heemskerk

Abstract

Intravital fluorescence microscopy is increasingly used to measure experimental arterial thrombosis in large and small arteries of mice in vivo. This chapter describes protocols for applying this technology to detect and measure thrombi formed by: (1) ultrasound-induced rupture of an atherosclerotic plaque in the carotid artery of adult *Apoe–/–* mice; (2) $FeCl_3$ or ligation in the carotid artery of nonatherosclerotic mice; and (3) $FeCl_3$ in the mesenteric venules and arterioles of young mice. In addition, we describe a protocol using two-photon laser scanning microscopy for intraluminal scanning of thrombi formed in the carotid artery. These approaches provide important information that cannot be obtained with ex vivo methods and thus are likely to lead to new insights into the complex process of thrombosis.

Key words: Animal models, Arterial thrombus formation, Atherosclerosis, Coagulation, Two-photon microscopy, Plaque rupture, Platelet activation

1. Introduction

Rupture of an atherosclerotic plaque and subsequent athero-thrombosis results in myocardial infarction or stroke and, hence, is a life-threatening event (1). In man, this process is difficult to follow, since the precise time of plaque rupture cannot be predicted. However, in the last two decades murine thrombosis models have been developed, where the acute formation of an arterial thrombus can be followed in real time. This measurement is possible with experimentally damaged microvessels in the mouse mesenteric tissue and cremaster muscle. Intravital microscopic imaging has been of pivotal importance to demonstrate the dynamics of the thrombotic process in the vessels (2–4). Furthermore, large mouse

Jonathan M. Gibbins and Martyn P. Mahaut-Smith (eds.), *Platelets and Megakaryocytes: Volume 3, Additional Protocols and Perspectives*, Methods in Molecular Biology, vol. 788, DOI 10.1007/978-1-61779-307-3_1, © Springer Science+Business Media, LLC 2012

vessels, such as the carotid and femoral arteries, can also be used for intravital imaging, provided that thrombus formation is detected using preinjected fluorescent labels. In most experimental models, thrombus formation in the arteries is driven by platelet activation and triggering of the coagulation system, although the relative contribution of these two processes may vary. Key platelet agonists in arterial thrombosis are collagen, acting via the collagen receptor glycoprotein VI (GPVI), ADP (acting via the $P2Y_1$ and $P2Y_{12}$ receptors) and thrombin (operating by PAR receptors). Coagulation can be triggered by either the extrinsic (tissue factor) or intrinsic (factor XII) pathway.

In the majority of these thrombosis models, healthy blood vessels are damaged rather than diseased, atherosclerotic vessels, which cause thrombosis in patients. To overcome this limitation, a new model has been developed, in which atherosclerotic lesions in the carotid artery of $Apoe^{-/-}$ mice are ruptured by targeted ultrasound treatment (5). Also in this model, platelet-collagen interaction via GPVI and simultaneous thrombin generation appear to trigger the formation of subocclusive thrombi (5, 6). Recently, a role of ADP/ $P2Y_{12}$-dependent platelet activation in the stability of these thrombi was also established (7). The present chapter explains in detail the common intravital microscopic techniques to measure thrombus formation in healthy and plaque-containing mouse arteries. Furthermore, following thrombus formation with one of the above in vivo models, it describes the use of two-photon laser scanning microscopy (TPLSM) to further assess the vessel wall damage and thrombus composition within the intact arteries after their isolation ex vivo.

2. Materials

2.1. Reagents and General Equipment

1. Carboxyfluorescein diacetate succinimidyl ester (CFSE, also abbreviated as DCF) (Molecular Probes, Eugene, OR, USA).

2. Syto41 (Invitrogen, Leiden, The Netherlands).

3. Ketamine and xylazine (Eurovet, Bladel, The Netherlands).

4. Unfractionated heparin (Sigma Aldrich, St. Louis, MI, USA).

5. D-Phenylalanyl-L-prolyl-L-arginine chloromethyl ketone (PPACK) (VWR International, Amsterdam, The Netherlands).

6. 0.9% (w/v) NaCl.

7. $FeCl_3$ (Merck Darmstadt, Germany): 500 mM in saline.

8. Phosphate-buffered saline (PBS): 136 mM NaCl, 6.5 mM Na_2HPO_4, 2.7 mM KCl, and 1.5 mM KH_2PO_4.

9. Tyrode's solution: 130 mM NaCl, 25 mM $NaHCO_3$, 13.1 mM sucrose, 11.1 mM glucose, 5.6 mM KCl, 2.2 mM $CaCl_2$, 1.2 mM NaH_2PO_4, 0.56 mM $MgCl_2$, and saturated with 95%

N_2 and 5% CO_2; pH 7.35 at 37°C (pH set by CO_2:HCO_3 buffer; further titration not required). To reduce bowel movement, 1 µg/ml isoprenaline is added to this buffer.

10. Hank's balanced salt solution (HBSS): 144 mM NaCl, 14.9 mM Hepes, 5.5 mM glucose, 4.7 mM KCl, 2.5 mM $CaCl_2$, 1.2 mM KH_2PO_4 and 1.2 mM $MgSO_4$, pH 7.4 at 37°C.

11. 30 gauge needle (Microlance, Becton Dickinson, Breda, The Netherlands) and polyethylene-10 catheter (0.28 mm ID, 0.61 mm OD from Rubber B.V., Hilversum, The Netherlands).

12. Tissue glue: Histoacryl glue from TissueSeal (Ann Arbor, MI, USA) or second glue Tipper from Bison (Goes, The Netherlands).

13. 37°C heated platform suitable for the microscope stage (Linkam, Tadworth, UK).

14. Ceramic heating lamp for maintaining animal temperature. The lamp is regulated by a thermo-analyzer system that receives information on body temperature from a rectal probe. Our system is manufactured by Maastricht Instruments, Maastricht University.

15. VibraCell VCX130 ultrasound processor (Sonics, Newtown, CT, USA), equipped with a titanium probe with 0.5 mm polished tip (5) for ultrasonic disruption of atherosclerotic plaques. The probe is mounted on a manipulator to allow positioning.

16. Dissection microscope.

17. Triangular piece of black plastic (5 × 10 mm) for enhancing contrast of dissected carotid arteries.

18. Prolene 5/0 ligature for inducing injury by ligation (Ethicon, Somerville, NJ, USA).

19. Digital analysis software: e.g., ImagePro software (Media Cybernetics, Silverspring, MD, USA).

20. Sphygmomanometer (e.g., from Riester, Jungingen, Germany) for the application of pressure to the isolated carotid artery in TPLSM.

21. Perfusion chamber for holding the isolated carotid artery in TPLSM: Built by Maastricht Instruments (Maastricht University, The Netherlands), containing two glass micropipettes (tip diameter 120–150 µm) (11).

22. Siliconized glass plate for the central opening of the intravital microscope (IVM) (used in studies of thrombus formation in the mesenteric vessels).

2.2. Animal Requirements

1. *Mice for measurement of thrombus formation in a plaque-containing carotid artery.* To provoke plaque rupture, *Apoe*−/− mice of either sex are used with a C57Bl/6 genetic background, e.g., available from Charles River (Maastricht, The Netherlands).

In order to allow the development of atherosclerotic plaques, mice from the age of 4 weeks are fed a Western-type diet containing cholesterol (0.15%) for a prolonged period of 18–20 weeks (5). At the age of 22–24 weeks, visible plaques are present near the bifurcation of one or both carotid arteries. For successful experimentation, the plaques should not completely surround the vessel. For plaque rupture experiments, labeled platelets are prepared from other mice of the same genotype. Preferentially, adult animals are used, which are not fed a high-fat diet, since high plasma lipid levels decrease the yield of platelets.

2. *Mice for measurement of thrombus formation in the healthy carotid artery.* Adult mice of either sex are used to assess ligation- or $FeCl_3$-induced thrombus formation in the healthy carotid artery, for example wild-type C57Bl/6 mice obtained from Charles River (8).

3. *Mice for measurement of thrombus formation in the mesenteric arterioles and venules.* Young mice of either sex, aged 4–5 weeks, are used to determine thrombus formation in the mesenteric arterioles and venules. The mesentery is one of the few tissues that is transparent enough to visualize intravessel thrombus formation with transillumination brightfield microscopy. Older mice are not used due to the accumulation of fat cells around these vessels. Wild-type mice with a C57Bl/6 or other genetic background are suitable, e.g., from Charles River (9, 10).

2.3. Intravital Fluorescence Microscope Requirements

State-of-the-art upright (nonconfocal) fluorescence microscopes from several manufacturers are suitable for in vivo measurements of thrombus formation, provided there is sufficient space on the microscopic stage to position an anesthetized mouse. Brightfield transillumination optics are suitable for monitoring thrombus formation in the vessels from transparent tissues, in particular the arterioles and venules in the mesenteric tissue (9, 10, 12) and the microvessels of the cremaster muscle (2). In these tissues, transillumination image recording can also be combined with epi-illumination fluorescence recordings, if fluorescent platelets or probes are preinjected. However, in the case of larger vessels with a thick wall, such as the carotid artery, transillumination is not possible and intravital microscopy of the thrombotic process is confined to the recording of fluorescence images (5, 13).

The precise microscopic optics depends on the vessel of interest and the camera system used. In general, for the various protocols described below, state-of-the-art 5–20× air objectives are suitable. For capturing transillumination images, a halogen white light source together with a conventional CCD camera are sufficient. For fluorescence imaging, a high intensity fluorescence source is needed (e.g., a 100 W Mercury lamp). To record fluorescein-like probes, an I3 or similar filter cube is inserted in the light path, and a sensitive CCD camera is used for image capture (e.g., the EM-CCD camera from Hamamatsu, Japan). The I3 filter cube contains a 450–490 nm band-pass excitation filter, a 510 nm dichroic mirror, and a >515 nm long-pass emission filter. Depending on the light source and the sensitivity of the camera, gray filters can be inserted in the illumination pathway to reduce fluorescence bleaching. In the protocols below, additional specifications for the microscope and camera setup are given.

For two-photon confocal imaging, we use a standard Bio-Rad 2100 MP multiphoton system (Zeiss/Bio-Rad, Hemel Hempstead, GB) (8, 11) coupled to a Nikon E600FN microscope (Nikon Corporation, Tokyo, Japan). The excitation source is a 140-fs-pulsed Ti:sapphire laser (Spectra Physics Tsunami, Mountain View, Calif., USA) tuned and mode-locked at 800 nm (for visualization of fluorescent probes). Laser light reaches the sample through the microscope objective (60× water dipping, numerical aperture 1.0, working distance 2 mm; 40× water dipping, numerical aperture 0.8, working distance 2 mm).

3. Methods

All experiments and operation procedures with animals must be performed in accordance with national legislations and local regulations, and with appropriate permissions from ethical or animal care committees. The surgery and all further animal handling must only be performed by personnel with appropriate certificates for animal experimentation. Researchers should be aware that the in vivo procedures described below also require training under guidance and will only be successful if the experimenter has gained sufficient expertise.

3.1. Preparation of Fluorescently Labeled Mouse Platelets for Injection (5, 9)

1. An adult donor wild-type or *Apoe*$^{-/-}$ mouse, as required, is anesthetized by subcutaneous injection of ketamine and xylazine (100 µg/g and 20 µg/g body weight, respectively).

2. Blood is drawn by the method of choice (see Note 1). A volume of 700 µl blood is collected into 300 µl anticoagulant solution that consists of 15 units heparin/ml and 120 µM PPACK in 0.9% NaCl. Hence, the final anticoagulant concentrations are 4.5 units heparin/ml and 35 µM PPACK. After blood collection, the animal is sacrificed.

3. Divide the blood into two Eppendorf tubes, each containing 500 µl. Centrifuge for 3 min at 290×g (no brake).

4. Take off the upper 350 µl, consisting of platelet-rich plasma (PRP) plus some red blood cells and transfer to two new tubes.

5. Centrifuge for 1 s at $625 \times g$ (no brake). Stop centrifugation once the required speed is reached.

6. Take off about 500 µl PRP totally from both tubes. Then, add PBS (see Note 2) to bring to a total volume of 3 ml.

7. Add 5 µg/ml CFSE (see Note 3), incubate 1–2 min in the dark, and centrifuge at $950 \times g$ for 10 min (50% brake).

8. Discard the supernatant, and resuspend the platelet pellet in 350 µl PBS. The resulting platelet suspension may be used for injection into two experimental mice.

9. Measure the number of platelets; the platelet count should be around $4–5 \times 10^8$/ml. Check that the suspended platelets swirl, which is an indication that they are not activated.

10. Use the labeled platelets, as described in Subheadings 3.2.2 or 3.2.3. For the study of thrombosis, 2×10^8 platelets are injected into the experimental adult mouse immediately before the experiment. This gives a labeling of 10% of all circulating platelets, assuming a blood volume of 3 ml and a platelet count of 7×10^8/ml. Note that it will not be possible to visualize the formation of a thrombus if the number of circulating labeled platelets is too low.

3.2. Thrombus Formation in Healthy and Atherosclerotic Mouse Carotid Arteries

3.2.1. Surgical Preparation of Mouse Carotid Arteries (5, 8)

1. Anesthetize the wild-type or $Apoe^{-/-}$ mouse by subcutaneous injection of ketamine and xylazine (100 µg/g and 20 µg/g body weight, respectively). The anesthesia is maintained by re-administration of ketamine and xylazine (30 µg/g and 6.7 µg/g body weight) every 30 min.

2. Shave the throat of the mouse, and place the animal in a stable position on a 37°C heated platform. The body temperature is maintained at 37°C using a ceramic heating lamp, which is controlled by a thermo-analyzer system connected to a rectal probe.

3. Insert subcutaneously into the neck a 30 gauge needle with polyethylene-10 catheter for administration of anesthetics. Fix to the skin with tissue glue (see Note 4).

4. Insert into a tail vein a 30 gauge needle with polyethylene-10 catheter for administration of labeled platelets and inhibitors. Fix also to the skin with tissue glue (see Note 4). Alternatively, the jugular vein can be catheterized for this purpose.

5. Dissect both carotid arteries free from surrounding tissue using a dissection microscope. Carefully remove the surrounding fat tissue without touching the arteries. Note that the presence of fat disturbs the clarity of microscopic images.

6. In $Apoe^{-/-}$ mice, the atherosclerotic plaques are visible around the carotid bifurcation as white spots (Fig. 1a).

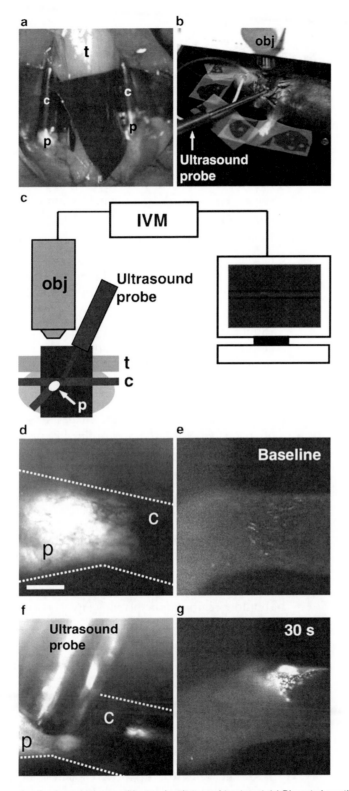

Fig. 1. Rupture of atherosclerotic plaque in the carotid artery by ultrasound treatment. (**a**) Dissected carotid arteries (c) from an *Apoe*[−/−] mouse next to trachea (t). Note the presence of white plaques (p) near the bifurcations. (**b**) Prepared mouse on the stage of the intravital microscope. Indicated are the air objective (obj) and the tip of the ultrasound probe. (**c**) Schematic representation of the setup of the intravital microscope (IVM); *c* carotid artery, *obj* objective, *p* plaque, and *t* trachea. (**d**) Brightfield image of a 0.5 mm plaque near the artery bifurcation (bar = 250 μm). (**e**) Baseline fluorescence image of circulating CFSE-labeled platelets. (**f**) Brightfield image of tip of ultrasound probe placed at plaque shoulder. (**g**) Fluorescence image of CFSE-labeled platelets trapped in a thrombus at 30 s after plaque rupture. Reproduced from ref. 5 with permission from John Wiley and Sons.

7. For enhanced optical contrast, place a triangular piece of black plastic (5×10 mm) under the dissected artery at the site, where the vessel will be damaged.

8. Keep the carotid arteries wet with saline ($37°C$) until start of the thrombosis experiment.

9. Proceed as described in Subheadings 3.2.2 or 3.2.3.

3.2.2. Ultrasound-Induced Plaque Rupture in the Carotid Artery and Intravital Recording of Thrombus Formation (5, 7)

1. Mount the platform with anesthetized mouse onto the IVM stage (Fig. 1b). For intravital brightfield and fluorescence recordings, use a 5–10× air objective (Fig. 1c).

2. Remove the surplus of saline from the dissected carotid arteries.

3. Select a suitable plaque near the bifurcation in one of the carotid arteries using brightfield illumination. Record brightfield images from the plaque area with the CCD camera (Fig. 1d).

4. Slowly inject a suspension of 2×10^8 CFSE-labeled platelets (Subheading 3.1) via the tail or jugular vein over a period of about 30 s. Circulating fluorescently labeled platelets are immediately visible by camera recording in the fluorescence mode. Note that injected platelets will not circulate when activated (aggregated).

5. Capture baseline fluorescence images for 1 min (Fig. 1e). Use appropriate color filters and optimal camera settings to obtain a high signal-to-noise ratio. Preselect the camera exposure time, gain, and sensitivity; and use the same camera settings for all experiments in a series. Record digital images at a frequency of 10 Hz or higher. Continuously control the focus position (manually or automatically).

6. Position the ultrasound probe (attached to a micromanipulator) near the plaque area, avoiding damage of the vessel. Under brightfield illumination use the micromanipulator to position the tip of the probe at an angle of approximately 30° at the shoulder region of the plaque (Fig. 1f). Then, gently press the tip against the vessel.

7. If needed, inject inhibitors or antagonists via the tail or jugular vein.

8. Add saline around the carotid artery and the sonicator tip in order to adsorb the heat that will be generated by ultrasound application.

9. Prepare the camera and other settings for immediate capturing of fluorescence images. Change filters, if needed.

10. With the sonicator tip placed against the plaque shoulder, induce rupture of the plaque by applying ultrasound irradiation for 10 s at a frequency of 6 kHz. Check that the tip does not vibrate away from the vessel during the sonication procedure. Successful rupture immediately causes adhesion of fluorescent platelets.

11. Quickly remove the sonicator probe and the saline. Switch to record fluorescence images using the preselected camera settings. Record formation of the fluorescent thrombus, while focusing, for at least 10 min (Fig. 1g). Do not change the camera settings during an experiment, since this makes it difficult to interpret the data.

12. Position the other carotid artery under the objective for a second experiment, if required.

13. After finishing the last experiment, draw blood from the vena cava for ex vivo analysis, if needed and sacrifice the animal.

3.2.3. Ligation- or FeCl₃-Induced Damage of the Carotid Artery and Intravital Recording of Thrombus Formation (8)

1. If ligation-induced damage is to be applied, place a prolene 5/0 ligature around the dissected carotid artery. Make a loose knot without tightening it.

2. Mount the platform with anesthetized mouse onto the stage of the intravital microscope. For recording fluorescence with the CCD camera, use a 5–10× air objective.

3. Remove the surplus of saline from the dissected artery.

4. Select a suitable segment of the common carotid artery under bright-field illumination.

5. Slowly inject a suspension of 2×10^8 CFSE-labeled platelets (Subheading 3.1) via the tail or jugular vein over a period of about 30 s. Circulating fluorescently labeled platelets are immediately visible by camera recording in fluorescence mode. Note that injected platelets will not circulate when activated (aggregated).

6. Record baseline fluorescence images with the camera for a period of 1 min. Use appropriate color filters and optimal camera settings to obtain a high signal-to-noise ratio. Preselect the camera exposure time, gain, and sensitivity; and use the same camera settings for all experiments in a series. Record digital images at a frequency of 10 Hz or higher. Continuously control the focus position (manually or automatically).

7. Inject desired inhibitors or antagonists through the tail or jugular vein.

8. Apply damage to the selected part of the common carotid artery in one of the following ways:

   (a) Ligate the artery by vigorously tightening the knot of the ligature. After 5 min loosen the knot. Immediately start recording of fluorescence images, refocusing if necessary.

   (b) Apply a filter paper (1 × 2 mm) soaked with 500 mM FeCl₃ onto the artery (see Note 7). After 5 min, remove the filter paper, rapidly rinse with saline, and remove the surplus. Immediately start recording of fluorescence images, refocusing if necessary.

9. Monitor the thrombus-forming process with the CCD camera at a capture rate of 10 Hz for 10 or more minutes.

10. Position the other carotid artery under the objective for a second experiment, if required.

11. After finishing the last experiment, draw blood from the vena cava for ex vivo analysis, if needed, and sacrifice the animal.

3.2.4. Off-Line Image Analysis of Thrombus Formation in the Carotid Artery (5)

1. Recorded time-series of fluorescence images from thrombi formed in (atherosclerotic) carotid arteries can be processed with various software packages. Analysis of complete video series shows a spiking pattern of increased fluorescence intensity, which is due to continuous movement of the carotid artery during the experiment (Fig. 2a). Usually, the images can be converted into a format (e.g., TIFF), which can be analyzed by the freely available software package, Image J (http://rsbweb.nih.gov/ij/). This allows calculation of the integrated pixel intensities at the thrombus site above background.

2. For more precise measurements, only in-focus single digital images taken from the time series are analyzed (Fig. 2b), e.g., with ImagePro software. Quantitative analysis of images taken at 0.5, 1, 2, 5, and 10 min after plaque rupture or vessel damage gives a good impression of the fluorescence accumulation in the formed thrombus. A common procedure is to draw two similar regions-of-interest, one representing the thrombus area and an adjacent area for the background (Fig. 2b). A threshold level is then set to eliminate all pixels with an intensity lower than that in 99% of the pixels (gray levels) of the background region. In the resulting mask of the thrombus, the integrated pixel intensity is calculated, presenting the accumulated fluorescent platelets.

3.3. Two-Photon Laser Scanning Microscopy

Using TPLSM, the lumen of a carotid artery can be scanned from outside the vessel due to the high penetration depth of this technology (100–200 μm), in contrast to single photon confocal microscopy which scans tissues of no more than 50 μm thick (14). However, the relatively slow scanning rate of the first generation of TPLSM in combination with the rapid movements of the carotid in living mice hampers the use of this technology under in vivo conditions. The following protocol describes how TPLSM is used to scan the thrombotic site within an isolated carotid artery after thrombus formation (e.g., using one of the methods described in Subheading 3.2).

3.3.1. Preparation of Mouse Carotid Artery for Analysis by TPLSM (11)

1. After thrombus formation using one of the protocols described in Subheading 3.2, remove the carotid artery (including bifurcation in case of atherosclerotic plaques) from the experimental mouse.

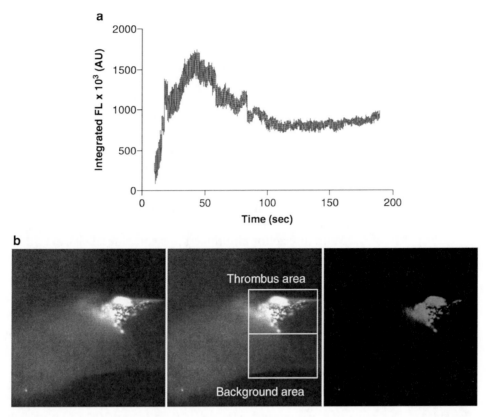

Fig. 2. Analysis of thrombus formation on ruptured atherosclerotic plaques. (**a**) Time series of integrated fluorescence of thrombus formation of CFSE-labeled platelets following ultrasound treatment of a plaque, analyzed with Hamamatsu software. The spiking pattern is caused by continuous in- and out-of-focus movements of the carotid artery. (**b**) Raw fluorescence images after 30 s (*left*); regions-of-interest, representing thrombus area and adjacent background area (*middle*); background-subtracted mask image representing CFSE-labeled platelets (*right*). Modified from ref. 5 with permission from John Wiley and Sons.

2. Using a dissection microscope, the artery is mounted in a perfusion chamber. Both ends are placed at each site on a glass micropipette (tip 120–150 μm) and tied with 17 μm nylon threads. For the bifurcation, attach the internal branch to a micropipette, and ligate the external branch and other side branches of the bifurcation.

3. The following procedures are all carried out at room temperature in the absence of luminal flow.

4. The perfusion chamber with mounted vessel is carefully filled with HBSS.

5. Attach a sphygmomanometer to one of the outlets of the chamber and apply an intraluminal pressure of 40–80 mmHg (depending on the pressure that the ligations can withstand). If needed, the distance between the two pipettes is adjusted, until the mounted artery is straight.

6. If required, carefully remove residual blood by gently flushing the vessel with Hank's solution. Perfuse with red fluorescent label(s) to stain desired cells or structures within the thrombus or vessel wall (e.g., the collagen probe CNA35 tagged with Oregon Green (5, 11) or quantum dot (QD) 585). Counter-stain the nuclei of vascular cells by external application of Syto41 (2.5 µM) for 10 min. The mounted vessel is viable for up to 4 h.

3.3.2. TPLSM Multicolor Analysis of the Carotid Artery Ex Vivo (5, 8)

A microscope setup used for the imaging of mounted arteries has been previously described in detail (14).

1. Scan the mounted vessel near the plaque-lumen interface (Fig. 3a) for the fluorescent probes using two-photon excitation at 800 nm (100 fs pulse-width, suitable laser intensity). With three photomultipliers present, blue labels can be detected at 450–470 nm emission wavelengths, green at 500–550 nm and red at 560–600 nm.

2. CFSE-labeled platelets are visualized in green and collagen labeled with CNA35 in red before (Fig. 3b) and after (Fig. 3c, d) counter-staining of the nuclei of vascular cells with Syto41 in blue.

3. Analysis of all confocal images and 3D reconstructions (Fig. 3e, f) are performed with Image Pro software.

3.4. Thrombus Formation in the Mouse Mesenteric Arterioles and Venules

3.4.1. Surgical Procedure for Exposure of Mesenteric Tissue (10)

1. Anesthetize a mouse of 4–5 weeks by subcutaneous injection of ketamine and xylazine (100 µg/g and 20 µg/g body weight, respectively). Anesthesia is maintained by injecting ketamine and xylazine (30 µg/g and 6.7 µg/g body weight) every 30 min (see Note 5).

2. Shave the throat of the mouse, and place the animal in a stable position on the 37°C heated platform. Body temperature is maintained at 37°C using a ceramic heating lamp, which is controlled by a thermo-analyzer system connected to a rectal probe.

Fig. 3. Cross-sectional analysis of damaged carotid artery and luminal thrombus by TPLSM. Ultrasound-treated and control carotid arteries were isolated from *Apoe*[−/−] mice, mounted in a perfusion chamber and scanned with TPLSM for monitoring fluorescence at deep penetration. CFSE-labeled platelets (*green*) were injected prior to plaque rupture; no fixation was applied. (**a**) Schematic representation of optical sections through plaque-containing artery. (**b**) Image of cross-section close to plaque shoulder of an ultrasound-treated vessel, postperfused with collagen-binding QD585-CNA35 (*red*). Note the aggregated CFSE-labeled platelets (*green*) close to the site of collagen staining (*red*). (**c, d**) Cross-sections through plaque (p) shoulder and lumen (l) of an ultrasound-treated vessel and a control vessel, postperfused with QD585-CNA35 and nuclear stain Syto41 (*blue*). (**c**) *Treated vessel:* ultrasound treatment provoked exposure of a collagen network (*red*) with attached platelets (*green*) in the shoulder adjacent to the vessel lumen (*black*). Note that normal alignment of smooth muscle nuclei (*blue*) and incidental endothelial nucleus (*arrow*). (**d**) *Control vessel:* undamaged vessel wall with weak staining for collagen (*red*) and monolayer of endothelial nuclei (*blue, arrow*) near the vessel lumen (*black*). Image sizes, 150 × 150 µm. (**e, f**) Three-dimensional overview of stained ultrasound-treated (179 × 179 × 84 µm) and control (179 × 179 × 55 µm) vessels from the luminal side. Note the locations of plaque (p) shoulder, and the marked exposed collagen network (*red*) in the treated vessel. From ref. 5 with permission from John Wiley and Sons.

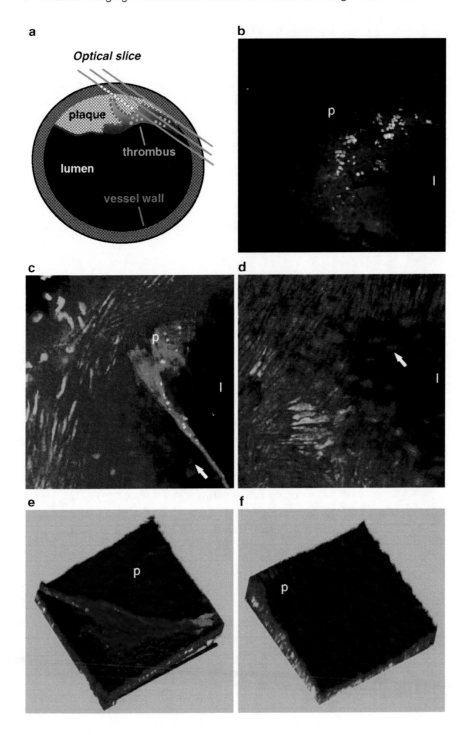

3. Insert subcutaneously into the neck a 30 gauge needle with polyethylene-10 catheter for the administration of anesthetics. Fix this to the skin with tissue glue (see Note 4).

4. Insert into a tail vein a 30 gauge needle with polyethylene-10 catheter for administration of labeled platelets and inhibitors. Fix to the skin with tissue glue. Alternatively, catheterize the jugular vein for this purpose.

5. Apply a total of 1.5 ml saline subcutaneously in both flanks to prevent tissue dehydration during the experiment.

6. Position the mouse with its abdominal surface downward and make a right-side incision of about 1 cm near the abdomen. The animal is now ready for exteriorization of the mesenteric tissue (Subheading 3.4.2).

3.4.2. Intravital Measurement of Thrombus Formation by FeCl$_3$- Induced Damage of Mesenteric Arteries and Veins (9, 10)

1. Position the mouse on the heated stage of the IVM (Subheading 3.4.1). The central opening of the stage is covered with a siliconized glass plate for transillumination. Select a 25× air or similar objective.

2. Place the head of the mouse on a cushion to prevent drowning.

3. From the abdominal incision, exteriorize an ileum segment with mesenteric tissue.

4. Spread the mesentery over the siliconized glass plate. Continuously superfuse the mesenteric tissue with buffered Tyrode's solution. Use wet gauzes to keep the exteriorized ileum still and moist.

5. Select a fat-free venule with adjacent arterioles close to the ileum using transillumination microscopy (Fig. 4a).

6. Inject fluorescently labeled platelets, other labels and/or inhibitors through the tail or jugular vein, as required for the specific experiment (see Note 6).

7. Record baseline transillumination or fluorescence images with the CCD camera for a period of 1 min. Use appropriate color filters and optimal camera settings to achieve a high signal-to-noise ratio. Preselect camera exposure time, gain, and sensitivity; use these settings for all experiments in a series. Record the images at a frequency of 10 Hz or higher while continuously controlling the focus (manually or automatically).

8. Stop superfusion of the mesentery and remove the surplus fluid.

9. Fill a pipette tip with 30 µl of 500 mM FeCl$_3$ (see Note 7).

10. Locally apply the FeCl$_3$ solution on the selected tissue site by placement of the pipette tip in the spotlight. Note that the arterioles constrict by about 20% while the diameter of venules does not change (10).

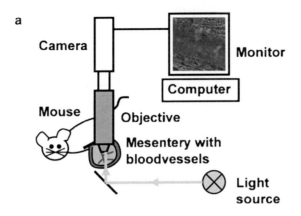

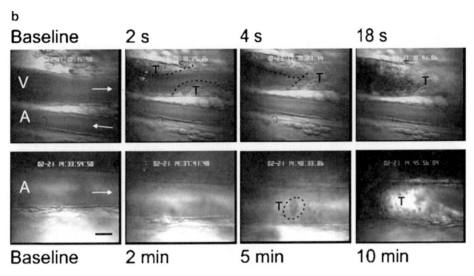

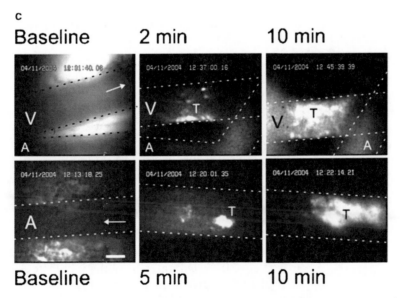

Fig. 4. FeCl$_3$-induced thrombus formation in mesenteric arterioles and venules. (a) Schematic representation of intravital microscope setup in transillumination mode. (b) Transillumination images showing progression of thrombus (T) formation at indicated times after FeCl$_3$ application in arterioles (A) and venules (V). *Upper panels* show near immediate formation of venous thrombus; lower panels show greatly delayed (5 min) formation of arterial thrombus. (c) Fluorescence images from labeled platelets in a thrombus at indicated times after FeCl$_3$ application. Again, more early accumulation in venules (*upper panels*) than in arterioles (*lower panels*). *Arrows* indicate direction of blood flow; *bars* represent 50 μm.

11. Continue recording of transillumination or fluorescence images to monitor thrombus formation, for up to 15 min after $FeCl_3$ application. In this model, thrombi are rapidly formed in venules and only after a delay of several minutes in arterioles (Fig. 4b, c).

12. Select another part of the mesenteric tissue for a second experiment, if required.

13. After finishing experimentation, draw blood from the vena cava for ex vivo analysis, if needed, and sacrifice the animal.

3.4.3. Off-Line Image Analysis of Thrombus Formation in mesenteric Arteries and Veins (9, 10)

1. Recorded movies of both transillumination and fluorescence images are analyzed off-line to determine key parameters of the $FeCl_3$-induced thrombus formation in arterioles and adjacent venules. First, measure the diameter of the vessels by comparison with a calibrated scale. In addition, determine the following lag times: (1) time to start of thrombus formation (i.e., of a platelet aggregate with height of 10 μm), (2) the time to half or full occlusion. The number of emboli shed in a specific time interval can also be counted. From individual transillumination images at 2, 5, and 10 min after $FeCl_3$ application, the en face thrombus height perpendicular to the vessel wall can be measured (expressed relative to the vessel diameter). This image analysis is possible with the freely available software package Image J.

2. Individual fluorescence images can be processed to give integrated fluorescence intensities above background, as described in Subheading 3.2.4. Select in-focus images for this analysis. Movies of fluorescence images additionally enable the measurement of the lag time to first platelet adhesion.

4. Notes

1. According to international accepted recommendations, the preferred methods of blood collection are by orbital puncture or by drawing from the vena cava (15). Cardiac puncture can result in high exposure to tissue factor, which triggers blood coagulation.

2. Use filtered, bacterial-free buffers and media for platelet isolation. Keep these at room temperature.

3. Dissolve 6.0 mg CFSE into 1.0 ml DMSO. Aliquots can be frozen at –20°C. Dilute 25 μl of the CFSE stock solution into 225 μl PBS to give a concentration of 0.6 mg/ml. The final concentration of CFSE in PRP is 5 μg/ml.

4. Catheters can be fixed onto the skin with tissue glue. Suitable sources are the more expensive histoacryl glue from TissueSeal

(Ann Arbor, MI, USA) and the cheaper second glue Tipper from Bison (Goes, The Netherlands).

5. Young mice of 4–5 weeks are very sensitive to anesthesia thus be cautious not to overdose.

6. In these transparent vessels, thrombus formation can easily be visualized by recording of transillumination images. However, in the presence of fluorescent labels, specific processes can also be monitored from epi-fluorescence images.

7. Comparisons between laboratories indicate that the most effective concentration of $FeCl_3$ can depend on the precise surgical procedure and the experimenter. It is advisable to initially test a dose range of 250–700 mM $FeCl_3$.

References

1. Virmani R, Burke AP, Farb A, Kolodgie FD (2006) Pathology of the vulnerable plaque. J Am Coll Cardiol 47: C13–18.

2. Falati S, Gross P, Merrill-Skoloff G, Furie BC, Furie B (2002) Real-time in vivo imaging of platelets, tissue factor and fibrin during arterial thrombus formation in the mouse. Nat Med 8: 1175–1181.

3. Westrick RJ, Winn ME, Eitzman DT (2007) Murine models of vascular thrombosis (Eitzman series). Arterioscler Thromb Vasc Biol 27: 2079–2093.

4. oude Egbrink MGA, van Gestel MA, Broeders MD, Tangelder GJ, Heemskerk JWM, et al. (2005) Regulation of microvascular thromboembolism in vivo. Microcirculation 12: 287–300.

5. Kuijpers MJE, Gilio K, Reitsma S, Nergiz-Unal R, Prinzen L, et al. (2009) Complementary roles of platelets and coagulation in thrombus formation on plaques acutely ruptured by targeted ultrasound treatment: a novel intravital model. J Thromb Haemost 7: 152–161.

6. Hechler B, Eckly A, Magnenat S, Freund M, Cazenave JP, et al. (2009) Localized arterial thrombosis on ruptured atherosclerotic plaques of ApoE–/– mice after vascular injury with ultrasound. J Thromb Haemost 7: Abstract OC-TH-033.

7. Nergiz-Unal R, Cosemans JMEM, Feijge MAH, van der Meijden PEJ, Storey RF, et al. (2010) Stabilizing role of platelet $P2Y_{12}$ receptors in shear-dependent thrombus formation on ruptured plaques. PLoS One 2010; 5: e10130.

8. Munnix IC, Kuijpers MJ, Auger J, Thomassen CM, Panizzi P, et al. (2007) Segregation of platelet aggregatory and procoagulant microdomains in thrombus formation: regulation by transient integrin activation. Arterioscler Thromb Vasc Biol 27: 2484–2490.

9. Kuijpers MJ, Pozgajova M, Cosemans JM, Munnix IC, Eckes B, et al. (2007) Role of murine integrin alpha2beta1 in thrombus stabilization and embolization: Contribution of thromboxane A(2). Thromb Haemost 98: 1072–1080.

10. Kuijpers MJE, Munnix ICA, Cosemans JMEM, Van Vlijmen BJ, Reutelingsperger CPM, et al. (2008) Key role of platelet procoagulant activity in tissue factor- and collagen-dependent thrombus formation in arterioles and venules in vivo. Differential sensitivity to thrombin inhibition. Microcirculation 15: 269–282.

11. Megens RT, Oude Egbrink MG, Cleutjens JP, Kuijpers MJ, Schiffers PH, et al. (2007) Imaging collagen in intact viable healthy and atherosclerotic arteries using fluorescently labeled CNA35 and two-photon laser scanning microscopy. Mol Imaging 6: 247–260.

12. Denis C, Methia N, Frenette PS, Rayburn H, Ullman Cullere M, et al. (1998) A mouse model of severe von Willebrand disease: defects in hemostasis and thrombosis. Proc Natl Acad Sci USA 95: 9524–9529.

13. Massberg S, Gawaz M, Gruner S, Schulte V, Konrad I, et al. (2003) A crucial role of glycoprotein VI for platelet recruitment to the injured arterial wall in vivo. J Exp Med 197: 41–49.

14. van Zandvoort M, Engels W, Douma K, Beckers L, Oude Egbrink M, et al. (2004) Two-photon microscopy for imaging of the (atherosclerotic) vascular wall: a proof of concept study. J Vasc Res 41: 54–63.

15. Jirouskova M, Shet AS, Johnson GJ (2007) A guide to murine platelet structure, function, assays, and genetic alterations. J Thromb Haemost 5: 661–669.

Chapter 2

Assessment of Platelet Aggregation Responses In Vivo in the Mouse

Christopher Moore and Michael Emerson

Abstract

Platelet aggregation responses are conventionally assessed in cuvette-based systems using either isolated platelets or whole blood. Unfortunately, in vitro aggregometry poorly predicts in vivo functionality, since mediators derived from the vascular endothelium are major regulators of platelet function. There is a need, therefore, for functional assays that assess platelet responsiveness in vivo in the presence of an intact and functional vascular endothelium. We have developed methodology for monitoring aggregation responses of freely circulating radiolabelled platelets using external detection probes in the anaesthetised mouse. Intravenous injection of platelet agonists induces reversible, dose-dependent aggregation responses that are sensitive to anti-platelet therapies and modification of the vascular endothelium. The technique provides a means of determining the effects of pharmacological and genetic manipulation upon platelet function in vivo.

Key words: Animal model, Indium Oxine, Mouse, Radiolabelled, Vascular endothelium

1. Introduction

Platelets are regulated not only by autologous signals, but also by mediators originating from the environment external to the platelet, such as the vascular endothelium and other tissues (1). Assessing platelet function in isolation limits studies to basic investigations of platelet behaviour that poorly reflect the responses of circulating platelets in intact animals and man (2). In particular, the contributions made by mediators derived from the vascular endothelium, such as nitric oxide and prostanoids, are not taken into account.

Models of thrombosis involving damage of the vascular endothelium by the application of chemicals, laser injury, or mechanical damage replicate elements of the thrombotic disease

Jonathan M. Gibbins and Martyn P. Mahaut-Smith (eds.), *Platelets and Megakaryocytes:*
Volume 3, Additional Protocols and Perspectives, Methods in Molecular Biology, vol. 788,
DOI 10.1007/978-1-61779-307-3_2, © Springer Science+Business Media, LLC 2012

process, such as vascular injury and dysfunction. Such models are driven by platelet activation but have additional haemostatic, vascular, and neurological determinants (3) and do not functionally isolate the platelet (4). Furthermore, these models are inappropriate for the evaluation of the role of endothelial products in regulating platelet functionality since the vascular endothelium is damaged as part of the process for inducing thrombus formation such that the role of endothelial products is likely to be variable and underestimated. This is evidenced by confusion in published literature concerning the role of endothelial nitric oxide synthase in regulating thrombotic responses (5, 6) and the failure of conventional models of thrombosis to predict the thrombotic tendencies of COX-2 selective antagonists (7).

We therefore established experimental procedures for assessing platelet aggregation responses in intact mice in the presence of an intact vascular endothelium (1). This approach may also be applied to larger species, such as rabbits (8). We have demonstrated that our model is sensitive to conventional anti-platelet therapy (1) and to pharmacological and genetic manipulation of the vascular endothelium (6).

2. Materials

2.1. Blood Collection, Platelet Isolation, and Radio-labelling

1. Ca^{2+} and Mg^{2+}-free Tyrode's solution (CFTS): 138 mM NaCl, 2.6 mM KCl, 5.5 mM glucose, 12 mM $NaHCO_3$, and 0.2 mM $NaHPO_4$ (see Note 1).

2. Acidified citrate solution (ACD): 74.8 mM trisodium citrate and 41.6 mM citric acid (see Note 1).

3. Prostaglandin E1 (PGE1, Sigma, Cat# P5515) stock: 1 mg/ml in ethanol can be frozen for 6 months.

4. CFTS/ACD solution is a 10:1 mixture of CFTS:ACD, with 0.34 µl/ml PGE1 stock: e.g. 18 ml CFTS, 1.8 ml ACD, and 6.7 µl PGE1 stock (see Note 1).

5. 1 ml syringes and needles (26-gauge × 10 mm) for intraperitoneal (ip) anaesthetic administration.

6. 1 ml syringes and needles (25-gauge × 16 mm) for blood collection by cardiac puncture.

7. Urethane (25% w/v) anaesthetic (Sigma). This is stable at room temperature for up to 2 months.

8. 1.5 ml Eppendorfs.

9. [111]Indium oxine (supplied as 37 MBq in 1 ml; GE Healthcare, Bucks, UK). This is gamma-emitting radioactive material and should be stored at 4°C in a lead-lined container.

Appropriate local licensing and permits should be obtained. The half-life of indium is 72 h and one pot will be usable for approximately 1 week after purchase.

10. Lead shielding to create a contained work space. In addition, you will require a lead-lined box for containment of radioactive waste, a containment tray for cell-labelling procedures, and lead-lined personal protective wear (see Note 2).

2.2. Platelet Monitoring

1. 2×1 cm Single Point Extended Area Radiation (SPEAR) detectors (eV Products, PA, US) for detection of radiation in pulmonary and abdominal regions.

2. Stands and clamps to hold detectors in position (see Fig. 1).

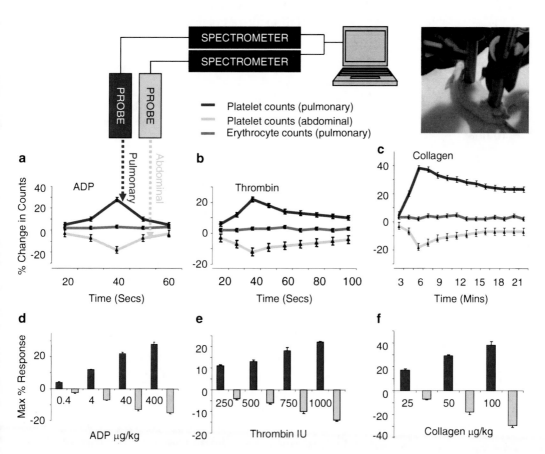

Fig. 1. Measurement of platelet aggregation responses in vivo in the mouse. The figure depicts the equipment used to obtain our data, the placement of probes in monitored mice and typical responses to common platelet agonists. (**a–c**) Time course of radio-labelled platelet responses in the pulmonary (*blue line*) and abdominal (*yellow line*) regions and erythrocyte responses in the pulmonary region (*red line*) following intravenous injection of (**a**) ADP, 400 µg/kg; (**b**) thrombin, 1,000 U/kg; and (**c**) collagen, 100 µg/kg. Data are plotted as the mean percentage change in radioactive counts between consecutive counting windows against time. (**d–f**) Radiolabelled platelet dose-responses following intravenous injection of bolus doses of (**d**) ADP, (**e**) thrombin, and (**f**) collagen. Results are presented as the mean ± SEM maximal percentage increase in radioactive counts above basal levels in the pulmonary (*blue bars*) and abdominal (*yellow bars*) regions ($n = 6$).

3. 2× UCS-20 spectrometer (Spectrum Techniques, TN, US; www.spectrumtechniques.com) for collection of data from radiation detectors.

4. Automated Isotope Monitoring System (Mumed Systems, London, UK): Software for data processing. Alternatively, free software from www.spectrumtechniques.com provides a more basic package.

5. Surgical equipment, including straight small scissors and two curved blunt forceps.

6. Urethane (25% w/v) from Sigma.

7. 1 ml syringes and needles (26 gauge × 10 mm) for ip anaesthetic administration.

8. 0.5 ml insulin syringes with needles (29 gauge × 13 mm, VWR) for administration of radiolabelled platelets and intravenous (iv) drugs.

9. Cotton buds.

10. Heat mat.

---

3. Methods

Appropriate licensing and ethical permissions (including Home Office Project and Personal Licenses in the UK) must be obtained before beginning animal procedures.

3.1. Blood Collection

1. Anaesthetise donor mice with urethane (25% w/v) at 10 μl/g ip (see Note 3 regarding number of animals required and Note 4 for further information on effective anaesthesia).

2. Bleed donor mice by cardiac puncture, using a 1 ml syringe loaded with 250 μl ACD for each animal. Between 0.7 and 1.0 ml of blood should be obtained per mouse.

3. Empty the contents of each syringe into individual 1.5 ml Eppendorfs. Add an extra 100 μl of ACD to each Eppendorf to prevent clotting.

3.2. Platelet Isolation

1. Centrifuge the blood for 3 min at $225 \times g$ to obtain platelet-rich plasma (PRP). Carefully transfer the PRP into a fresh tube using a pipette.

2. Add 300 μl CFTS/ACD solution to the remaining red blood cells (RBCs) and centrifuge again for 3 min at $225 \times g$ to acquire additional platelets. Remove the additional PRP and pool with the PRP obtained in step 1.

3. Repeat step 2 to obtain further PRP, thereby pooling PRP from three separate spins.

4. Remove contaminating RBCs by centrifuging the pooled PRP at $200 \times g$ for 2 min and transfer RBC-free PRP to fresh tubes.

5. Pellet the platelets by centrifugation at $640 \times g$ for 7.5 min to create a supernatant of platelet poor plasma (PPP). Discard the PPP and carefully resuspended the platelet pellet in ~200 μl CFTS/ACD solution and pool platelet suspensions from all donor mice for labelling. Do not exceed 800 μl, if there are more than four platelet pellets to pool, resuspend platelets in 150 μl or pool into an additional Eppendorf.

3.3. Radio-labelling of Platelets (and Erythrocytes)

1. Put on personal protective equipment and wear throughout labelling procedure and during monitoring experiments.

2. Incubate pooled platelets with 1.8 MBq [111]indium oxine (by adding an appropriate volume of stock solution) for 10 min behind lead shielding to minimise exposure to radiation.

3. Pellet the platelets by centrifugation at $640 \times g$ for 5 min (see Note 5).

4. Remove radioactive PPP and discard into a lead-shielded bin. Carefully wash the surface of the platelet pellet (i.e. without centrifugation) three times with CFTS to remove excess radiation.

5. Gently re-suspend the platelet pellet in CFTS to a volume of approximately 200 μl per mouse.

6. Obtain an approximate measure of the efficiency of platelet labelling by placing the platelet suspensions in front of a Geiger counter before and after washing. A labelling efficiency of at least 70% should be achieved; a lower level may indicate damage of platelets during the isolation process.

7. For labelling of erythrocytes, remove 50 μl of the erythrocyte-rich fraction resulting from step 1, Subheading 3.2 and make up to 200 μl with CFTS/ACD solution. Label erythrocytes with [111]indium oxine as described for platelets in steps 2–6, Subheading 3.3. This allows platelet responses to be compared with erythrocyte responses to confirm a platelet-specific observation rather than, e.g. blood redistribution (see Fig. 1).

3.4. Platelet Monitoring

1. Anaesthetise recipient mice with urethane (25% w/v) at 10 μl/g ip.

2. Expose a femoral vein and slowly infuse 200 μl radiolabelled platelets or erythrocytes using an insulin syringe and 29 gauge needle (see Notes 6–10). Allow to equilibrate for 15 min.

3. Once the platelets have been injected, place a cotton bud over the injection site and apply light pressure until bleeding stops. This closes the injection site. After a few minutes remove the bud and dispose of appropriately (all blood-contaminated

waste should be considered radioactive material). Use physiological saline-soaked tissue or cotton wool to cover the exposed femoral vein to prevent drying.

4. Expose a second femoral vein and place damp tissue over the exposed site (see Notes 6–10).

5. Place one SPEAR radiation detector over the pulmonary vascular bed (lungs). This is done by positioning the detector just above the diaphragm. The probe should make contact with the mouse but not impair breathing. Place a second control probe over the abdominal area to continuously record platelet levels in the peripheral circulation (see Fig. 1 for probe positioning).

6. Counts are continuously measured over fixed time periods (we recommend consecutive 4 s windows for ADP responses and 8 s windows for other agonists).

7. Administer platelet agonists and drugs via the newly exposed femoral vein using an insulin syringe and 29 gauge needle. If the vein becomes damaged, the original vessel (step 2) can be used as an alternative access site; this may involve repositioning of the animal.

8. Upon injection of platelet agonists, a rapid increase in counts should be observed in the pulmonary probe with a concomitant fall in the abdominal probe. This occurs as platelet aggregates form and become trapped in the pulmonary vascular bed so that peripheral platelet counts fall (1). Counts then return to baseline in a time period that varies with dose and between platelet agonists (see Fig. 1 for typical responses to commonly used platelet agonists and Note 11 for further information).

9. It may be possible to induce reproducible responses within an individual mouse (see Fig. 2 and Note 12) and so reduce mouse use.

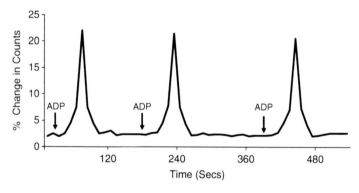

Fig. 2. Reproducible platelet responses within an individual mouse. Consecutive radiolabelled platelet aggregation responses following repeated administration of ADP (40 μg/kg). Data are presented as percentage change in radioactive counts following three injections of ADP within an individual mouse.

4. Notes

1. CFTS and ACD should be stored at 4°C and are stable for 1 week; however, the CFTS/ACD solution should be mixed fresh daily to avoid loss of PGE1 activity. All compounds can be purchased from Sigma.

2. A Radiation Protection Officer, or equivalent, should be consulted and radiation monitoring (of personnel and work spaces) should be performed in line with local regulations.

3. Approximately one donor mouse per monitored recipient is required. A typical experiment uses 9 mice per treatment group to obtain 4 sets of responses and detect changes of 15% between control and experimental groups.

4. Mice are left for approximately 10 min after ip administration of urethane to achieve full general anaesthesia. If there is no response to a paw pinch, then mice are considered fully anaesthetised and ready to be bled or undergo vessel exposure. If animals respond to a paw pinch, then an additional 40–50 μl of urethane may be administered to achieve effective anaesthesia. Care should be taken when administering anaesthetics since overdose can cause depression of breathing and death. There may be variations in tolerance to anaesthesia between strains and in genetically modified lines, requiring adjustment to the protocol.

5. Upon pelleting of the platelets in step 3 of Subheading 3.3, the platelet may be visibly contaminated with RBCs since it is difficult to obtain a completely clean platelet pellet. We have shown that RBCs will not affect the platelet experiment but obviously the platelet response is maximised by obtaining a pure platelet preparation.

6. Prior to exposure of the femoral vein, tape the mouse out on a heat mat with surgical tape so that limbs are extended, this will make it easier to expose a larger length of vein.

7. An alternative technique for administering radiolabelled platelets is via the tail vein in a conscious mouse. Place the mouse in a restraining tube and warm the tail by placing it in warm water (37°C). This should make the tail vein more clearly visible to facilitate injection of platelets. The mouse should then be housed in a lead shielded cage and anaesthetised as required.

8. When injecting either radio-labelled cells or drug solutions into the femoral vein of the mouse, bending the needle to 45° (after taking up the drug/radio-labelled cells) will facilitate injections. A characteristic blanching of the vein will indicate a successful intravenous injection.

9. In order to have a stable syringe/needle, it is recommended to hold the syringe in one hand (between index and middle finger with thumb on plunger) and with the other hand stabilise the injecting hand to avoid unnecessary movement of the needle.

10. Observing the vein through a dissection microscope and with an additional light source can facilitate injections.

11. We have obtained dose-dependent responses to platelet agonists as follows: ADP (0.4–400 μg/kg); thrombin (250–1,000 IU/kg); and collagen (10–100 μg/kg). These are applicable in both C57BL/6 and BALB/c mouse strains but may vary in other mouse lines. Dosing ranges may need to be established for other strains.

12. We have found that multiple reproducible responses can be obtained to ADP (up to four responses with 10 min intervals between responses). Collagen and thrombin responses are also reproducible (up to three responses with 15 min intervals) at low doses, but at higher doses responses may not return to baseline and desensitisation can occur.

Acknowledgements

Support for the work described in this chapter was provided by the National Centre for the Replacement, Refinement and Reduction of Animals in Research (www.nc3rs.org.uk).

References

1. Tymvios, C., et al., Real-time measurement of non-lethal platelet thromboembolic responses in the anaesthetized mouse. Thromb Haemost, 2008. **99**: p. 435–440.

2. Morley, J. and C.P. Page, Platelet aggregometry in vivo. TiPS, 1984. **5**: p. 258–60.

3. Nieswandt, B., et al., Platelets in atherothrombosis: lessons from mouse models. J Thromb Haemost, 2005. **3**(8): p. 1725–36.

4. Bodary, P.F. and D.T. Eitzman, Animal models of thrombosis. Curr Opin Hematol, 2009. **16**(5): p. 342–6.

5. Naseem, K.M. and R. Riba, Unresolved roles of platelet nitric oxide synthase. J Thromb Haemost, 2008. **6**(1): p. 10–9.

6. Tymvios, C., et al., Platelet aggregation responses are critically regulated in vivo by endogenous nitric oxide but not by endothelial nitric oxide synthase. Br J Pharmacol, 2009. **158**(7): p. 1735–42.

7. Mitchell, J.A. and T.D. Warner, COX isoforms in the cardiovascular system: understanding the activities of non-steroidal anti-inflammatory drugs. Nat Rev Drug Discov, 2006. **5**(1): p. 75–86.

8. May, G.R., et al., Radioisotopic model for investigating thromboembolism in the rabbit. J Pharmacol Methods, 1990. **24**(1): p. 19–35.

Chapter 3

Chapter 3

Focal Cerebral Ischemia

Stefan Braeuninger, Christoph Kleinschnitz, Bernhard Nieswandt, and Guido Stoll

Abstract

Rodent models of focal cerebral ischemia have been extremely useful in elucidating pathomechanisms of human stroke. Most commonly, a monofilament is advanced through the internal carotid artery of rodents to occlude the origin of the middle cerebral artery thus leading to critical ischemia in the corresponding vascular territory. The filament can be removed after different occlusion times allowing reperfusion (transient middle cerebral artery occlusion (MCAO) model) or is left permanently within the internal carotid artery (permanent MCAO model) both mimicking clinical thromboembolic stroke in which the occluding clot may resolve spontaneously or after thrombolysis, or may persist. Overall, the occlusion time determines the extent of ischemic brain damage, but infarcts still grow during reperfusion, a process involving complex interactions between platelets, endothelial cells, immune cells, and the coagulation system.

Key words: Experimental stroke, Monofilament, Middle cerebral artery occlusion, MCAO, Mice, Murine

1. Introduction

Ischemic stroke is a devastating disease and the second leading cause of death worldwide (1). It is in most cases caused by thromboembolic occlusion of major intracranial arteries with the occluding thrombus regularly originating from the heart (cardioembolic) or large extracranial arteries, such as the aorta or carotid arteries (arterioembolic). As the only currently approved therapeutic measure in acute stroke, recombinant tissue plasminogen activator (rt-PA) can be infused in selected patients within 4.5 h after the onset to resolve the vessel-occluding thrombus (thrombolysis) and to restore cerebral blood flow. Success rates, however, are moderate at best, since even recanalization does not guarantee salvage of brain

Jonathan M. Gibbins and Martyn P. Mahaut-Smith (eds.), *Platelets and Megakaryocytes:*
Volume 3, Additional Protocols and Perspectives, Methods in Molecular Biology, vol. 788,
DOI 10.1007/978-1-61779-307-3_3, © Springer Science+Business Media, LLC 2012

tissue, a phenomenon referred to as reperfusion injury or the no-reflow phenomenon. The pathophysiology of reperfusion injury during acute stroke is incompletely understood, but involves complex cell–cell interactions and probably thrombus formation in the microvascular bed (2–5). In the clinic, antiplatelet drugs like acetylsalicylic acid or clopidogrel are regularly used in acute stroke, but the small benefit in stroke outcome is mainly attributed to prevention of recurrent thromboembolism rather than blocking of thrombus formation within the microvasculature.

Various animal models in different species have been developed to mimic acute ischemic stroke in humans (6, 7). We exclusively focus on the focal cerebral ischemia model of middle cerebral artery occlusion (MCAO) in mice using an intraluminal filament while other paradigms, such as global hypoxic brain damage, are not considered. This is on the one hand because the monofilament MCAO model is probably the most frequently applied experimental stroke model worldwide. On the other hand, the clinical situation of vessel occlusion followed by resolution of clots and reperfusion can be easily reflected by withdrawing the intraluminal filament. Finally, transient MCAO, in contrast to other stroke models like cerebral photothrombosis (8), has proven to be useful for evaluating basic mechanisms of platelet function and thrombus formation in the downstream cerebral vasculature (5).

The MCAO model using an intraluminal occluding monofilament was originally developed in rats (9) and has subsequently been modified (10, 11) and adapted to mice (12). Craniotomy is not required, which renders this model of focal ischemia technically relatively simple. It has been firmly established that final infarct size depends on the prior occlusion time of the middle cerebral artery (6): ischemic periods of less than 30 min lead to infarctions of the caudate and putamen (basal ganglia) and only partly affect the neocortex because ischemic cortical tissue is salvaged by collateral blood supply and reperfusion. If reperfusion is delayed further, however, the size of neocortical infarctions will increase because the surrounding penumbra is subsequently involved in the definite infarct area.

2. Materials

2.1. Experimental Setup

1. Operating microscope.
2. Heating device.
3. Anesthesia unit.

2.2. Surgical Instruments and Materials

1. Scalpel (optional).
2. Scissors (e.g., delicate curved sharp/blunt iris scissors, Fine Science Tools Inc., Foster City, CA).

3. Spring scissors (e.g., Vannas spring scissors straight with 3 mm blade, Fine Science Tools).

4. Forceps: Two pairs of splinter forceps (e.g., 90 mm curved splinter forceps, Aesculap AG, Tuttlingen, Germany), and two pairs of delicate angled forceps (e.g., delicate angled forceps designed for eye surgery, Geuder AG, Heidelberg, Germany).

5. Vessel clip and clip applying forceps (e.g., microserrefine clips and microserrefine clip applying forceps, Fine Science Tools).

6. Needle holder (e.g., Halsey Micro Needle Holder, Fine Science Tools).

7. Occluding thread or monofilament (e.g., commercially available silicon rubber-coated monofilaments of 6–0 filament size and 20 mm length, Doccol Corporation, Redlands, CA).

8. Standard consumables used for surgery, such as sterile suture material (e.g., 3–0, 4–0, and 7–0 silk suture), swabs, and gloves. Eye ointment (e.g., dexpanthenol eye ointment).

Instruments and materials needed for murine transient MCAO using the intraluminal filament method are depicted in Fig. 1.

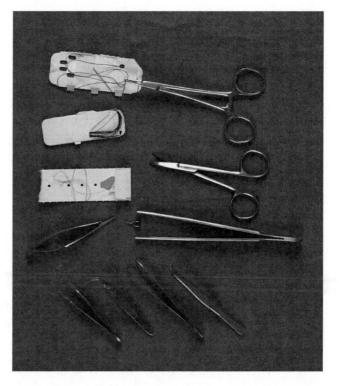

Fig. 1. Instruments and materials needed for murine intraluminal filament middle cerebral artery occlusion.

2.3. TTC Staining

1. Mouse brain matrix (e.g., mouse brain slicer matrix with 1 mm coronal slice intervals, Zivic Instruments, Pittsburgh, PA).

2. Razor blade (e.g., SIH1 razor blades, Hartenstein Laborbedarf, Wuerzburg, Germany).

3. Phosphate-buffered saline (PBS).

4. 2,3,5-triphenyltetrazolium chloride (e.g., Sigma-Aldrich, St. Louis, MO).

3. Methods

3.1. General Considerations

For the intraluminal thread MCAO model of experimental stroke, laboratory animals (commonly mice or rats) are used requiring an appropriate animal experimentation facility. It is also obvious that animal experiments have to be conducted in accordance with ethical standards to minimize animal discomfort, legal requirements (approval by the appropriate authorities), and institutional guidelines. To ensure maximal benefit from experimental studies in terms of translation from rodents to man, recommendations have been developed to improve the quality of preclinical stroke research (13, 14) that should be considered when designing experiments.

3.2. SetUp

The experimental setup needed for the intraluminal filament MCAO model of focal cerebral ischemia in mice is similar to the setup needed for other surgical experiments in laboratory animals. An example is shown in Fig. 2. Due to the small size of the brain-supplying arteries in mice, however, an operating or dissecting microscope is essential to magnify the operating field. Animals must be operated under appropriate anesthesia and analgesia. The choice of the anesthetic and analgesic agents is not trivial in preclinical stroke experiments because any type of anesthesia can alter stroke outcome (15). In general, inhalation anesthesia seems to be superior to intraperitoneal or intravenous administration of anesthetics. Some experimentalists even recommend endotracheal intubation and mechanical ventilation to achieve less variability of physiological parameters (16). We feel, however, that inhalation anesthesia in spontaneously breathing mice is acceptable if operation time per animal is kept below 15 min. Brain temperature may also have a profound influence on experimental stroke outcome (17). During the surgical procedure, a heating pad or similar preferably servo-controlled heating device for rodents is strongly recommended to record and maintain body temperature of the laboratory animals. Monitoring of additional physiological parameters, such as blood pressure, pulse rate, blood glucose levels, arterial blood gases, and pH may be considered at least in subgroups of animals.

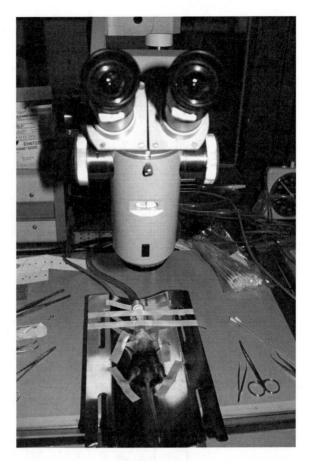

Fig. 2. Example of an experimental setup. Note placement of the mouse under an operating microscope and on a heating device. Inhalation anesthesia is applied.

3.3. Surgical Procedure

1. The anesthetized laboratory animal is placed under the operating microscope lying in supine position (see Fig. 2). The forelimbs and hindlimbs are immobilized by adhesive bandages. An appropriate ointment should be applied to the eyes of the animal prior to surgery to prevent drying out.

2. After rechecking depth of anesthesia by applying a painful stimulus, a midline skin incision is made in the neck (see Fig. 3). For this, a scalpel or scissors can be used in mice.

3. Following midline skin incision, the thyroid gland is exposed. The right and left lobes of the thyroid gland are mobilized by blunt dissection using two pairs of splinter forceps (see Note 1), separated at the isthmus and moved to the right and left. Underneath, the trachea comes into view. This is illustrated in Fig. 4.

4. The common carotid artery is located lateral to the trachea. It can be found in the carotid triangle bounded by the sternocleidomastoid muscle, the stylohyoid muscle and the posterior

Fig. 3. Midline skin incision.

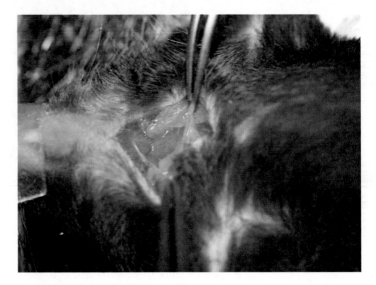

Fig. 4. Dissection of the thyroid gland.

belly of the digastric muscle, and the omohyoid muscle. In principle, the common carotid artery of either side can be operated on, though most right-handed experimentalists may find it easier to perform right MCAO. The common carotid artery can now be removed from adjacent connective tissue by blunt dissection using two pairs of delicate angled forceps (see Note 1). Special care should be taken to separate the vagus nerve (in the carotid sheath, lateral to the artery) from the common carotid artery (see Figs. 5 and 6).

5. By following the common carotid artery in the cranial direction and dissecting this vessel from the surrounding tissue, the point of division into the external and internal carotid arteries

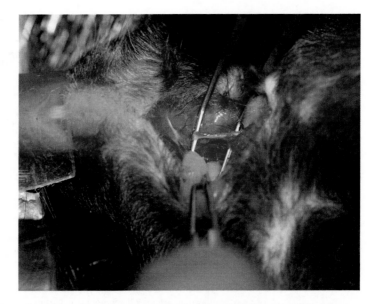

Fig. 5. Dissection of the common carotid artery. Note the adjacent whitish vagus nerve.

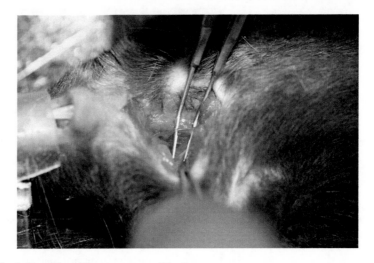

Fig. 6. Dissection of the common carotid artery.

(carotid bifurcation) becomes visible. The superior thyroid artery emerges as first branch of the external carotid artery but, unlike the situation in humans, the pterygopalatine artery emerges from the internal carotid artery. The proximal external and internal carotid arteries are also dissected from surrounding tissue. The anatomy following dissection of the carotid arteries is shown in Fig. 7.

6. A permanent ligature is tied around the proximal common carotid artery and another ligature around the external carotid artery (see Note 2). A 7–0 silk suture can be used. The ligatures are presented in Figs. 8 and 9.

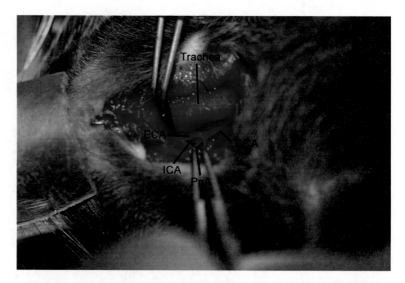

Fig. 7. Anatomy of the carotid arteries after dissection (*CCA* common carotid artery, *ECA* external carotid artery, *ICA* internal carotid artery, *PpA* pterygopalatine artery).

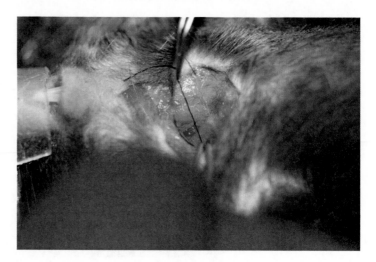

Fig. 8. Ligature of the proximal common carotid artery.

7. A tourniquet, e.g., a 4–0 silk suture, is loosely tied around the distal common carotid artery (see Fig. 10).

8. A vessel clip is applied to the distal common carotid artery or proximal internal carotid artery using a clip applying forceps. Skilled experimentalists, however, may find use of a vessel clip dispensable in mice, since the tourniquet suture (when tension is applied) can be used in lieu and is sufficient to prevent bleeding from the vessel hole proximal to the tourniquet.

9. Between the proximal common carotid artery ligature and the vessel clip, a stretch of the vasculature is cut off from blood circulation. There, the vessel is now incised using spring scissors (arteriotomy, see Fig. 11).

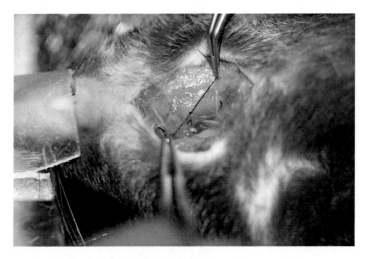

Fig. 9. Ligature of the external carotid artery.

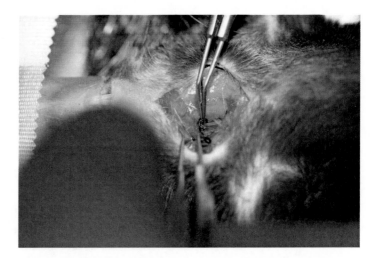

Fig. 10. Applying of a tourniquet suture around the distal common carotid artery.

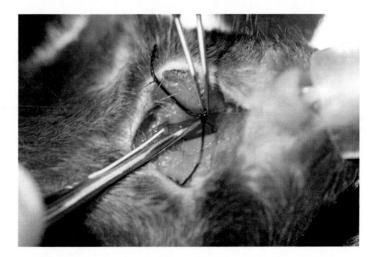

Fig. 11. Arteriotomy of the common carotid artery.

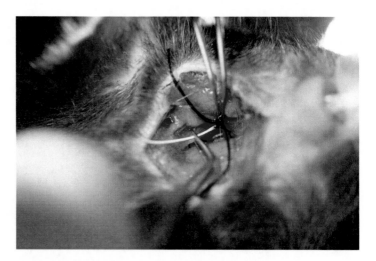

Fig. 12. Insertion of the occluding monofilament into the vessel lumen.

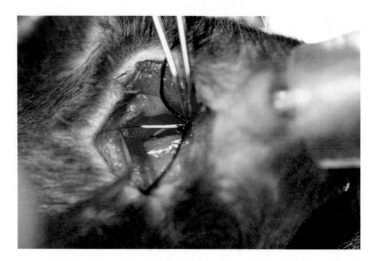

Fig. 13. Advancement of the intraluminal monofilament into the internal carotid artery.

10. The monofilament is inserted through the incised vessel wall into the common carotid artery lumen (see Fig. 12).

11. The vessel clamp is removed and the monofilament is further advanced. It should be visually controlled that the monofilament has been correctly placed into the internal carotid artery (and not into the pterygopalatine artery) (see Note 3). The monofilament is carefully pushed up through the internal carotid artery until a gentle resistance is felt (see Fig. 13). Then, the end of the monofilament should still protrude from the vessel hole and is now secured by tightening the tourniquet suture prepared before to prevent dislocation during the ischemia period. In our experience, the distance between the carotid bifurcation and the origin of the middle cerebral artery

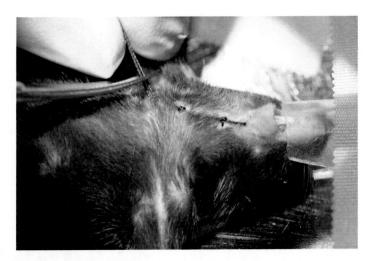

Fig. 14. Skin closure.

is usually 0.9 cm in normal-sized C57Bl/6 mice. Hence, when using a 20-mm long occluding filament, a distance of 11 mm from the carotid bifurcation to the end of the monofilament indicates appropriate placement. Given these premises, the tip of the monofilament should now be located intracranially at the origin of the ipsilateral middle cerebral artery thereby interrupting blood flow. Excessive advancement of the filament may lead to perforation of the vessel wall and thus cause subarachnoid hemorrhage while insufficient advancement may not induce adequate ischemia.

12. The occluding monofilament may be left in place permanently to induce permanent cerebral ischemia or may be removed after a desired time interval to model reperfusion. For transient MCAO, the tourniquet suture is loosened again, the occluding monofilament is removed, and the tourniquet is then secured by knots and thus changed into a ligature to prevent bleeding from the incised vessel.

13. Closure is achieved by standard skin suture using a needle holder and, e.g., 3–0 silk suture material with attached needle (see Fig. 14). Total time for surgery (if no additional measurements of physiological variables, such as cerebral blood flow, are performed) should usually not exceed 15 min.

3.4. Outcome

Outcome analysis after MCAO can include infarct size, histology, mortality rate, and functional (behavioral, motor, and cognitive) scores (14).

A common method to assess infarct size is the triphenyltetrazolium chloride (TTC) method. For this, a solution of 2% (w/v) 2,3,5-triphenyltetrazolium chloride (TTC) in PBS is prepared. Brains are harvested and coronal slices of typically 1 or 2 mm

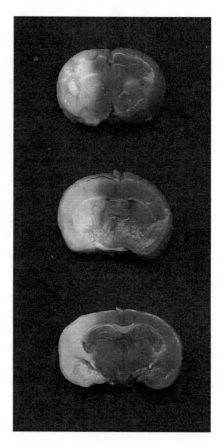

Fig. 15. 2-mm thick mouse brain slices stained with TTC. Note the pale infarct area.

thickness are cut in a mouse brain matrix using a razor blade. The brain slices are immersed in 2% TTC (w/v) solution until the non-infarcted tissue is stained brick red (see Note 4). The infarct area remains pale (see Fig. 15). The slices may be stored (e.g., in form-aldehyde solution) and, given the known slice thickness, infarct volumes can be calculated by obtaining planimetric measurements of infarcted areas.

A commonly used functional outcome score is the neurologic score originally developed by Bederson et al. in rats (18). Animals subjected to MCAO are scored according to a grading system essentially assessing hemiparesis symptoms: 0, no observable deficit; 1, forelimb flexion; 2, decreased resistance to lateral push (and forelimb flexion) without circling; 3, as for 2, plus circling. A modification results in the following scoring system: 0, no deficit; 1, forelimb flexion; 2, as for 1, plus decreased resistance to lateral push; 3, unidirectional circling; 4, longitudinal spinning or seizure activity; 5, no movement.

Another functional score is the string test, also known as grip test (19). Here, the mouse is placed midway on a string (or rod)

Fig. 16. Mouse tested in the string test.

between two vertical supports (see Fig. 16) and rated as follows: 0, falls off; 1, hangs onto string by two forepaws; 2, as for 1, and attempts to climb onto string; 3, hangs onto string by two forepaws plus one or both hindpaws; 4, hangs onto string by all four paws plus tail wrapped around string; 5, escape (to the supports).

4. Notes

1. Some researchers use electrocauterization, which, however, has been associated with complications and is unnecessary if surgery is carefully performed by a skilled experimentalist.

2. Above, we have described a straightforward technique inserting the occluding monofilament into the common carotid artery. In the original description of the intraluminal filament method in murine MCAO (12), however, the external carotid artery has been used for insertion of the monofilament. For this technical variant, the external carotid artery is coagulated distal to the bifurcation and the monofilament is inserted through the resulting external carotid artery stump.

3. When trying to advance the monofilament into the internal carotid artery, it often undesirably ends up in the pterygopalatine artery. This can be prevented by ligature of this artery or simply by using the forceps to occlude the origin of the pterygopalatine artery while advancing the filament.

4. TTC and TTC solutions have to be shielded from light. The immersed brain slices should be turned occasionally to ensure even staining. Warm TTC solution stains faster than fridge-cold solution.

Acknowledgments

The authors wish to thank Mrs. A. Götz for making the illustrating MCAO photographs.

References

1. Lopez, A. D., Mathers, C. D., Ezzati, M., Jamison D. T., and Murray, C. J. L. (2006) Global and regional burden of disease and risk factors, 2001: systematic analysis of population health data. *Lancet.* **367**, 1747–1757.

2. Kleinschnitz, C., Stoll, G., Bendszus, M., Schuh, K., Pauer H. U., Burfeind, P., Renné, C., Gailani, D., Nieswandt, B., and Renné, T. (2006) Targeting coagulation factor XII provides protection from pathological thrombosis in cerebral ischemia without interfering with hemostasis. *J. Exp. Med.* **20**, 513–518.

3. Kleinschnitz, C., Pozgajova, M., Pham, M., Bendszus, M., Nieswandt, B., and Stoll, G. (2007) Targeting platelets in acute experimental stroke: impact of glycoprotein Ib, VI, and IIb/IIIa blockade on infarct size, functional outcome, and intracranial bleeding. *Circulation.* **115**, 2323–2330.

4. Kleinschnitz, C., De Meyer, S. F., Schwarz, T., Austinat, M., Vanhoorelbeke, K., Nieswandt, B., Deckmyn, H., and Stoll, G. (2009) Deficiency of von Willebrand factor protects mice from ischemic stroke. *Blood.* **113**, 3600–3603.

5. Stoll, G., Kleinschnitz, C., and Nieswandt, B. (2008) Molecular mechanisms of thrombus formation in ischemic stroke: novel insights and targets for treatments. *Blood.* **112**, 3555–3562.

6. Carmichael S. T. (2005) Rodent models of focal stroke: size, mechanism, and purpose. *NeuroRx.* **2**, 396–409.

7. Traystman R. J. (2003) Animal models of focal and global cerebral ischemia. *ILAR J.* **44**, 85–95.

8. Kleinschnitz, C., Braeuninger, S., Pham, M., Austinat, M., Nölte, I., Renné, T., Nieswandt, B., Bendszus, M., and Stoll, G. (2008) Blocking of platelets or intrinsic coagulation pathway-driven thrombosis does not prevent cerebral infarctions induced by photothrombosis. *Stroke* **39**, 1262–1268.

9. Koizumi, J., Yoshida, Y., Nazakawa, T., and Ooneda, G. (1986) Experimental studies of ischemic brain edema: A new experimental model of cerebral embolism in rats in which recirculation can be introduced in the ischemic area. *Jpn. J. Stroke.* **8**, 1–8.

10. Longa, E. Z., Weinstein, P. R., Carlson, S., and Cummins, R. (1989) Reversible middle cerebral artery occlusion without craniectomy in rats. *Stroke.* **20**, 84–91.

11. Belayev, L., Alonso, O. F., Busto, R., Zhao, W., and Ginsberg, M. D. (1996) Middle cerebral artery occlusion in the rat by intraluminal suture. Neurological and pathological evaluation of an improved model. *Stroke.* **27**, 1616–1623.

12. Clark, W. M., Lessov, N. S., Dixon, M. P., and Eckenstein, F. (1997) Monofilament intraluminal middle cerebral artery occlusion in the mouse. *Neurol. Res.* **19**, 641–648.

13. Stroke therapy academic industry roundtable (STAIR). (1999) Recommendations for standards regarding preclinical neuroprotective and restorative drug development. *Stroke* **30**, 2752–2758.

14. Braeuninger, S., and Kleinschnitz, C. (2009) Rodent models of focal cerebral ischemia: procedural pitfalls and translational problems. *Exp. Transl. Stroke Med.* **1**: 8.

15. Kirsch, J. R., Traystman, R. J., and Hurn, P. D. (1996) Anesthetics and cerebroprotection: experimental aspects. *Int. Anesthesiol. Clin.* **34**, 73–93.

16. Zausinger, S., Baethmann, A., and Schmid-Elsaesser, R. (2002) Anesthetic methods in rats determine outcome after experimental focal cerebral ischemia: mechanical ventilation is required to obtain controlled experimental conditions. *Brain Res. Brain Res. Protoc.* **9**, 112–121.

17. Busto, R., Dietrich, W. D., Globus, M. Y., Valdés, I., Scheinberg, P., and Ginsberg, M. D. (1987) Small differences in intraischemic brain temperature critically determine the extent of ischemic neuronal injury. *J. Cereb. Blood Flow Metab.* **7**, 729–738.

18. Bederson, J. B., Pitts, L. H., Tsuji, M., Nishimura, M. C., Davis, R. L., and Bartkowski, H. (1986) Rat middle cerebral artery occlusion: evaluation of the model and development of a neurologic examination. *Stroke* **17**, 472–476.

19. Moran, P. M., Higgins, L. S., Cordell, B., and Moser, P. C. (1995) Age-related learning deficits in transgenic mice expressing the 751-amino acid isoform of human β-amyloid precursor protein. *Proc. Natl. Acad. Sci. USA* **92**, 5341–5345.

Chapter 4

Extracorporeal Assays of Thrombosis

Lina Badimon, Teresa Padro, and Gemma Vilahur

Abstract

Platelet deposition, adhesion/aggregation, to the damaged vessel wall or atherosclerotic plaque components has shown to play a major role in hemostasis, thrombosis, and the development of atherosclerosis. Platelet-vessel wall interaction and thrombus formation is driven by blood flow rheology/hemodynamics (changes in local flow conditions), the nature of the flowing blood, and the characteristics of the triggering substrate (lesion type). An extracorporeal perfusion system (the Badimon chamber) was developed to investigate the dynamics of platelet deposition and thrombus formation on: (a) different surfaces (biological and synthetic); (b) under controlled blood flow conditions with varying degrees of stenosis mimicking various vascular conditions (patent and stenotic arteries); and (c) with varying perfusing blood treatments.

In the following chapter, we thoroughly describe this experimental approach that has helped to improve the understanding of the pathophysiology of the acute coronary syndromes (Badimon et al., Arteriosclerosis 6:312–320, 1986; Badimon et al., J Lab Clin Med 110:706–718, 1987; Badimon et al., Blood 73:961–967, 1989; Mailhac et al., Circulation 90:988–996, 1994) and is a useful tool for the study and screening of new antithrombotic and platelet-inhibitory compounds (Badimon et al., Thromb Haemost 71:511–516, 1994; Vilahur et al., Circulation 110:1686–1693; Vilahur et al., Thromb Haemost 92:191–200, 2004; Vilahur et al., Cardiovasc Res 61:806–816; Vilahur et al., Thromb Haemost 98:662–669, 2007; Vilahur et al., Thromb Haemost 97:650–657, 2007; Zafar et al., J Thromb Haemost 5:1195–1200, 2007; Lev et al., J Am Coll Cardiol 43:966–971) and for the evaluation of the thrombogenicity associated to synthetic/prosthetic surfaces (Badimon et al., J Biomater Appl 5:27–48, 1990; Badimon et al., ASAIO Trans 33:621–625, 1987) and/or plasma components (cholesterol, glucose levels, etc.) (Badimon et al., Arterioscler Thromb 11:395–402, 1991; Osende et al., J Am Coll Cardiol 38:1307–1312; 2001).

Key words: Platelets, Atherothrombosis, Pig animal model of thrombosis

1. Introduction

Thrombosis is a process that involves various blood cells and proteins, the nature of the blood flow, and surface characteristics (1, 2). Several perfusion chambers have been developed to study platelet-vessel wall

Jonathan M. Gibbins and Martyn P. Mahaut-Smith (eds.), *Platelets and Megakaryocytes:*
Volume 3, Additional Protocols and Perspectives, Methods in Molecular Biology, vol. 788,
DOI 10.1007/978-1-61779-307-3_4, © Springer Science+Business Media, LLC 2012

interactions under controlled rheological conditions; yet, the majority of these studies have been performed in laminar flow conditions (3). In atherosclerotic vessels, however, laminar flow conditions may not be maintained since stenotic narrowing induces flow disturbances that modify cell–cell and cell–vessel interactions and the local concentration of fluid-phase chemical mediators necessary for cell interaction. *H.R. Baumgartner* developed the first popular annular perfusion chamber that helped to advance the knowledge and understanding of platelet adhesion to the subendothelium (4). Briefly, the Baumgartner chamber consists of a central rod surrounded by a cylinder. The arterial segment (usually reverted rabbit aorta) is mounted on the rod, and the gap between the subendothelial surface and the cylinder allows blood to pass over the surface (4). Indeed, a wide range of shear rates can be obtained by varying the width of the gap and/or varying the blood flow rate. However, the Baumgartner technique is laborious, not easily amenable to use for nonvascular materials, and requires a preselection of vessels for appropriate dimensional characteristics. Subsequently, other ex vivo chamber systems were developed for investigating prosthetic surfaces (5); however, in counterpart, they do not allow exposure of biological substrates to blood over a broad range of flow conditions.

The Badimon perfusion chamber is a bioreactor designed to retain the cylindrical shape typical of the vasculature (6), flexible enough to accept a variety of biological and prosthetic materials (7, 8), and capable of simulating a broad range of pathophysiological flow conditions, including laminar and nonparallel streamline flows (9–11). Thrombus formation can be sensitively and quantitatively assessed by estimating the amount of radiolabelled platelets and fibrinogen deposited on the tested surface, by histological computer-assisted morphometry and by immunofluorescence.

2. Materials

2.1. Badimon Perfusion Chamber

The Badimon perfusion chamber consists of a Plexiglass block through which a cylindrical hole of 0.2 or 0.1 cm diameter has been machined in order to mimic the tube-like shape of the vascular system (Fig. 1). The upper face of the plastic block, parallel to the axis of the cylindrical hole, has been milled to create a channel. Surfaces, either biologic or prosthetic, are placed on the channel, and a pressure plate, also constructed from Plexiglass, is placed on the nonexposed side of the test surface. A 0.2 or 0.1 cm thick ridge in the face of the pressure plate contacts the surface around the periphery of the channel gap and pressure imposed on the back of the pressure plate by a screw compresses the test surface to ensure a leak-free system when blood is passed through the cylindrical tube.

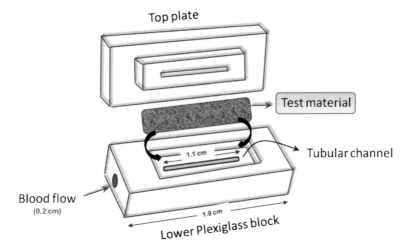

Fig. 1. Side view of an unassembled Plexiglass Badimon chamber.

Table 1
Flow characteristics of perfusion chambers

Tubular diameter (cm)	Blood flow rate (ml/min)	Blood wall shear rate (1/s)
0.2	5	105
	10	212
	20	425
	30	640
	40	850
0.1	5	840
	10	1,680
	20	3,360
	30	5,040
	40	6,720

Shear conditions calculated in this table are calculated from the expression for shear rate given for a Newtonian fluid in tube flow

Thus, the test material serves as its own gasketing seal, although for noncompressible surfaces a gasketing material can also be used. As mentioned above, two chambers of different internal diameters (0.1 or 0.2 cm in width) were constructed in order to obtain a broad range of blood velocities and wall shear rates on the substrate with moderate changes in average blood flow (Table 1) (6, 9).

Furthermore, a second generation of chambers with varying degrees of stenosis (ranging from 0 to 80%) has been used to mimic stenotic vessels and study nonparallel streamlines under controlled rheological conditions (9).

46   L. Badimon et al.

2.2. Test Surfaces

Different biological substrates and biomaterials from different origin can be placed within the chamber. Seeded proteins and cells on substrates or matrices can also be placed in the chamber (see Notes 1 and 2).

2.2.1. Vascular Substrates

Exposure of deendothelialized vessel mimicking mild type II vascular injury

Briefly, type II vascular injury includes endothelial denudation and intimal injury with intact internal elastic lamina and media (Fig. 2). Such lesions can occur as a result of toxic products released by macrophages in the early stages of atherogenesis, in the syndromes of accelerated atherosclerosis as a result of surgical manipulation and the relatively high pressure in the coronary vein graft within the first postoperative year, or as a result of immune or other injuries in coronary arteries of patients after heart transplantation (12). As to the degree of platelet attachment upon exposure of subendothelium, this is a mild thrombogenic material that mainly induces the deposition of a platelet monolayer. Indeed, we have observed that under low shear rate conditions there is a direct attachment, or adhesion, of platelets to the vascular surface that is apparently irreversible. In contrast, platelet–platelet interactions (aggregates) formed at high wall shear rates are dislodged or embolized by the flow for the same exposure time (13) (see Note 3).

In order to produce deendothelialization of the vessels, fresh aortas (e.g., obtained in-house or in the slaughter house) must be rinsed with phosphate-buffered saline (PBS), cleaned of the surrounding connective tissue, and immediately frozen. This procedure induces a selective endothelial injury without damaging the basement membrane assessed by "en face" staining with silver nitrate (13).

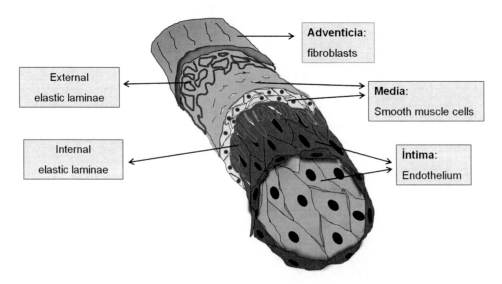

Fig. 2. Image depicting the different vessel layers.

Exposure of tunica media fibrillar collagen that mimics a deep type III injury

Type III injury encompasses endothelial denudation with damage to both intima and media. This surface, consisting of fibrillar native collagens (types III and I), proteoglycans, glycosaminoglycans, elastin, and smooth muscle cells, is exposed in the ruptured atherosclerotic wall, although probably in different relative proportions with respect to the normal vessel wall. We have observed that this lesion is highly thrombogenic and induces thrombus formation at any shear rate and perfusion time. In fact, high shear rates are unable to dislodge significant quantities of platelet thrombi, and deposits continue to increase with exposure time (13).

In order to expose tunica media, the intima with a thin portion of the subjacent media is lifted up and peeled off starting from one corner (by retracting a raised corner) leaving the substrate with the major portion of the remaining media.

Exposure of atherosclerotic vessels

Interestingly, human atherosclerotic specimens may also be placed within the extracorporeal perfusion system. We have previously described that the different components of human atherosclerotic plaques result in different degrees of thrombus formation (14). For instance, taking into consideration the following macroscopic classification, including (1) fatty streaks; (2) sclerotic plaques; (3) fibrolipid plaques; and (4) atheromatous plaques, the atheromatous core, and more precisely tissue factor within the plaque is the most thrombogenic component (15, 16).

2.2.2. Biological Materials

Exposure of isolated proteins

Isolated proteins or collagen seeded in substrates/plastic slides cut to fit the chamber trough have been exposed in the perfusion chamber to perfusing blood. For instance, in order to obtain collagen-coated plastic slides, plastic slides (30 × 10 mm) are coated with fibrillar collagen of bovine Achilles tendon at a concentration of 16 µg/ml. Then, the plastic slides are kept in a humid atmosphere overnight, washed, and blocked with 3% bovine serum albumin for 2 h. After washing with PBS solution, collagen-coated surfaces can be placed in the perfusion chamber as we have previously reported (17).

Exposure of cell cultures

Cell cultures (e.g., endothelial cells, smooth muscle cells) grown on special slides fitted to the chamber trough have been successfully studied.

2.2.3. Prosthetic materials, such as polytetrafluoroethilene (Goretex)

Prosthetic materials can be placed within the perfusion chamber although a gasket might be necessary to ensure proper sealing of the chamber (7, 8).

2.3. Platelet Labeling   Platelets are labeled with Indium-Oxine[111] (Amersham Biosciences). Platelets labeled with Indium-Oxine[111] deposited in the triggering substrate are assessed by gamma-counting (Beckman Coulter, Gamma 800).

2.4. Arteriovenous Shunt Animal Model   The extracorporeal circuit (animal/chamber) is composed of a nonthrombogenic material (tygon tubing; Clay Adams PE 200, Cole-Paimer). The placement of three-way switch valves (Discofix® C, Braun) within the circuit allows change from PBS to blood and vice versa avoiding stasis within the perfusion chamber. A peristaltic pump (Masterflex Model 7013) is required to ensure blood recirculation to the animal. Before placing the extracorporeal perfusion system (carotid artery-jugular vein), animals should be anticoagulated with heparin (Hospira Prod. Farm y Hosp, SL). Heparin plasma levels are measured by using Coatest, heparin [S-2222]; (KabiVitrum).

2.5. Morphological Measurements of Formed Thrombus   *Perfused substrates* are sectioned and stained with hematoxilin–eosin (HE) and combined Masson's trichrome elastin (CME) for the presence of thrombus. Evaluation of total thrombus area is performed on six sections taken from each perfusion chamber by computer-assisted planimetry using Image-Pro Plus software (v4.1, Media Cybernetics, Silver Spring, MD).

2.6. Solutions

2.6.1. EDTA-Anticoagulant   1. Resuspend 5 g of EDTA in 100 ml of milli-Q water.

2.6.2. Acid-Citrate-Dextrose-Anticoagulant   This anticoagulant preserves the platelet membrane and thus protects against platelet activation.
   Dissolve in 100 ml of milliQ water:

1. Citric acid (monohydrate): 0.8 g.

2. Trisodium citrate (dihydrate): 2.5 g.

3. Anhydrous glucose (dextrose): 1.2 g.

2.6.3. Saline (0.9%)   1. Dissolve 0.9 g of sodium chloride in 100 ml of milli-Q water.

2.6.4. ACD-Saline   This buffer is used to wash the platelets.

1. Dissolve 14.4 ml of acid-citrate-dextrose (ACD)-anticoagulant in 100 ml of saline.

2. Adjust pH to 6.5 in order to avoid platelet activation.

2.6.5. Phosphate-Buffered Saline   1. Sodium chloride: 137 mM.

2. Potassium chloride: 2.7 mM.

3. Sodium phosphate (dibasic): 10 mM.

4. Potassium phosphate (monobasic): 2 mM.

5. pH of 7.4.

2.6.6. Sodium Citrate at 3.8%

1. Dissolve in 100 ml of milli-Q water 3.8 g of sodium citrate (anhydrous).

2.6.7. Sucrose 30% in PBS

1. Dissolve 30 g of sucrose to a total volume of 100 ml in PBS.

2.6.8. Paraformaldehyde at 4% (w/v) (see Note 4)

1. Dissolve 4 g paraformaldehyde in 100 ml PBS in a conical flask. Cover with parafilm and transfer to the fume hood: shake thoroughly – take care not to splash paraformaldehyde – it is a rapid fixative and is toxic.

2. Place flask on top of the hotplate/stirrer inside the fume cupboard and heat with moderate stirring. Allow the solution to warm up – it will turn from being cloudy to clear when ready. Inspect regularly to avoid overheating and consequent spilling.

3. When the paraformaldehyde has dissolved, switch off the heat but leave to stir: do not handle for safety reasons. Allow to cool.

4. When cooled, transfer the fixative to a 4°C refrigerator.

3. Methods

3.1. Experimental Perfusion Systems

Two types of ex vivo perfusion systems can be applied with the Badimon perfusion chamber.

3.1.1. Recirculating Experiments

As depicted in Fig. 3, a carotid artery-jugular vein shunt is established in the animal model of interest. This experimental design is usually performed in animals of medium to large size (mainly pig and less frequently rabbits).

1. Animals are anesthetized by injecting a mixture of ketamine (25 mg/kg) and xylazine (1 mg/kg), and the carotid artery and contralateral jugular vein are cannulated.

2. Blood is collected for baseline determination of hematocrit, platelet number, prothrombin time, and activated partial thromboplastin time (APTT) (see Note 5).

3. The animals are intravenously heparinized (120 U/kg bolus plus continuous infusion of 100 U/kg per h for the duration of the study) (see Note 6).

4. The cannulated carotid artery is connected by polyethylene tubing (20 cm in length, Clay Adams PE 200, Cole-Paimer) to the input of the Plexiglass chamber.

5. The output of the chamber is connected to a peristaltic pump (Masterflex Model 7013) so that blood that passes through the chamber is recirculated back into the animal by the contralateral jugular vein.

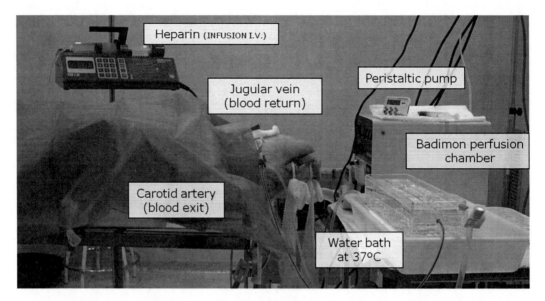

Fig. 3. Scheme of the extracorporeal perfusion system. Substrates/segments are placed in the Badimon perfusion chamber and exposed to flowing blood from carotid artery of pigs at different shear rates. Blood returns to the jugular vein by a peristaltic pump. For further details of the dimensions of the chambers, please refer to Badimon et al. (6).

6. The substrate of interest is placed within the Badimon chamber as detailed in Fig. 4a–e.

7. Blood flows from the carotid artery through the chamber, which is immersed in a 37°C water bath to avoid blood temperature fluctuations (Fig. 4f). Two or three chambers can be placed in line to simultaneously evaluate several thrombogenic conditions by varying the thrombogenic substrate and shear rate (chamber internal diameter).

8. The specimens are perfused with PBS solution, at 37°C for 60 s before perfusing blood to ensure the wash-out of nondesired elements and equilibration of the system.

9. After the 60 s preperfusion period, blood enters the chamber at a preselected flow rate and for predefined perfusion times.

10. At the termination of blood flow, buffer is again passed for 30 s to eliminate the remaining nonattached cells (see Note 7).

11. Blood samples are regularly collected (every two perfusion experiments) for APTT and heparin plasma level measurements (Coatest, heparin [S-2222]; KabiVitrum).

12. At the end of the perfusion, the substrate is carefully removed from the chamber and fixed in 4% paraformaldehyde in PBS to avoid platelet detachment (Fig. 4g–i).

13. The perfused segments are then counted in a gamma counter (Wizard, Wallac, USA) for quantification of deposited platelets (Fig. 4j).

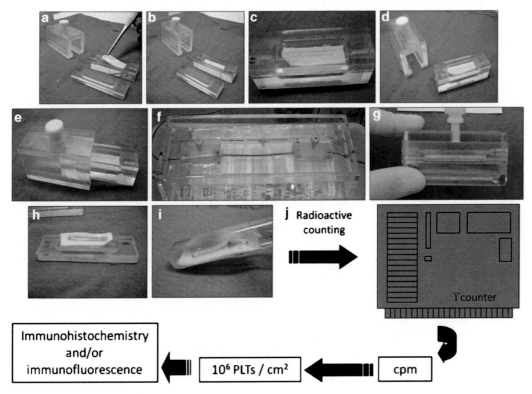

Fig. 4. (**a–f**) The detailed sequence of substrate placement within the Badimon perfusion chamber and perfusion system. (**g–h**) Once the perfusion is finished, thrombus may be detected macroscopically in the cylindrical channel that has been directly exposed to the flowing blood. (**i**) Perfused substrates, carefully removed, are immersed in paraformaldehyde 4% to avoid platelet detachment and then placed in a gamma-counter to measure counts per minute (cpm). Cpm values are then converted to platelets/cm² (**j**).

3.1.2. Nonrecirculating Experiments

This modified Badimon perfusion system is also appropriate for studies conducted in humans, where blood withdrawn is not recirculated. Such approach offers the possibility of performing homologous perfusion studies in humans (14, 17).

Nonanticoagulated blood is obtained from a 21-gauge catheter in the subjects' antecubital vein. Blood may then follow two pathways depending on the purpose of the study:

1. *Ex vivo treatment (Fig. 5a)*. Blood is directly introduced into a mixing device made in-house containing a magnetic stirrer at a constant flow rate (ex vivo treatment) in which medication or saline (control) is infused from an infusion pump through a second inlet. A magnetic stirrer thoroughly mixes the drug/saline with the blood before passing the mixture to the perfusion chamber. The changes from blood to buffer and vice versa are achieved by a switch valve without the introduction of stasis in the chamber. The blood enters the chambers at a preselected flow rate, and then it is collected for additional testing and discarded (18, 19).

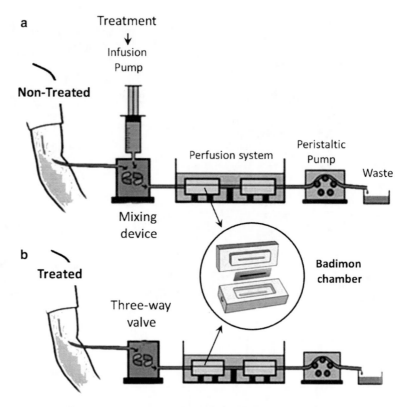

Fig. 5. Graphic illustration of the modified Badimon ex vivo perfusion system with: (**a**) a special mixing device connected to an infusion pump for ex vivo mixing of medications with nonanticoagulated blood; and (**b**) three way valves directly connected to the inlet of the first perfusion chamber.

2. *In vivo treatment (Pretreated human/patient) (Fig. 5b)*. Blood directly flows, by means of a standard three-way switch valve to the inlet of the perfusion chamber, and then it is collected for additional testing and discarded (20–22).

3.2. Evaluation of Thrombus Formation in the Vessel Segments (animal studies)

Following each perfusion study, the substrates are fixed in 4% (w/v) paraformaldehyde for 30 min and then directly counted in a gamma-well counter for quantification of deposited platelets (without washing).

3.2.1. Quantification of Platelet Deposition by the Gamma Counter

The number of platelets deposited on each specimen is calculated from the platelet count and the Indium[111]-activity on the perfused area and in blood (6).

Labeling of Platelets
with Indium[111]

Platelets are isolated and labeled according to a modified method of Dewanjee et al. (23) (see Note 8). This procedure should be performed as follows:

1. Collect blood from the donor in ACD-anticoagulant. Perform a platelet count before manipulating the sample.

2. Centrifuge, ($400 \times g$, room temperature, 10 min) the whole blood to obtain platelet rich plasma (PRP). Perform a platelet count on the PRP (please refer to volume 1 Chap. 3 for further details on platelet counting methodology).

3. Centrifuge ($1,100 \times g$, room temperature, 10 min) the PRP to obtain platelet poor plasma (PPP). Perform a platelet count which should be close to zero. Keep PPP for step 9.

4. Cautiously resuspend platelet pellets in ACD-saline which has been previously warmed to room temperature. Perform another platelet count of the resuspension. It is important to highlight that platelets are susceptible to aggregation upon inadequate handling.

5. Spin platelet suspension at $1,100 \times g$ for 15 min at room temperature.

6. Resuspended platelet pellet in 2 ml ACD-saline and perform a platelet count.

7. Add 250 μCi of indium-oxine[111] solution to the 2 ml of ACD-saline containing the platelet resuspension. Incubate for 20 min at room temperature and check indium dose (dose-1) by counting the counts per minute (cpm) of the sample.

8. Centrifuge the platelet suspension at $800 \times g$ for 10 min at room temperature (Tanetzki T32C) and discard supernatant. Check cpm counts (indium dose-2). Resuspended platelet pellet in PPP (from step 3). Take small volume of this resuspension in order to check the amount of lysed platelets (see step 12 below). In addition, take 25 μl of the resuspension to perform the last platelet count (prediluted).

9. Charge delivery syringe through a cannula and check cpm count (dose-3).

10. Slowly inject into the animal via the marginal ear vein.

11. Calculate:

    (a) The % of platelet lysis (indium not incorporated within the platelets and thus released to the supernatant) by using the following equation: supernatant counts (cpm)/pellet counts (cpm). This should be less than 1%. If not, most of the indium has not been adequately incorporated within the platelets and labeling should be repeated.

    (b) Calculate the indium loading efficiency (dose-3/dose-1 × 100). This should be higher than 80%.

3.2.2. Biodistribution of Indium-¹¹¹-Labeled Platelets

At the end of the perfusion experiment, animals are euthanized and the heart, lungs, liver, spleen, kidneys, and blood are isolated and weighed. The radioactivity of weighed portions of these organs is measured and the organ biodistribution determined.

Average of indium-¹¹¹ organ distribution in a large animal should be around 27% liver, 15% spleen, 10%, lung 1% heart, 0.3% kidney, and 47% blood to ensure that excess platelets have not been retained in the organs and are largely circulating in the blood stream (13).

3.2.3. Morphometry Analysis

Substrates fixed in 4% (w/v) paraformaldehyde are processed for morphometric analysis. Tissue slides are serially cut using a cryostat and stained with hematoxilin–eosin or with Combined Masson trichrome Elastin. Evaluation of total thrombus area can be performed by acquiring several images of different sections from each perfusion chamber and analyzing them by computer-assisted planimetry.

3.2.4. Immunofluorescence

Serially cut sections may also be processed for immunohistochemical analysis of fibrin(ogen) and platelet detection (Fig. 6).

3.2.5. Evaluation of Effluents

Platelet activation can be evaluated by biomarkers of activation in the effluent blood.

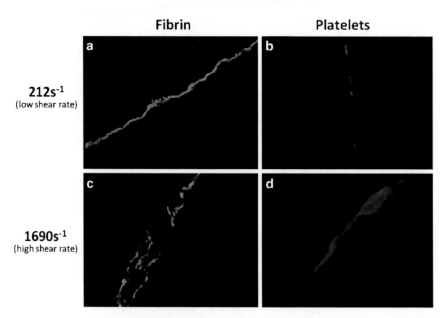

Fig. 6. Representative immunophotomicrographs of fibrin (*green*; **a, c**) and platelets (*red*; **b, d**) deposition triggered by severely damaged vessel wall perfused at low shear rate (**a, b**) and high shear rate (**c, d**).

3.3. Choice of Animal Model

The ideal animal model for the extracorporeal Badimon perfusion system is the commercial swine (body weight around 40 kg) (6). Diverse animal species, including cats, rabbits, rats, and dogs may also be utilized (24). In our group, we are interested in swine because of their similarities to humans in cardiovascular anatomy and physiology (26–34). Cats and rabbits can only be used for a few perfusions because of blood volume limitations, whereas dogs present low platelet deposition levels under a high thrombogenic stimulus (25).

4. Notes

1. When obtaining the vascular/prosthetic segments, in all cases, special care has to be taken to avoid irregularities on the surface and, in the case of natural surfaces, substrates must be stored frozen at –70°C until the day of the experiment.

2. All segments (natural, seeded, seeded biological matrices, and biomaterials) must be 3 cm × 0.8 cm to perfectly fit into the chamber.

3. Blood flow conditions directly influence the arrival and removal rates of platelets at the surface. In our perfusion system, the flow conditions are well-defined and platelets deposit on both deendothelialized vessel wall and collagen fibrils in quantities that increase with local shear rate.

4. Paraformaldehyde has to be manipulated with gloves, gown, and mask and should be performed in a fume hood. In order to ensure proper solute dilution, an "agitator" and heat should be used (never exceed 80°C).

5. Cell wall encounters arise from the continual shear-induced collisions between red cells that lead to lateral displacements of their paths as well as those of white cells and platelets. Thus, cells flowing in the periphery of vessels are frequently displaced to the wall. The hematocrit has, therefore, a very important role in cell–wall interaction. Hematocrit must be similar in all the experiments so that it cannot be considered an interassay variable that may have influenced platelet deposition results at the different flow conditions.

6. With appropriate heparinization the average prothrombin time ratio during the study should be approximately 1.5.

7. Within the extracorporeal perfusion system, changes from buffer to blood and vice versa must be achieved by a switch valve without the introduction of stasis in the chamber.

8. Indium labeling procedure should not take longer than 2 h to avoid platelet alterations and maintain the properties of fresh platelets.

Acknowledgments

This article has been possible thanks to support from PNS 2006–10091 (to L.B.) from the Spanish Ministry of Science; Lilly Foundation (to L.B.) and CIBER$_{OBN}$06 (to L.B.). We thank Fundacion Juan Serra, Barcelona, for their continuous support. G.V. is recipient of a grant from the Spanish Ministry of Science and Innovation (RyC, MICINN).

References

1. Badimon L, Vilahur G, Padro T. (2009) Lipoproteins, platelets and atherothrombosis. *Rev Esp Cardiol.* **62**, 1161–1178.

2. Badimon L, Badimon JJ, Vilahur G, et al. (2002) Pathogenesis of the acute coronary syndromes and therapeutic implications. *Pathophysiol Haemost Thromb.* **32**, 225–231.

3. Sakariassen KS, Aarts PA, de Groot PG, et al. (1983) A perfusion chamber developed to investigate platelet interaction in flowing blood with human vessel wall cells, their extracellular matrix, and purified components. *J Lab Clin Med.* **102**, 522–535.

4. Baumgartner HR. (1973) The role of blood flow in platelet adhesion, fibrin deposition, and formation of mural thrombi. *Microvasc Res.* **5**, 167–179.

5. Martines E, McGhee K, Wilkinson C, et al. (2004) A parallel-plate flow chamber to study initial cell adhesion on a nanofeatured surface. *IEEE Trans Nanobioscience.* **3**, 90–95.

6. Badimon L, Turitto V, Rosemark JA, et al. (1987) Characterization of a tubular flow chamber for studying platelet interaction with biologic and prosthetic materials: deposition of indium 111-labeled platelets on collagen, subendothelium, and expanded polytetrafluoroethylene. *J Lab Clin Med.* **110**, 706–718.

7. Badimon L, Badimon JJ, Turitto VT, et al. (1990) Platelet interaction to prosthetic materials--role of von Willebrand factor in platelet interaction to PTFE. *J Biomater Appl.* **5**, 27–48.

8. Badimon L, Badimon JJ, Turitto VT, et al. (1987) Role of von Willebrand factor in platelet interaction with an expanded PTFE surface. *ASAIO Trans.* **33**, 621–625.

9. Badimon L, Badimon JJ. (1989) Mechanisms of arterial thrombosis in nonparallel streamlines: platelet thrombi grow on the apex of stenotic severely injured vessel wall. Experimental study in the pig model. *J Clin Invest.* **84**, 1134–1144.

10. Badimon L, Badimon JJ, Turitto VT, et al. (1987) Thrombosis: studies under flow conditions. *Ann N Y Acad Sci.* **516**, 527–540.

11. Badimon L, Badimon JJ, Rand J, et al. (1987) Platelet deposition on von Willebrand factor-deficient vessels. Extracorporeal perfusion studies in swine with von Willebrand's disease using native and heparinized blood. *J Lab Clin Med.* **110**, 634–647.

12. Fuster V, Ip JH, Badimon L, et al. (1991) Importance of experimental models for the development of clinical trials on thromboatherosclerosis. *Circulation.* **83**, IV15–25.

13. Badimon L, Badimon JJ, Galvez A, et al. (1986) Influence of arterial damage and wall shear rate on platelet deposition. Ex vivo study in a swine model. *Arteriosclerosis.* **6**, 312–320.

14. Mailhac A, Badimon JJ, Fallon JT, et al. (1994) Effect of an eccentric severe stenosis on fibrin(ogen) deposition on severely damaged vessel wall in arterial thrombosis. Relative contribution of fibrin(ogen) and platelets. *Circulation.* **90**, 988–996.

15. Toschi V, Gallo R, Lettino M, et al. (1997) Tissue factor modulates the thrombogenicity of human atherosclerotic plaques. *Circulation.* **95**, 594–599.

16. Badimon JJ, Lettino M, Toschi V, et al. (1999) Local inhibition of tissue factor reduces the thrombogenicity of disrupted human atherosclerotic plaques: effects of tissue factor pathway inhibitor on plaque thrombogenicity under flow conditions. *Circulation.* **99**, 1780–1787.

17. Vilahur G, Duran X, Juan-Babot O, et al. (2004) Antithrombotic effects of saratin on human atherosclerotic plaques. *Thromb Haemost.* **92**, 191–200.

18. Zafar MU, Vilahur G, Choi BG, et al. (2007) A novel anti-ischemic nitric oxide donor (LA419) reduces thrombogenesis in healthy human subjects. *J Thromb Haemost.* **5**, 1195–1200.

19. Chelliah R, Lucking AJ, Tattersall L, et al. (2009) P-selectin antagonism reduces thrombus formation in humans. *J Thromb Haemost.* 7, 1915–1919.

20. Sarich TC, Osende JI, Eriksson UG, et al. (2003) Acute antithrombotic effects of ximelagatran, an oral direct thrombin inhibitor, and r-hirudin in a human ex vivo model of arterial thrombosis. *J Thromb Haemost.* 1, 999–1004.

21. Zafar MU, Vorchheimer DA, Gaztanaga J, et al. (2007) Antithrombotic effects of factor Xa inhibition with DU-176b: Phase-I study of an oral, direct factor Xa inhibitor using an ex-vivo flow chamber. *Thromb Haemost.* 98, 883–888.

22. Dangas G, Badimon JJ, Coller BS, et al. (1998) Administration of abciximab during percutaneous coronary intervention reduces both ex vivo platelet thrombus formation and fibrin deposition: implications for a potential anticoagulant effect of abciximab. *Arterioscler Thromb Vasc Biol.* 18, 1342–1349.

23. Dewanjee MK, Rao SA, Didisheim P. (1981) Indium-111 tropolone, a new high-affinity platelet label: preparation and evaluation of labeling parameters. *J Nucl Med.* 22, 981–987.

24. Alfon J, Pueyo Palazon C, Royo T, et al. (1999) Effects of statins in thrombosis and aortic lesion development in a dyslipemic rabbit model. *Thromb Haemost.* 81, 822–827.

25. Badimon L, Fuster V, Chesebro JH, et al. (1983) New "ex vivo" radioisotopic method of quantitation of platelet deposition - studies in four animal species. *Thromb Haemost.* 50, 639–644.

26. Badimon L, Badimon JJ, Turitto VT, et al. (1989) Role of von Willebrand factor in mediating platelet-vessel wall interaction at low shear rate; the importance of perfusion conditions. *Blood.* 73, 961–967.

27. Badimon JJ, Weng D, Chesebro JH, et al. (1994) Platelet deposition induced by severely damaged vessel wall is inhibited by a boroarginine synthetic peptide with antithrombin activity. *Thromb Haemost.* 71, 511–516.

28. Vilahur G, Segales E, Salas E, et al. (2004) Effects of a novel platelet nitric oxide donor (LA816), aspirin, clopidogrel, and combined therapy in inhibiting flow- and lesion-dependent thrombosis in the porcine ex vivo model. *Circulation.* 110, 1686–1693.

29. Vilahur G, Baldellou MI, Segales E, et al. (2004) Inhibition of thrombosis by a novel platelet selective S-nitrosothiol compound without hemodynamic side effects. *Cardiovasc Res.* 61, 806–816.

30. Vilahur G, Casani L, Badimon L. (2007) A thromboxane A2/prostaglandin H2 receptor antagonist (S18886) shows high antithrombotic efficacy in an experimental model of stent-induced thrombosis. *Thromb Haemost.* 98, 662–669.

31. Vilahur G, Pena E, Padro T, et al. (2007) Protein disulphide isomerase-mediated LA419-NO release provides additional antithrombotic effects to the blockade of the ADP receptor. *Thromb Haemost.* 97, 650–657.

32. Lev EI, Hasdai D, Scapa E, et al. (2004) Administration of eptifibatide to acute coronary syndrome patients receiving enoxaparin or unfractionated heparin: effect on platelet function and thrombus formation. *J Am Coll Cardiol.* 43, 966–971.

33. Badimon JJ, Badimon L, Turitto VT, et al. (1991) Platelet deposition at high shear rates is enhanced by high plasma cholesterol levels. In vivo study in the rabbit model. *Arterioscler Thromb.* 11, 395–402.

34. Osende JI, Badimon JJ, Fuster V, et al. (2001) Blood thrombogenicity in type 2 diabetes mellitus patients is associated with glycemic control. *J Am Coll Cardiol.* 38, 1307–1312.

Chapter 5

Platelet Life Span and Apoptosis

Emma C. Josefsson, Michael J. White, Mark R. Dowling, and Benjamin T. Kile

Abstract

Like many nucleated mammalian cells, the life and death of the anucleate platelet is regulated by Bcl-2 family proteins. Platelets depend on Bcl-x_L for survival. Bcl-x_L maintains platelet viability by restraining the killer protein Bak. When Bak is unleashed, it triggers classical intrinsic apoptosis by causing mitochondrial damage. The latter leads to caspase activation and phosphatidylserine (PS) exposure. Platelet apoptosis can be blocked by caspase inhibitors, or by genetic deletion of Bak and its close relative Bax. Perturbations in the platelet apoptosis program lead to changes in platelet life span in vivo. Here, we describe methods to determine platelet life span, enumerate young platelets, and measure hallmarks of platelet apoptosis, such as PS exposure, caspase activation, and mitochondrial dysfunction.

Key words: Apoptosis, Platelet life span, Caspase activity, Mitochondrial function, ABT-737, Phosphatidylserine, Bcl-x_L, Bak

1. Introduction

Apoptosis is a highly conserved molecular program of cell death, which ensures that aged, dysfunctional, or infected cells do not accumulate in the body but, instead, are removed in a swift and immunologically silent manner. The intrinsic apoptosis pathway is molecularly regulated by the balance between pro-survival and pro-apoptotic proteins within the Bcl-2 family (1). If not restrained by pro-survival proteins, pro-apoptotic Bak/Bax induce mitochondrial damage, which initiates a caspase cascade through activation of the apical caspase, caspase-9. In platelets, pro-survival Bcl-x_L is the key player. It is required to restrain pro-apoptotic Bak in order to maintain platelet viability (2). Induction of the intrinsic apoptosis pathway, either by pharmacological inhibition of Bcl-x_L

Jonathan M. Gibbins and Martyn P. Mahaut-Smith (eds.), *Platelets and Megakaryocytes: Volume 3, Additional Protocols and Perspectives*, Methods in Molecular Biology, vol. 788, DOI 10.1007/978-1-61779-307-3_5, © Springer Science+Business Media, LLC 2012

with the BH3-mimetic compound ABT-737 (3), or genetic mutation of Bcl-x$_L$, results in platelet death and clearance from the circulation (2, 4).

Assays designed to measure features of platelet apoptosis in vitro, such as mitochondrial damage or dysfunction, caspase activation, and phosphatidylserine (PS) externalization, should be interpreted with caution. For example PS, the archetypal "eat me" signal on nucleated cells undergoing apoptosis (5–7) is also externalized during agonist-induced platelet activation, and treatment with calcium ionophore (8, 9). Evidence suggests, however, that the processes and pathways involved in mediating PS exposure in each of these cases are distinct (9, 10). Other apoptotic signatures, such as caspase activation, have been observed in platelets treated with agonists of platelet activation (11, 12), suggesting that the apoptotic machinery may play a role in platelet functional responses. Therefore, no one assay alone can be used to ascertain whether a platelet is undergoing classical intrinsic apoptosis. Equally, many of the assays we describe measure parameters that are affected by a broad range of nonapoptotic processes, and, as such, are useful tools for examining platelet production and function in any setting.

2. Materials

2.1. Platelet Life Span Analysis

2.1.1. In Vivo Biotinylation of Mouse Platelets

1. (+) Biotin N-hydroxysuccinimide ester (NHS-biotin) (Sigma Aldrich).

2. Dimethyl sulfoxide (DMSO) (Sigma Aldrich).

3. Saline for injection of NHS-biotin: 145 mM NaCl, unbuffered.

4. Antibodies: Rat anti-mouse CD41 conjugated to phycoerythrin (PE), clone MWReg30 (BD Biosciences, San Jose, CA), streptavidin conjugated to allophycocyanin (APC) (BD Biosciences).

5. Aster Jandl (13) citrate-based anticoagulant (AJ): 85 mM sodium citrate dihydrate, 69 mM citric acid, 20 mg/ml glucose, pH 4.6.

6. Platelet wash buffer: 140 mM NaCl, 5 mM KCl, 12 mM sodium citrate, 10 mM glucose, 12.5 mM sucrose, pH 6.0.

7. Platelet buffer: 10 mM HEPES, 140 mM NaCl, 3 mM KCl, 0.5 mM MgCl$_2$ hexahydrate, 0.5 mM NaHCO$_3$, 10 mM glucose, pH 7.4. Titrated with NaOH.

8. Needles 25 gauge 5/8 in. (BD Biosciences), syringes 1 ml (Terumo, Tokyo, Japan).

9. Heat lamp.

10. Flow cytometer (e.g., FACSCalibur, BD Biosciences, or similar).

2.1.2. Transfusion of Biotinylated Mouse Platelets (Adoptive Platelet Transfer)

1. Sphero blank calibration particles (3.5–4.0 μm) (Spherotech Inc., Libertyville, IL).
2. See in vivo biotinylation of platelets (Subheading 2.1.1, items 1–10).

2.1.3. In Vivo Platelet Double Labeling (Cohort and Population Label)

1. X488, a DyLight488-labeled rat IgG derivate (Emfret Analytics, Eibelstadt, Germany) 0.2 mg/ml.
2. See in vivo biotinylation of platelets (Subheading 2.1.1, items 1–10).

2.1.4. Reticulated Platelet Labeling

1. Heparinized capillaries and EDTA-coated tubes (Microvette, Sarstedt, Numbrecht, Germany) for blood collection.
2. Thiazole orange (TO) powder (Sigma Aldrich): Dissolved in methanol at 1 mg/ml and stored in aliquots at −20°C in the dark.
3. Antibody: Rat anti-mouse CD41 conjugated to PE, clone MWReg30 (BD Biosciences).
4. 1% paraformaldehyde (PFA) in phosphate-buffered saline (PBS).

2.2. Hallmarks of Platelet Apoptosis

2.2.1. Flow Cytometric Analysis of Phosphatidylserine Externalization

1. DMSO (Sigma Aldrich).
2. ABT-737 (Abbott Laboratories, Abbott Park, IL): Dissolved in DMSO at 10 mM.
3. Aster Jandl anticoagulant (AJ): 85 mM sodium citrate dihydrate, 69 mM citric acid, 20 mg/ml glucose, pH 4.6.
4. Platelet wash buffer: 140 mM NaCl, 5 mM KCl, 12 mM sodium citrate, 10 mM glucose, 12.5 mM sucrose, pH 6.0.
5. Platelet buffer: 10 mM HEPES, 140 mM NaCl, 3 mM KCl, 0.5 mM $MgCl_2$ hexahydrate, 0.5 mM $NaHCO_3$, 10 mM glucose, pH 7.4. Titrated with NaOH.
6. Needles 25 gauge 5/8 in. (BD Biosciences), syringes 1 ml (Terumo).
7. Sphero blank calibration particles (3.5–4.0 μm) (Spherotech).
8. Flow cytometer (FACSCalibur, BD Biosciences).
9. Fluorophore-conjugated Annexin-V protein (BD Biosciences).
10. Calcium ionophore A23187 hemicalcium salt powder (Sigma Aldrich): Dissolved in DMSO at 10 mM.
11. Annexin-V buffer (Invitrogen, Carlsbad, CA).

2.2.2. Analysis of Caspase Activity: Determining Caspase Expression and Processing by Western Blot

1. Materials as described (Subheading 2.2.1, items 1–8).
2. Quinoline-Val-Asp-CH_2-O-Ph (Q-VD-OPh) powder (Alexis, Lausen, Switzerland): dissolved in DMSO at 25 mM.
3. NP40 3× lysis buffer: 3% nonidet P40, 450 mM NaCl, 150 mM Tris–HCl, pH 7.4 (14).

4. Complete protease inhibitor (Roche, Basel, Switzerland).

5. Ethylene glycol-bis(β-aminoethyl ether)- N, N, N', N'-tetraacetic acid (EGTA) tetrasodium salt (200 mM).

6. Precast 4–12% Bis–Tris polyacrylamide gel (Invitrogen).

7. Modified Laemmli (15): 167 mM Tris, 26.7% (v/v) glycerol, 2% (w/v) SDS, 20% (v/v) 2-mercaptoethanol, 26.8 μg/ml bromophenol.

8. 2-N-morpholino ethanesulfonic acid (MES) buffer (Invitrogen).

9. Polyvinylidene fluoride (PVDF) membrane (Micron Separation, Westborough, MA).

10. Kaleidoscope precision protein standard (Bio-Rad, Hercules, CA).

11. Gel running cell unit (Bio-Rad).

12. 10× NuPage transfer buffer (Invitrogen).

13. Methanol (MeOH).

14. Gel transfer tank (Bio-Rad).

15. T-TBS blocking buffer: 1% BSA (w/v), 137 mM NaCl, 20 mM Tris, 0.2–0.5% (v/v) Tween 20, pH 7.4. Titrated with HCl.

16. T-TBS buffer without BSA: 137 mM NaCl, 20 mM Tris, 0.2–0.5% (v/v) Tween 20, pH 7.4. Titrated with HCl.

17. Primary antibody.

18. Secondary antibody (horseradish peroxidase conjugated).

19. Chemiluminescence detection system reagents (Millipore, Billerica, MA).

20. X-ray film cassette.

2.2.3. Analysis of Caspase Activity: Luminescent Assay for Quantifying Caspase Activity

1. Materials as described in Subheading 2.2.1, items 1–8.

2. Quinoline-Val-Asp-CH$_2$-O-Ph (Q-VD-OPh) powder (Alexis): Dissolved in DMSO at 25 mM.

3. Caspase-Glo-3/7, Caspase-Glo-9, Caspase-Glo-8 (Promega, Madison, WI).

4. 96-well U-bottom Maxisorp plate (Nunc, Rochester, NY).

5. LumiSTAR Galaxy luminometer (BMG Labtech, Offenburg, Germany).

2.2.4. Luminescent Assay for Assessing Mitochondrial Function

1. Materials as described in Subheading 2.2.1, items 1–8.

2. CellTiter-Glo (Promega).

3. 96-well U-bottom Maxisorp plate (Nunc).

4. LumiSTAR Galaxy luminometer (BMG Labtech).

3. Methods

3.1. Platelet Life Span Analysis

In the early days, platelet life span measurements were made using radioisotopes, and today this method is still being used to a small degree in humans. More recent methods to determine platelet life span in animals involve nonradioactive compounds, such as biotin (16) and fluorescent dyes or Ig derivates (17–21). Fluorescent dyes became popular from 1996 for ex vivo platelet labeling before transfusion. We have also developed a new method of double labeling platelets that allows us to simultaneously follow the survival of both the entire population of platelets and a young cohort born within a 24 h period (22). This involves injecting a fluorescent label specific for platelets (developed by Emfret, Germany), followed by the standard technique of in vivo biotinylation 24 h later.

3.1.1. In Vivo Biotinylation of Mouse Platelets (2, 16)

1. Dissolve NHS-biotin powder (new batch) in DMSO to give a stock of 30 mg/ml.

2. Dilute the fresh NHS-biotin stock 1/10 in 145 mM NaCl at room temperature (RT). Ensure that there are no precipitates (see Note 1).

3. Inject 200 μl of NHS-biotin in 145 mM NaCl into the tail vein of each mouse (600 μg NHS-biotin). We typically analyze 6–8 adult mice, between 8 and 12 weeks of age, from each genotype. All animal work should be conducted in accordance with local and national guidelines and under appropriate ethical approval.

4. Bleed mice from the tail vein 2 h after injection. Dilate tail veins under heat lamp and take 5 μl blood into 125 μl anticoagulant mix (25 μl AJ, 100 μl platelet buffer). Mix gently (see Note 2).

5. Centrifuge samples at $125 \times g$ at RT, and collect 100 μl from the supernatant. This is the platelet fraction. Discard the red cell pellet.

6. Mix CD41-PE antibody and streptavidin-APC (both 1/100 dilution) in platelet buffer. Add 100 μl of this mixture to 100 μl of platelets and incubate for 40 min at RT in the dark.

7. Add 500 μl of platelet wash buffer to each sample and spin for 6 min at $860 \times g$. Discard supernatant and resuspend the platelet pellet (invisible) in 300 μl platelet buffer. Samples are now ready for flow cytometry.

8. Acquire 30,000–50,000 events in the platelet gate on a FACSCalibur. The platelet population can be identified by forward and side scatter. Platelets are positive for CD41-PE (FL2).

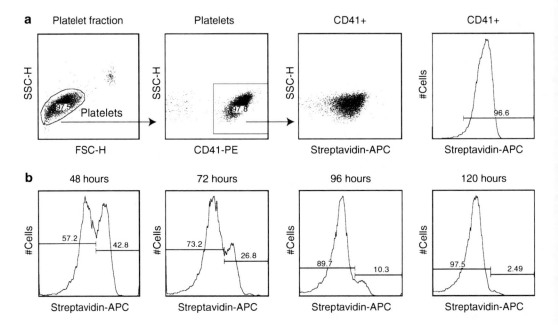

Fig. 1. In vivo biotinylation of platelets. (**a**) Gating of the platelet population from the prepared platelet fraction by forward and side scatter. Furthermore, platelets are identified as CD41 positive (CD41+), and the percent biotinylated platelets (streptavidin positive) are determined 2 h after biotin injection. (**b**) An example of the percent biotinylated CD41+ platelets 48–120 h postbiotin injection. Note the two peaks in the histogram (negative to the left and positive to the right).

Determine the percent platelets positive for biotin (streptavidin-APC positive) in FL4. See Fig. 1 for more details.

9. Repeat bleeds daily for a further 5 days to provide data of percent labeled platelets at 24, 48, 72, 96, and 120 h in a wild type mouse.

3.1.2. Transfusion of Biotinylated Mouse Platelets (Adoptive Platelet Transfer) (2, 16, 19)

1. See in vivo biotinylation in Subheading 3.1.1, steps 1–3.

2. 2 h after injection, heart-bleed mice into 0.1 volume of Aster Jandl anticoagulant. We typically recover ~ 1 ml of blood per mouse. Transfer the blood: anticoagulant mixture into an Eppendorf tube. To avoid shear stress, remove needle when expelling blood into the tube.

3. Obtain diluted platelet rich plasma (PRP) by centrifuging blood at $125 \times g$ for 8 min (RT) (see Note 3). Take the top layer plus ~ 200 µl red cells. Add 300 µl platelet wash buffer and spin again at $125 \times g$.

4. Wash the platelets by taking the top layer (transparent, slightly cloudy), adding 300 µl wash buffer, and spinning at $860 \times g$, 5 min. Resuspend the pellet in 500 µl wash buffer, and centrifuge again at $860 \times g$, 5 min.

5. Resuspend the pellet in a small volume of platelet buffer. Pool platelets from mice with the same genotype, and determine platelet counts using Sphero blank calibration particles on an

FACSCalibur flow cytometer. By flow cytometry you will also be able to tell if platelets are resting (nonactivated) by forward and side scatter. In addition, by eye, a solution of resting platelets will appear "swirly" (see Note 4).

6. Adjust platelet counts with platelet buffer or PBS and inject 200 µl intravenously per mouse. Assuming normal posttransfusion recovery, injection of 2×10^8 platelets/ml in 200 µl (a total of 4×10^7 cells) will result in a labeled platelet fraction of approximately 3–5% in the recipient mouse. Inject the same number of platelets if comparing different genotypes or platelet treatments. Higher percentages of transfused platelets can be achieved by using 5×10^8–1×10^9 platelets/ml in 200 µl.

7. See in vivo biotinylation, Subheading 3.1.1 steps 4–9. In order to determine the transfusion recovery, the initial bleed can be done earlier than 2 h.

3.1.3. In Vivo Platelet Double Labeling (Cohort and Population Label) (2, 16, 19, 21, 22)

1. Inject mice intravenously with 0.15 µg (0.75 µl) X488 per gram body weight in a suitable volume (50–200 µl) of PBS.

2. Check the labeling efficiency by flow cytometry. X488 will label ~90% of the platelets within minutes postinjection. X488 is visualized in FL1 on the FACSCalibur. Take 5 µl blood from the tail vein into 125 µl anticoagulant mix (25 µl AJ, 100 µl platelet buffer). Dilute whole blood further before flow cytometry or separate the PRP and run immediately on FACS. Use CD41-PE positivity to confirm the platelet gate.

3. 24 h after X488 injection, inject the same mice intravenously with 600 µg NHS-biotin. Follow the instructions given in platelet in vivo biotinylation in Subheading 3.1.1, steps 1–5.

4. Mix CD41-PE antibody and streptavidin-APC (both 1/100 dilution) in platelet buffer. Add 100 µl of this mixture to 100 µl of platelets and incubate for 40 min at RT in the dark.

5. Add 500 µl of platelet wash buffer to each sample and spin for 6 min at $860 \times g$. Discard the supernatant and resuspend the platelet pellet (invisible) in 300 µl platelet buffer. The samples are then ready for flow cytometry.

6. Acquire 30,000–50,000 events in the platelet gate on a FACSCalibur. Platelets can be identified by forward and side scatter and positivity for CD41-PE (FL2). Thereafter the percentage of platelets positive for X488 (FL1) and streptavidin-APC (FL4) is determined. The "population label" includes those positive for X488. The "cohort label" (newly produced platelets) comprises those positive for streptavidin-APC and negative for X488.

3.1.4. Reticulated Platelet Labeling (2, 23)

Young platelets are also referred to as reticulated platelets, due to their RNA content. It can be useful to determine the age profile of circulating platelets by assessing the percentage of reticulated platelets in different mouse models and disease states, as it can provide information about production and or clearance defects. The following protocol fluorescently labels reticulated platelets using thiazole orange.

1. Obtain anti-coagulated mouse blood by retro-orbital bleeding through heparinized capillaries into EDTA coated tubes (see Note 5).

2. Prepare TO in PBS by taking a TO aliquot from the freezer and diluting it to 0.1 μg/ml in PBS.

3. Prepare antibody solution by diluting CD41-PE 1/40 in PBS.

4. Mix 50 μl TO in PBS, 9 μl antibody solution, and 1 μl blood. Pipette gently, and incubate for 15 min at room temperature. Avoid light exposure.

5. Add 1 ml 1% PFA in PBS to each sample. Do not breathe in PFA, as it is toxic. Cover with foil and place samples on ice.

6. Run samples immediately on a FACSCalibur. Gate for platelets on FSC vs. CD41-PE (FL2).

7. TO staining is seen in FL1. There should be two peaks in the histogram. Reticulated platelets are TO+ while mature platelets are TO−.

8. Determine the percentage of platelets that are TO+.

3.2. Hallmarks of Platelet Apoptosis

3.2.1. Flow Cytometric Analysis of Phosphatidylserine Externalization

1. Obtain washed platelets as described in Subheading 3.1.2, steps 3–4 and resuspend the pellet in a small volume of platelet buffer.

2. Determine platelet count using Sphero blank calibration particles by flow cytometry, according to the manufacturer's specifications, and adjust the count with platelet buffer to 1×10^8 platelets/ml.

3. To induce Bak/Bax-mediated intrinsic apoptosis, incubate 100 μl of platelet preparation with 1 μM ABT-737 in 1% DMSO (v/v) at 37°C for 90 min. As a positive control for phosphatidylserine externalization, incubate 100 μl of platelet preparation with 1 μM calcium ionophore A23187 at room temperature for 15 min.

4. Following incubation, transfer 10 μl of platelet preparation to a fresh tube containing 100 μl of fluorophore-conjugated Annexin-V protein [1/50 (v/v) in Annexin-V buffer] and incubate in the dark at room temperature for 20 min.

5. Dilute samples by adding 600 μl of Annexin-V buffer and analyze immediately by flow cytometry (see Fig. 2 for more details).

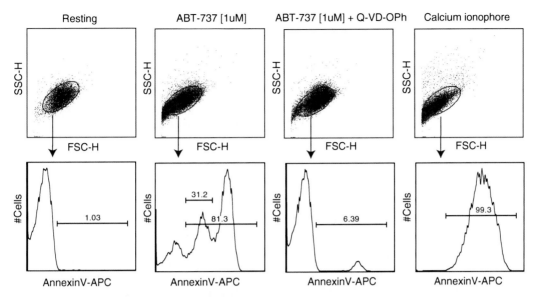

Fig. 2. Flow cytometric analysis of platelet phosphatidylserine exposure. Platelets were treated with ABT-737 (1 μM), calcium ionophore A23187 (1 μM), or left untreated (resting) followed by incubation with fluorescently conjugated Annexin-V protein. The platelet population was gated and the percentage of platelets with externalized phosphatidylserine quantified by flow cytometry. Preincubation with the caspase inhibitor Q-VD-OPh (100 μM) was included as a negative control.

3.2.2. Analysis of Caspase Activity: Determining Caspase Expression and Processing by Western Blot

These instructions are easily adapted to other formats and with antibodies preferred by the experimentalist.

1. Obtain washed platelets as described (Subheading 3.1.2, steps 3–4) and resuspend the pellet in a small volume of platelet buffer.

2. Determine platelet counts and adjust platelet count as described (Subheading 3.2.1, step 2) but make up platelets to $5 \times 10^8 - 1 \times 10^9$ cells/ml.

3. As a negative control for caspase activity, pre-incubate a sample of the platelet preparation with the cell-permeable caspase inhibitor Q-VD-OPh (100 μM) for 30 min at room temperature.

4. To activate caspases via the induction of intrinsic apoptosis, incubate a sample of the platelet preparation as described in Subheading 3.2.1, step 3.

5. During incubation, dissolve one complete protease inhibitor tablet in 2 ml distilled water (to make 25×). Prepare the lysis reagent by mixing 5 ml NP40 lysis buffer with 200 μl of the protease inhibitor solution and 75 μl of EGTA stock.

6. Mix this lysis reagent with platelet samples (one volume of platelet sample to two volumes of lysis reagent) and transfer to ice. Remove insoluble fraction by centrifugation at $1,400 \times g$ for 10 min and immediately transfer to ice or store at –80°C. Protein concentration may then be determined by Bio-Rad Protein Assay according to manufacturers specifications (we recommend

a total platelet protein abundance for Western blot analysis of 30–60 μg per lane).

7. Add Laemmli, containing reducing reagent, to the platelet protein samples (one volume of Laemmli to 3 volumes of platelets) and incubate at 100°C for 5 min.

8. Carefully remove the comb from a precast 4–12% Bis–Tris polyacrylamide gel and rinse wells with MES buffer to remove excess gel fragments.

9. Add MES buffer to the chambers of the gel unit and load 30–60 μg of total protein sample to each well. Include one well for the protein standard.

10. Connect the gel unit to a power supply and run at ~100–150 V until the blue lines have moved to the bottom of the gel (~2 h).

11. While the gel is running, prepare 1 L of 1× NuPage transfer buffer by mixing 100 ml of 10× NuPage transfer buffer with 200 ml MeOH and 700 ml distilled water (additional 1× NuPage transfer buffer will be required if running multiple gels).

12. After the platelet protein samples have been separated by SDS-PAGE, transfer to a membrane by electrophoresis. This method assumes the use of a Bio-Rad transfer tank. Cut a piece of PVDF membrane just larger than the gel and lay on the surface of the gel, surrounded by filter paper and sponges in a transfer tank cassette, to allow transfer of the protein in the samples to the membrane.

13. Presoak filter papers and soft white sponges in 1× NuPage transfer buffer.

14. Activate PVDF membrane by presoaking in MeOH for at least 5 min.

15. Assemble the gel transfer tank and place the cassette inside the tank, orientated in such a way that the PVDF membrane is between the gel and the anode.

16. Connect to the power supply and run at ~30–100 V for 1.5–2 h to allow transfer.

17. When transfer is complete, remove the cassette and disassemble the apparatus. The protein standard should be clearly visible on the membrane.

18. Incubate the membrane in 5 ml T-TBS blocking buffer for 1 h at room temperature or at 4°C overnight on a rocking platform.

19. Add primary antibody, at the desired concentration, in T-TBS blocking buffer and incubate overnight at 4°C on a rocking platform.

20. Discard the primary antibody and wash the membrane three times for 30 min each with T-TBS (without BSA), discarding T-TBS between each wash.

21. Add the secondary antibody [1/5,000–1/10,000 (v/v)] in T-TBS blocking buffer and incubate for 1 h at room temperature on a rocking platform.

22. Discard the secondary antibody and wash the membrane three times for 30 min each with T-TBS buffer without BSA, discarding the T-TBS between each wash and after the final wash.

23. Mix chemiluminscent substrate reagents together and immediately add to the membrane for 3 min.

24. Remove the membrane from the substrate reagents and place between sheets of clear plastic film.

25. Place the clear plastic films, containing the membrane, in an X-ray film cassette.

26. In a darkroom, place an X-ray film with the membrane in the cassette for an appropriate exposure time (typically ranging from seconds to minutes).

3.2.3. Analysis of Caspase Activity: Luminescent Assay for Quantifying Caspase Activity

1. Obtain washed platelets as described in Subheading 3.1.2, steps 3–4 and resuspend the pellet in a small volume of platelet buffer.

2. Adjust platelet count as described in Subheading 3.2.1, step 2.

3. For the induction of intrinsic apoptosis, incubate platelet preparation as described in Subheading 3.2.1, step 3. To assess basal caspase activity in untreated samples, include a negative control for caspase activity as described in Subheading 3.2.2, step 3.

4. Following incubations, transfer 50 µl of platelet preparation to each well of a 96-well Maxisorp plate.

5. Add an equal volume (50 µl) of Caspase-Glo reagent, previously reconstituted according to the manufacturer's specifications, and allow reaction to proceed at room temperature for 30 min.

6. Immediately measure light signal output, proportional to the caspase enzymatic activity, with the LumiSTAR Galaxy luminometer.

3.2.4. Luminescent Assay for Assessing Mitochondrial Function

1. Obtain washed platelets as described in Subheading 3.1.2, steps 3–4 and resuspend the pellet in a small volume of platelet buffer.

2. Adjust platelet count as described in Subheading 3.2.1, step 2.

3. To induce intrinsic apoptosis, incubate platelet preparation as described in Subheading 3.2.1, step 3.

4. Following incubations, transfer 50 µl of platelet preparation to each well of a 96-well Maxisorp plate.

5. Add an equal volume (50 µl) of CellTiter-Glo reagent, previously reconstituted according to the manufacturer's specifications, and allow reaction to proceed at room temperature for 30 min.

6. Immediately measure light signal output, proportional to total cellular ATP content, with the LumiSTAR Galaxy luminometer.

4. Notes

1. In order to avoid precipitates when diluting NHS-biotin/ DMSO solution in saline, take a small volume of the biotin solution into the full volume of saline and mix. Repeat until all the biotin is in suspension.

2. Bring all solutions to RT before use. Make sure that solutions are sterile. To minimize shear stress to the platelets, avoid excessive pipeting.

3. In order to maximize in vivo recovery after transfusion, it is crucial to move quickly with the platelet preparation and not to have the platelets sitting on the bench for any longer than absolutely necessary.

4. This step also allows the user to assess the biotin platelet labeling efficiency before the sample is used for transfusion. Stain a small aliquot of platelets with CD41-PE and streptavidin-APC, and analyze by flow cyometry. Include a negative control.

5. When investigating the percentage of reticulated platelets, it may be useful to include some blood from a mouse with an increased proportion of reticulated platelets as a positive control. For example, a mouse treated with anti-platelet serum (APS) will incur severe thrombocytopenia followed by recovery and rebound thrombocytosis. During the recovery phase (we typically take blood 3 days after APS treatment), a very high proportion of the platelet population will be reticulated. A negative control can be obtained by treating the platelets with RNAse (23).

Acknowledgments

We thank Chloé James, Katya Henley, Philip Hodgkin, David Huang, Marina Carpinelli, and Kylie Mason for helpful discussions and collaborative input. This work was supported by Project Grants (516725, 575535), Fellowships (M.R.D., B.T.K.), and an Independent Research Institutes Infrastructure Support Scheme Grant (361646) from the Australian National Health and Medical Research Council, a Fellowship from the Sylvia and Charles Viertel Charitable Foundation (B.T.K.), a Fellowship from the Leukemia & Lymphoma Society (E.C.J), a Fellowship from the Leukemia Foundation of Australia (M.J.W.), and a Victorian State Government Operational Infrastructure Support grant.

References

1. Youle RJ, Strasser A (2008) The BCL-2 protein family: opposing activities that mediate cell death. Nat Rev Mol Cell Biol 9: 47–59.

2. Mason KD, Carpinelli MR, Fletcher JI, Collinge JE, Hilton AA, et al. (2007) Programmed anuclear cell death delimits platelet life span. Cell 128: 1173–1186.

3. Oltersdorf T, Elmore SW, Shoemaker AR, Armstrong RC, Augeri DJ, et al. (2005) An inhibitor of Bcl-2 family proteins induces regression of solid tumours. Nature 435: 677–681.

4. Zhang H, Nimmer PM, Tahir SK, Chen J, Fryer RM, et al. (2007) Bcl-2 family proteins are essential for platelet survival. Cell Death Differ 14: 943–951.

5. Miyanishi M, Tada K, Koike M, Uchiyama Y, Kitamura T, et al. (2007) Identification of Tim4 as a phosphatidylserine receptor. Nature 450: 435–439.

6. Park SY, Jung MY, Kim HJ, Lee SJ, Kim SY, et al. (2008) Rapid cell corpse clearance by stabilin-2, a membrane phosphatidylserine receptor. Cell Death Differ 15: 192–201.

7. Wang X, Wu YC, Fadok VA, Lee MC, Gengyo-Ando K, et al. (2003) Cell corpse engulfment mediated by C. elegans phosphatidylserine receptor through CED-5 and CED-12. Science 302: 1563–1566.

8. Dachary-Prigent J, Freyssinet JM, Pasquet JM, Carron JC, Nurden AT (1993) Annexin V as a probe of aminophospholipid exposure and platelet membrane vesiculation: a flow cytometry study showing a role for free sulfhydryl groups. Blood 81: 2554–2565.

9. Schoenwaelder SM, Yuan Y, Josefsson EC, White MJ, Yao Y, et al. (2009) Two distinct pathways regulate platelet phosphatidylserine exposure and procoagulant function. Blood 114: 663–666.

10. Jobe SM, Wilson KM, Leo L, Raimondi A, Molkentin JD, et al. (2008) Critical role for the mitochondrial permeability transition pore and cyclophilin D in platelet activation and thrombosis. Blood 111: 1257–1265.

11. Shcherbina A, Remold-O'Donnell E (1999) Role of caspase in a subset of human platelet activation responses. Blood 93: 4222–4231.

12. Leytin V, Allen DJ, Mykhaylov S, Lyubimov E, Freedman J (2006) Thrombin-triggered platelet apoptosis. J Thromb Haemost 4: 2656–2663.

13. Aster RH, Jandl JH (1964) PLATELET SEQUESTRATION IN MAN. I. METHODS. J Clin Invest 43: 843–855.

14. Josefsson EC, Gebhard HH, Stossel TP, Hartwig JH, Hoffmeister KM (2005) The macrophage alphaMbeta2 integrin alphaM lectin domain mediates the phagocytosis of chilled platelets. J Biol Chem 280: 18025–18032.

15. Laemmli UK (1970) Cleavage of structural proteins during the assembly of the head of bacteriophage T4. Nature 227: 680–685.

16. Heilmann E, Friese P, Anderson S, George JN, Hanson SR, et al. (1993) Biotinylated platelets: a new approach to the measurement of platelet life span. Br J Haematol 85: 729–735.

17. Michelson AD, Barnard MR, Hechtman HB, MacGregor H, Connolly RJ, et al. (1996) In vivo tracking of platelets: circulating degranulated platelets rapidly lose surface P-selectin but continue to circulate and function. Proc Natl Acad Sci USA 93: 11877–11882.

18. Baker GR, Sullam PM, Levin J (1997) A simple, fluorescent method to internally label platelets suitable for physiological measurements. Am J Hematol 56: 17–25.

19. Hoffmeister KM, Felbinger TW, Falet H, Denis CV, Bergmeier W, et al. (2003) The clearance mechanism of chilled blood platelets. Cell 112: 87–97.

20. Bergmeier W, Piffath CL, Cheng G, Dole VS, Zhang Y, et al. (2004) Tumor necrosis factor-alpha-converting enzyme (ADAM17) mediates GPIbalpha shedding from platelets in vitro and in vivo. Circ Res 95: 677–683.

21. Pleines I, Eckly A, Elvers M, Hagedorn I, Eliautou S, et al. (2010) Multiple alterations of platelet functions dominated by increased secretion in mice lacking Cdc42 in platelets. Blood 115: 3364–3373.

22. Dowling MR, Josefsson EC, Henley KJ, Hodgkin PD, Kile BT (2010) Platelet senescence is regulated by an internal timer, not damage inflicted by hits. Blood 116: 1776–1778.

23. Matic GB, Chapman ES, Zaiss M, Rothe G, Schmitz G (1998) Whole blood analysis of reticulated platelets: improvements of detection and assay stability. Cytometry 34: 229–234.

Chapter 6

A Live Cell Micro-imaging Technique to Examine Platelet Calcium Signaling Dynamics Under Blood Flow

Warwick S. Nesbitt, Ian S. Harper, Simone M. Schoenwaelder, Yuping Yuan, and Shaun P. Jackson

Abstract

The platelet is a specialized adhesive cell that plays a key role in thrombus formation under both physiological and pathological blood flow conditions. Platelet adhesion and activation are dynamic processes associated with rapid morphological and functional changes, with the earliest signaling events occurring over a subsecond time-scale. The relatively small size of platelets combined with the dynamic nature of platelet adhesion under blood flow means that the investigation of platelet signaling events requires techniques with both high spatial discrimination and rapid temporal resolution. Unraveling the complex signaling processes governing platelet adhesive function under conditions of hemodynamic shear stress has been a longstanding goal in platelet research and has been greatly influenced by the development and application of microimaging-based techniques. Advances in the area of epi-fluorescence and confocal-based platelet calcium (Ca^{2+}) imaging have facilitated the in vitro and in vivo elucidation of the early signaling events regulating platelet adhesion and activation. These studies have identified distinct Ca^{2+} signaling mechanisms that serve to dynamically regulate activation of the major platelet integrin $\alpha_{IIb}\beta_3$ and associated adhesion and aggregation processes under flow. This chapter describes in detail a ratiometric calcium imaging protocol and associated troubleshooting procedures developed in our laboratory to examine live platelet Ca^{2+} signaling dynamics. This technique provides a method for high-resolution imaging of the Ca^{2+} dynamics underpinning platelet adhesion and thrombus formation under conditions of pathophysiological shear stress.

Key words: Platelets, Calcium, Shear stress, Adhesion, Imaging

1. Introduction

Ca^{2+} signaling mechanisms control a vast array of cellular functions ranging from short-term responses, such as contraction, secretion, and cellular adhesion, to long-term responses such as the regulation of cellular growth and proliferation. In the case of blood platelets, Ca^{2+} signal transduction is a key regulator and integrator of a

Jonathan M. Gibbins and Martyn P. Mahaut-Smith (eds.), *Platelets and Megakaryocytes: Volume 3, Additional Protocols and Perspectives*, Methods in Molecular Biology, vol. 788, DOI 10.1007/978-1-61779-307-3_6, © Springer Science+Business Media, LLC 2012

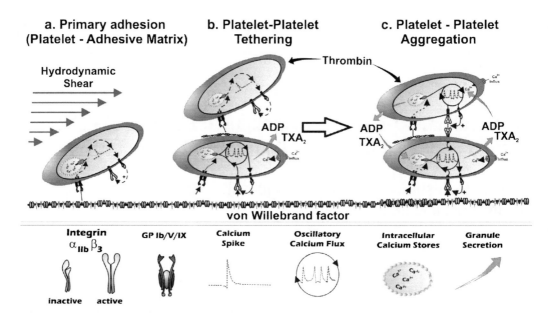

a. Primary adhesion
(Platelet - Adhesive Matrix)

b. Platelet-Platelet
Tethering

c. Platelet - Platelet
Aggregation

Fig. 1. Platelet Ca^{2+} signaling under high shear blood flow. (**a**) Primary adhesion. Initial outside-in signaling events mediated by GPIb/V/IX engagement of surface-immobilized vWF triggers the initiation of an elementary calcium event. Subsequent integrin $\alpha_{IIb}\beta_3$ engagement of the matrix initiates elevated and oscillatory calcium flux, driving further rounds of integrin activation and stationary adhesion at the vWF surface (9). (**b**) Platelet–platelet tethering. Platelets within the bulk flow interact with a primary adherent platelet via GPIb/V/IX binding to surface expressed vWF. The nucleating (adherent) platelet presents a surface bearing active integrin $\alpha_{IIb}\beta_3$ and locally secretes ADP and TXA_2.(8). (**c**) Aggregation. Integrin $\alpha_{IIb}\beta_3$ engagement of platelet-expressed vWF and/or fibrinogen triggers further ADP release and stimulates platelet calcium signaling events throughout the local platelet population. ADP engagement of $P2Y_{12}$ potentiates integrin $\alpha_{IIb}\beta_3$ activation and calcium signaling, thereby promoting platelet aggregation in the shear field (8). Thrombin generation further drives the Ca^{2+} signaling response leading to full thrombus consolidation.

diverse array of adhesion and soluble agonist-mediated activation mechanisms leading to cytoskeletal remodeling and integrin activation under blood flow (Fig. 1) (1–12). Platelets demonstrate a number of Ca^{2+} signal transduction mechanisms that are dependent on the form of stimulation (both mechanical and chemical) and the magnitude or concentration of the activating agent. As is the case for nonexcitable cells in general, activation of platelets results in a Ca^{2+} flux that comprises two key components: (1) transient release of Ca^{2+} predominantly from the dense tubular system (DTS), with a minor contribution suggested to be derived from acidic (2,5-di-(t-butyl)-1,4-hydroquinone (TBHQ) sensitive) intracellular stores, and (2) sustained influx of Ca^{2+} across the plasma membrane (13–17). Stimulation of the battery of platelet adhesion and soluble agonist receptors leads to the specific activation of various phospholipase C (PLC) isoforms (PLC β, γ) leading to the generation of both inositol-1,4,5-trisphosphate (IP_3) and 1,2-diacylglycerol (DAG) at the plasma membrane (12, 18–21). Subsequent IP_3 binding and direct activation of IP_3-receptor isoforms (IP_3-R1 and IP_3-R2) in the platelet DTS, results in transient elevations of platelet cytosolic Ca^{2+} $((Ca^{2+})_c)$ and concomitant Ca^{2+} store depletion (22, 23). In most nonexcitable cells including platelets, intracellular Ca^{2+} store depletion triggers the influx of extracellular Ca^{2+} down its concentration gradient.

While the mechanisms of intracellular Ca^{2+} release have been known for some time, the key components mediating Ca^{2+} influx have only recently started to emerge. Ca^{2+} influx associated with platelet activation can be categorized into: (1) store operated calcium entry (SOCE) mechanisms and (2) receptor-operated Ca^{2+} entry that is not dependent upon the release of Ca^{2+} from internal stores (non-SOCE mechanisms). According to the current model of SOCE, IP_3-dependent discharge of DTS Ca^{2+} stores leads to the rapid redistribution and clustering of the protein *stromal interaction molecule 1* (STIM1) in the dense tubule system(24, 25). Interaction of STIM1, either directly or indirectly, with the pore forming subunit *Ca^{2+}-release activated Ca^{2+} modulator 1* (CRACM1 aka Orai1) at the plasma membrane triggers *store operated channel* (SOC) pore formation and Ca^{2+} influx (26). In addition to this predominant SOCE mechanism, activation of *transient receptor potential channel 6* (TRPC6) by DAG and/or PIP_2 depletion, and ATP-dependent activation of the platelet purinergic receptor operated Ca^{2+} channel $P2X_1$ along with reversed mode Na^+/Ca^{2+} exchange activity at the plasma membrane (non-SOCE mechanisms) also contribute to global Ca^{2+} influx (27–31). Overall, the nature and contribution of both intracellular store release and Ca^{2+} influx pathways depend to a large extent on the nature of the initiating platelet stimulus.

Many studies that have led to the elucidation of the mechanics of platelet Ca^{2+} signaling, while informative, have been conducted under experimental conditions that do not take into consideration the effects of the mechanical shear environment on platelet signaling dynamics. Blood flow can have a significant effect on both soluble agonist and adhesion receptor-mediated platelet activation and adhesion events. In the case of soluble agonist signaling, the rapid clearance of agonists due to flow can significantly reduce overall agonist concentrations and therefore the magnitude of the platelet Ca^{2+} response. Hemodynamic shear and extensional forces can also have direct effects on adhesion receptor function and signaling. In the case of platelet-von Willebrand Factor (vWF) binding interactions, published studies from our laboratory and others suggest that downstream Ca^{2+} signaling dynamics may be sensitized to the rate of change in shear stress rather than the absolute magnitude of the applied shear (4, 32). Furthermore, this shear gradient sensitivity is critically dependent on concomitant integrin $\alpha_{IIb}\beta_3$ engagement, as blocking integrin binding blocks the responsiveness to rapid shear accelerations on vWF. More recent findings from our laboratory examining discoid platelet aggregation under localized shear microgradients, suggest that a distinct integrin-dependent Ca^{2+} signaling mechanism underpins aggregation at the growing face of developing thrombi (32).

The ability to examine platelet Ca^{2+} signaling dynamics under conditions of high or rapidly altering shear rates has been limited by the relatively small size (2 μm diameter) of platelets, detection device sensitivity and spatial resolution, and the high speed dynamic nature of hemodynamic flow. The application of microimaging-based techniques to the analysis of platelet Ca^{2+} signaling in combination with advances in flow-based technologies has greatly facilitated the investigation of the way in which Ca^{2+} flux regulates platelet function under both static (nonsheared) and flow conditions. This chapter describes in detail a dual-dye (pseudo-ratiometric) Ca^{2+} imaging protocol developed in our laboratory and addresses many of the important troubleshooting procedures necessary for the successful application of this method to flow-based studies of platelet function.

2. Materials

2.1. Blood Platelet Preparation

1. 7× Acid Citrate Dextrose (ACD) containing theophylline: 85 mM sodium citrate, 85 mM citric acid (Anhydrous), 110 mM D-glucose, 70 mM theophylline.

2. 10× Platelet Wash Buffer (PWB): 43 mM K_2HPO_4, 43 mM Na_2HPO_4, 243 mM NaH_2PO_4, 1.13 M NaCl, 55 mM D-glucose, 100 mM theophylline (see Note 1).

3. 1× PWB, pH 6.5, conductivity 13–15 ms/cm (50 ml): 10× PWB (5 ml), 0.5% BSA (2.5 ml 10% BSA), dH_2O 42.5 ml.

4. 10× Tyrode's Buffer: 120 mM $NaHCO_3$, 100 mM Hepes, 1.37 M NaCl, 27 mM KCl, 55 mM D-glucose, pH 7.2 (at room temperature) with NaOH.

5. 1× Tyrode's Buffer pH 7.2, conductivity 13–15 ms/cm.

6. Hirudin (Lepirudin) (Celgene) 50 mg/ml (800,000 IU) in ddH_2O, store at 4°C.

7. Apyrase (Grade VII, from potato (Sigma-Aldrich)): stock: 50 U/ml (ADPase activity) in Tyrode's Buffer. Store at –20°C.

8. 50 ml Falcon tubes.

9. Automated blood analyzer (Sysmex).

10. 1.0 ml syringes.

11. Clexane (enoxaparin sodium) (Sanofi Aventis) 100 mg/ml (10,000 IU) in ddH_2O, store at room temperature.

12. 1.2 ml microtubes (HydroLogix Specialty Tubes, catalog # REF 3492, Molecular BioProducts, Inc. San Diego, CA).

13. Bovine serum albumin (BSA).

14. 1.5 M probenicid (Sigma Aldrich) in dimethyl sulfoxide (DMSO), store at room temperature.

2.2. Calcium Probes and Reagents for Calibration

1. 2 mM Oregon Green 488 BAPTA-1 (OG), AM (Molecular Probes Inc.) in dimethyl sulfoxide (DMSO), store at −20°C.

2. 5 mM Fura Red (FR), AM (Molecular Probes Inc.) in DMSO, store at −20°C.

3. 50 mM 5,5-dimethyl-BAPTA (DM-BAPTA), AM (Molecular Probes), in DMSO, store at −20°C.

4. 5 mM 4-Bromo A23187 (Molecular Probes) in DMSO, store at −20°C.

5. 1 M CaCl$_2$ in unsupplemented Tyrode's buffer pH 7.2, store at room temperature.

6. 0.5 M EGTA in ddH$_2$O pH 8 at 37°C with KOH (1:3 w/w).

2.3. In Vitro Flow System

1. The reagents and equipment required to setup the in vitro flow apparatus utilized in our laboratory have been described in detail (33, Volume 1, chapter 15).

2.4. Wide-Field Epi-Fluorescence Imaging Setup

Standard Leica DMIRBE and/or Olympus IX81 invert microscopes are the instruments of choice for platelet calcium imaging experiments within our laboratory; however, any laboratory grade inverted microscope ideally fitted with differential interference contrast (DIC) optics and high magnification, high numerical aperture, plan, apochromatic (PlanApo) objectives can be utilized.

1. Objectives: Immersion objective lenses are specifically selected, as they provide both higher resolution as well as superior light capture for the same magnification compared to dry lenses. The 40×/1.30 PlanFL Apo (Oil), 60×/1.35 PlanS APO (Oil), 63×/1.20 W CORR Plan APO (water immersion) (Leica Confocal), and 100×/1.4 PlanS APO (Oil) have all been utilized in our laboratory for Ca^{2+} imaging experiments.

2. Alternate Objectives: Alternatively, we have also successfully utilized very high total internal reflection fluorescence (TIRF) quality objectives, which have NAs typically larger than 1.4: the PLAPO 60XOTIRFM/1.45 (oil) and APO100× OHR/1.65 (1.75 immersion liquid, Cargill) objectives have been used in the course of our Ca^{2+} imaging experiments with exceptional results. These objectives, although specialized for TIRF microscopy applications, are particularly useful for in vitro platelet flow experiments, where dye loading concentrations have to be minimized and the associated fluorescence emission is low and where high resolution platelet images are a prerequisite.

3. Excitation Source: Lambda DG-4 ultra high speed wavelength switcher fitted with a 175 W, prealigned, Xenon arc lamp with integrated shutter (Sutter Instruments). For dual-dye platelet Ca^{2+} imaging experiments with Oregon Green BAPTA-1 and Fura Red, the DG4 should be fitted with S484/15 excitation filter from the Chroma Technology Corp. 86013 filter set.

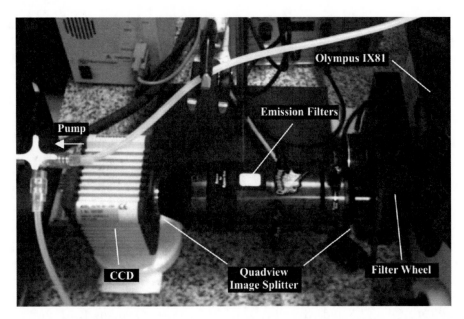

Fig. 2. Emission splitter to CCD layout. Overview of the imaging splitter (Quadview), Filter wheel to CCD (Orca–ER) arrangement attached to an IX81 epifluorescence microscope.

4. Dichroic Mirror : For dual dye ratiometric imaging using Oregon Green BAPTA-1 and Fura Red (FR) combinations, the IX81 microscope is fitted with a z488/543 or 86013bs dichroic mirror (Chroma & Olympus).

5. In the case of high speed Ca^{2+} imaging, such as that associated with high shear rate ($\geq$1,800 per s) platelet flow experiments, it is advisable to utilize an image field splitting system that divides the emission output onto the same camera CCD rather than the filter wheel assembly. We utilize the Quadview (Optical Insights) simultaneous imaging system fitted with ET520/40m, ET605/55 m, D700/10 m emission filters for Oregon Green BAPTA-1 and Fura Red and DIC emissions, respectively. In our case, the Quadview is fitted between a Sutter Instruments Lambda 10–2 optical filter changer (wheel) and the CCD allowing for greater experimental flexibility (Fig. 2).

6. CCD Camera: Although most microscopy grade cooled CCD cameras are acceptable, we have opted for the Orca-ER CCD (Hamamatsu Photonics) operating in low light mode for wide-field dual dye calcium imaging experiments. The low dark current and high sensitivity of this CCD have proven to be particularly useful for platelet calcium experiments in our hands. However, in the case of very low dye loading concentrations or where the available light budget is limiting, the use of a high gain electron multiplied EM-CCD camera is advisable and we have had good success using the QuantEM: 512SC EM-CCD (Photometrics).

2.5. Confocal Imaging Setup

For both in vitro and in vivo Ca^{2+} imaging experiments, we have utilized a Leica spectral confocal system with an electronically controlled inverted microscope (Leica SP with DM IRBE). Excitation of both Oregon Green BAPTA-1 and Fura Red is achieved via a 488 nm Argon ion laser. The pinhole is opened to maximum (3.0) to maximize light collection: confocal sectioning is not a priority for these experiments. Oregon Green BAPTA-1 and Fura Red emissions are detected using photomultipliers 1 (green emission) and 2 (orange/red emission) for dual channel imaging. Based on empirical observations of maximal and minimal fluorescence outputs of the respective dyes the PMT gain settings are set and not readjusted for all subsequent imaging: For example, on our system, we routinely use PMT gain settings (detector voltage) of 661 and 788 where the range is 100–1,000, and a PMT offset of 0 in both channels, respectively, for the two dyes. The confocal is equipped with an acousto optical tunable filter (AOTF) which regulates the intensity of the incoming laser beams, and this is routinely set to 50% to minimize photo-damage and image saturation which is especially important when imaging the maximal response of Oregon Green BAPTA-1 loaded platelets. The image (optical field) zoom is set to 1.3×. Imaging is carried out in XYT mode with speed of acquisition set to Medium 2.

2.6. Environmental Flow Hood

To ensure that experiments are performed at a physiological temperature (37°C), we use a custom-built microscope heater (model TX, Scientific Concepts, Victoria, Australia). A custom-made clear Perspex hood is also fitted around the microscope, allowing efficient heating of the flow chamber and blood/platelet samples. The air temperature inside the Perspex chamber is regulated by a precision temperature probe placed inside the hood at a point close to the glass perfusion chamber.

3. Methods

3.1. Collection of Whole Human Blood

1. Collect blood into ACD (at a ratio of 6/1, blood/ACD) via venesection of the antecubital vein of consenting healthy volunteers who have not ingested any antiplatelet medication (e.g., aspirin or ibuprofen) in the 2 weeks preceding the procedure. Use of butterfly needles smaller than 19-gauge is not recommended, as high shear forces are generated when drawing through smaller-sized needles. Withdraw blood at a reasonably slow and steady rate, taking care to avoid frothing, and immediately mix by gentle inversions of the syringe.

2. To avoid the effects of trace amounts of thrombin generated during venepuncture, discard the first 5 ml of blood taken.

3. Following collection, place the blood sample in a Falcon tube containing hirudin (lepirudin) to a final concentration of 800 U/ml and 0.005 U/ml apyrase, and allow to rest at 37°C for 15 min prior to the preparation of washed platelets (see Note 2).

3.2. Preparation of Washed Human Platelets

1. Centrifuge the equilibrated blood sample at $300 \times g$ for a time dependent on the volume of blood taken (Table 1). The braking phase of the centrifugation step should be set to the slowest rate of deceleration to prevent mixing of the erythrocyte phase with the platelet rich plasma (PRP) phase.

2. Transfer the PRP to a fresh Falcon tube (use of a syringe with slow suction to avoid shearing is recommended), and allow to rest for 10 min at 37°C in a water bath.

3. Centrifuge the PRP at $1,700 \times g$ for 7 min to separate the platelet pellet from the platelet poor plasma (PPP) fraction.

4. Gently aspirate all of the PPP as quickly as possible without disturbance of the platelet pellet.

5. Immediately resuspend the resulting platelet pellet via gentle aspiration with 1× PWB + hirudin (800 U/ml final) and apyrase (0.01 U/ml) to a volume equal to the original volume of PRP.

6. Assess the platelet count and level of erythrocyte contamination via an automated blood analyzer (Sysmex) and correct the count to 3×10^8/ml via removal or addition of 1× PWB (see Note 3).

7. Allow the platelet suspension to equilibrate at 37°C for 10 min in the presence of apyrase (0.01 U/ml) prior to Ca^{2+} dye loading.

3.3. Preparation of Washed Murine Platelets

1. Withdraw murine blood into a 1.0 ml syringe containing 100 µl of clexane (40 U/ml). Immediately after blood collection, add a volume of 7× ACD to obtain a final concentration of 1× ACD.

2. Carefully mix by inversion and allow to rest at 37°C for 10 min.

3. Carefully mix the blood with 300 µl of PWB containing 20 U/ml clexane, 0.02 U/ml apyrase, and 5 mg/ml BSA, and transfer to 1.2 ml microtubes.

Table 1
Blood volume to centrifugation speed (× g)

Blood volume (ml)		
50 ml tubes	**Centrifugation speed (× g)**	**Time (min)**
50	300	16
40	300	15
25	300	13

4. Centrifuge at $250 \times g$ for 2 min (1 min if blood vol. <0.5 ml) and carefully remove the PRP, without disturbing the erythrocyte pellet. This is best done in increments of 100 μl with a cut yellow pipette tip.

5. Repeat steps 2–4 twice, and pool the PRP.

6. Centrifuge PRP at $2,000 \times g$ for 1 min, discard the supernatant and gently resuspend the platelet pellet in PWB containing 20 U/ml clexane, 5.0 mg/ml BSA, 0.02 U/ml apyrase, and 1.5 mM probenecid, at a platelet count of 2.5×10^8/ml.

3.4. Ca²⁺ Dye Loading of Washed Platelet Suspensions

A range of fluorescent Ca^{2+} probes can and have been utilized to investigate platelet Ca^{2+} signaling events. Empirical judgement needs to be used when choosing the appropriate dye and loading concentration based on the known biological response being studied (see Note 4). The following generic platelet loading method has been applied to human platelet isolates in our laboratory using a range of AM ester probes including: Fura 2-AM (2.5 μM), Oregon Green BAPTA-1-AM (2.5 μM) and Fluo 3-AM (2.5 μM). However, the specific methodology described in this chapter has evolved for use with the dual-dye (pseudo-ratiometric) Ca^{2+} assay utilizing a combination of Oregon Green BAPTA-1-AM (1.25 μM) and Fura Red-AM (2.5 μM) co-loaded into isolated platelet suspensions (see Note 5).

1. Incubate 1 ml aliquots of the isolated platelet suspension (Human; 3×10^8/ml, Murine; 2.5×10^8/ml in 1× PWB) for 30 min at 37°C with Oregon Green BAPTA-1-AM (1.25 μM) and Fura Red-AM (2.5 μM) (see Note 6).

2. Centrifuge the dye-loaded platelet suspension in a bench-top Eppendorf centrifuge at $2,000 \times g$ for 1 min and gently aspirate the supernatant containing unincorporated dye.

3. Wash the platelet pellet once with an equivalent volume of 1× PWB containing apyrase (0.02 U/ml) and finally resuspend at a concentration of 3×10^8/ml (human) or 2.5×10^8/ml (murine) in 1× PWB containing apyrase (0.02 U/ml) (see Note 7).

4. Incubate the dye-loaded platelet suspension for a further 30 min at 37°C to allow for complete conversion of the AM ester form of the dyes to the free acid form (see Note 8).

3.5. Preparation of Reconstituted Blood

The generic platelet dye-loading method described (Subheading 3.4) requires the use of isolated platelet preparations. Published protocols from other laboratories have been described in which PRP (rather than isolated platelet preparations) is directly loaded with fluorescent Ca^{2+} probes. These protocols tend to utilize relatively high dye loading concentrations (8 μM) to overcome the effects of plasma protein binding interactions of the Ca^{2+} probes during the incubation period.

The specific dye combination of Oregon Green BAPTA-1 and Fura Red described (Subheading 3.4) preclude this approach as we have found that Fura Red AM-loading of platelets is extremely poor when the direct PRP approach is utilized. Because of this restriction, experiments must be carried out using reconstituted blood preparations in which autologous erythrocytes and plasma are recombined with the dye-loaded platelet suspension.

1. In reconstitution experiments, prepare platelets as described in Subheading 3.4 and resuspend with isolated erythrocytes in either modified Tyrode's buffer or PPP. When using PPP, platelets should be reconstituted at a final concentration of $50–300 \times 10^9/l$ no more than 10 min prior to experimentation.

2. Observation of reconstituted blood under physiological flow conditions has demonstrated that platelets under these conditions can efficiently tether and form adhesion contacts with immobilized matrix proteins, but are limited in their ability to form stable platelet aggregates due to the lack of soluble adhesive proteins, such as fibrinogen and vWF. Therefore, when examining thrombus growth using blood reconstituted with Tyrode's buffer, the suspension buffer should be supplemented with purified vWF (10–20 µg/ml) and/or fibrinogen (1–2 mg/ml) prior to perfusion.

3.5.1. Preparation of Autologous PPP

1. Collect 50 ml of blood (see Subheading 3.1).

2. Separate PRP from the erythrocyte pellet by centrifugation at $300 \times g$ for 16 min.

3. Carefully remove the PRP phase being careful not to disturb the buffy coat or erythrocyte phases.

4. Centrifuge the PRP at $2,000 \times g$ for 7 min (50 ml) and carefully remove the PPP phase without disturbing the platelet pellet.

5. Supplement the PPP with 800 U/ml hirudin and 0.02 U/ml apyrase and incubate for 30 min at 37°C prior to use.

3.5.2. Preparation of Washed Erythrocytes

1. Collect 50 ml blood in ACD (see Subheading 3.1).

2. Separate PRP from the erythrocyte pellet by centrifugation at $300 \times g$ for 16 min.

3. Carefully remove the PRP phase along with the buffy coat from the top of the erythrocyte phase.

4. Wash the erythrocyte phase with an equal volume of Tyrode's buffer (no BSA) and mix by gentle inversion.

5. Centrifuge at $2,000 \times g$ for 6 min (50 ml) to pack the erythrocyte phase.

6. Remove the supernatant and any remaining buffy coat.

7. Repeat steps 4–6.

8. Resuspend the erythrocyte pellet from step 7 in an equal volume of supplemented Tyrode's (+1 mM $CaCl_2$ + 0.5% BSA).

9. Repeat steps 5 and 6.

10. Supplement the packed erythrocyte suspension with 800 U/ml hirudin and 0.02 U/ml apyrase and incubate for 30 min at 37°C.

11. Immediately prior to reconstitution with the platelet suspension supplement the erythrocyte suspension with 0.02 U/ml apyrase to prevent secreted ADP from activating the isolated platelet suspension (see Notes 9 and 10).

12. Allow the reconstituted blood suspension to equilibrate for 10 min at 37°C prior to use.

3.6. Flow-Based Platelet Adhesion Assays

The reconstituted platelet flow assays used in conjunction with the Ca^{2+} imaging assays described in this chapter have been described in detail elsewhere (34).

3.7. Imaging of Platelet Ca^{2+} Flux

Real-time platelet Ca^{2+} flux is monitored via confocal or epifluorescence microscopy as the ratio of Oregon Green BAPTA-1 fluorescence over Fura Red fluorescence at emission wavelengths of 500–570 and 600–710 nm, respectively.

3.7.1. Calibration

In order to calibrate the dynamic range of both the imaging system and the probe loaded platelet preparations, the minimum (R_{min}) and maximal (R_{max}) fluorescence ratios (R) need to be determined as follows:

R_{min}: The mean fluorescence ratio determined from 200 to 400 cells preincubated with 50 µM DM-BAPTA-AM.

1. In order to determine R_{min}, resuspend an aliquot of the probe loaded platelet preparation (100 µl) in Tyrode's buffer supplemented with 5 mM EGTA and the membrane permeable Ca^{2+} chelator dimethyl BAPTA-AM (DM-BAPTA-AM) to a final solution concentration of 50 µM.

2. Incubate the 100 µl aliquot with DM-BAPTA-AM for 30 min at 37°C and wash as per Subheading 3.4.

3. Dilute (1/10 v/v) the 100 µl DM-BAPTA-loaded platelet sample in 1× Tyrode's buffer + 2 mM EGTA and pipette onto a clean microscope slide or coverslip.

4. Image ten random fields (equivalent to 200–400 cells) using an appropriate camera gain or PMT setup that produces an image of sufficient intensity without saturating the detector.

R_{max}: The mean fluorescence ratio determined from 200 to 400 cells suspended in Tyrode's buffer supplemented with 5 µM A23187 + 10 mM $CaCl_2$.

1. In order to determine R_{max} an aliquot of the probe loaded platelet preparation (100 µl) is resuspended in 1× Tyrode's buffer supplemented with 10 mM $CaCl_2$ and 5 µM A23187.

2. Incubate the platelet sample for 30 min at 37°C and image (as per R_{min} above).

$\Delta[Ca^{2+}]_c$ *Determination* (see Note 11).

1. Using standard image thresholding, apply a threshold that will demarcate the platelets in the acquired channels and then measure the average background-corrected fluorescence intensities for Oregon Green BAPTA-1 and Fura Red for each platelet in the field (see Note 12).

2. Fluorescence ratios (R) are calculated as OG/FR.

3. Determine the average R_{min} and R_{max} values for 200–400 platelets (as per R_{min} and R_{max} above) (see Note 13). These average values will be used for subsequent calculations of relative calcium flux.

4. Also determine the average F_{max} and F_{min} (determined from the R_{min} and R_{max} data sets) representing the mean fluorescence values (arbitrary units) of Oregon Green BAPTA-1 only for R_{max} and R_{min} measurements, respectively.

5. Convert fluorescence ratios to relative cytosolic Ca^{2+} concentration $(\Delta[Ca^{2+}]_c)$ according to (1) (see Notes 15–17),

$$\Delta\left[Ca^{2+}\right]_c = 170 \times (R - R_{min}) / (R_{max} - R) \times (F_{max} / F_{min}), \quad (1)$$

170 is the K_d value of Oregon Green BAPTA-1 Ca^{2+} binding (see Note 14); R represents the measured fluorescence ratio; R_{max} is the mean fluorescence ratio determined from 200 to 400 cells suspended in Tyrode's buffer supplemented with 5 µM A23187 + 10 mM $CaCl_2$; R_{min} is the mean fluorescence ratio determined from 200 to 400 cells preincubated with 50 µM DM-BAPTA-AM + 5 mM EGTA; F_{max} and F_{min} represents the mean fluorescence values (arbitrary units) of Oregon Green BAPTA-1 only for R_{max} and R_{min} measurements, respectively.

3.7.2. Real Time Ca²⁺ Flux Measurement

Real-time platelet Ca^{2+} flux images under conditions of blood flow (Subheading 3.6) are acquired using Metamorph (ver 6.2 (Fig. 3a)) or Leica confocal software (LCS). The first 3 min of platelet flow is captured. In the case of epifluorescence capture, a CCD exposure time of 150 ms (0 gain) is utilized resulting in a frame rate of 6.67 fps. In the case of confocal imaging (Subheading 2.5) a frame rate of 2 fps is achieved. Data is plotted as $\Delta[Ca^{2+}]_c$ as a function of time (Fig. 3b).

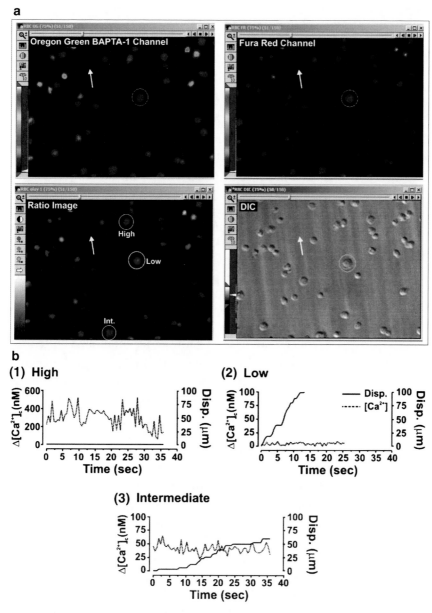

Fig. 3. Real-time epifluorescence-based ratiometric platelet Ca^{2+} flux under blood flow. (**a**) Single simultaneous snapshot images from a 3 min imaging stack showing Oregon Green BAPTA-1 + Fura Red dye-loaded human platelets reconstituted with autologous erythrocytes (40% Hematocrit) at the surface of immobilized von Willebrand Factor at an applied shear rate of 1,800 per second. Oregon Green BAPTA-1 Channel: Raw Oregon Green BAPTA-1 imaging data; Fura Red Channel: Raw Fura Red imaging data; Ratio Image: Pseudo ratio overlay of Oregon Green and Fura Red channels. *Red* low Δ[Ca^{2+}]$_c$; *Green* High Δ[Ca^{2+}]$_c$; *Int* intermediate Δ[Ca^{2+}]$_c$; *DIC* Concomitant Differential Interference Contrast imaging of platelet flow. *White arrows* show flow direction. (**b**) Δ[Ca^{2+}]$_c$ profiles and concomitant displacement versus time graphs for representative platelets undergoing interaction with vWF at an applied shear rate of 1,800 per second. (1) *High*: representative platelet undergoing highly oscillatory Δ[Ca^{2+}]$_c$ correlating with stationary adhesion formation; (2) *Low*: representative platelet displaying basal Δ[Ca^{2+}]$_c$ and rapid translocation on the vWF surface; (3) *Intermediate*: representative platelet displaying an intermediate level of Δ[Ca^{2+}]$_c$ undergoing slow translocation at the vWF surface.

4. Notes

1. Theophylline is insoluble at room temperature and the solution must be kept at $\geq 70°C$ to dissolve it prior to use.

2. From this step onward, blood/platelets and buffers should be kept at $37°C$.

3. The platelet count should not exceed 1×10^9/ml to avoid undue platelet activation.

4. Care needs to be taken when using any Ca^{2+} chelating probe in the context of live platelet studies due to potential Ca^{2+} buffering effects on platelet functional responses. It is recommended that platelet functional studies, including platelet aggregometry with low-dose ADP ($2.5 \mu M$) and FACS analysis of integrin $\alpha_{IIb}\beta_3$ activation (FITC-fibrinogen and/or PAC-1 binding) and P-selectin expression be undertaken as a function of Ca^{2+} probe concentration during the initial work-up phase to rule out Ca^{2+} buffering artifacts. Where appropriate and dependent on the sensitivity of the chosen microimaging apparatus, dye concentration can be lowered to minimize buffering effects.

5. For both the wide field and confocal imaging systems described in this chapter, the nominated dye loading concentrations have proven to result in minimal functional impairment, based on platelet aggregometry and measurements of platelet integrin $\alpha_{IIb}\beta_3$ activation.

6. As an alternative to reduce the dye loading concentration, the loading incubation time can be reduced; however, this may lead to an overall reduction in loading efficiency.

7. It is advisable to retain the remainder of the platelet preparation in $1\times$ PWB at $37°C$ and resuspend aliquots in modified Tyrode's buffer when needed, rather than maintaining the platelet sample in Tyrode's buffer for the duration of the experiment.

8. The AM ester form is membrane permeable and some loss of loaded probe from the platelet cytosol may occur during this final incubation period. This is especially problematic when dealing with platelets of murine origin which express relatively high plasma membrane anion exchange activity that can almost completely deplete the platelet of Oregon Green BAPTA-1 and Fura Red. When dealing with rodent platelet preparations, it is advisable to directly inhibit anion transport activity through co-treatment of the platelets with 1.5 mM probenecid.

9. Platelet interactions with thrombogenic surfaces under conditions of blood flow in vitro are influenced significantly by the presence and density of erythrocytes (hematocrit) in the bulk flow. In the case of reconstituted blood flow experiments, the % hematocrit directly impacts on the number and frequency of platelet-surface interactions and resulting adhesion.

10. Experiments involving both Ca^{2+} imaging and transmitted light imaging can be affected by hematocrit. Experience in our laboratory has demonstrated that DIC imaging is particularly sensitive to the presence and density of erythrocytes in reconstituted platelet preparations due to the reflection and refractive artifacts introduced by the relatively large erythrocyte component. The ideal hematocrit range for flow-based platelet Ca^{2+} measurements using the microcapillary apparatus utilized in our laboratory (Subheading 2.3) falls within the range of 20–50%. The decision on final hematocrit (within this range) is critically dependent on the choice of objective and imaging system along with the nature of transmitted light techniques that may be used in conjunction with the Ca^{2+} imaging technique. A balance needs to be achieved between the efficiency of platelet adhesion and image quality as a function of hematocrit.

11. The calculated Ca^{2+} flux values are designated $\Delta[Ca^{2+}]_c$ to indicate that all Ca^{2+} concentration estimates are relative to a zero point, set by DM-BAPTA Ca^{2+} chelation.

12. In our laboratory, Ca^{2+} imaging measurements are analyzed offline using either Leica Physiology Software (Leica TCS SP; Leica, Heidelberg, Germany) or Metamorph (version 6.2) (Universal Imaging). In all cases, background subtraction is applied based on a measured region of interest and objects less than or equal to 10 pixel diameters are routinely disregarded from further analysis. These small fluorescence objects may represent precipitated fluorescent probe or platelet microparticles.

13. Individual platelets in R_{min} image frames should demonstrate consistently low OG fluorescence levels with concomitantly elevated FR fluorescence levels. If heterogeneity is observed in the DM-BAPTA treated population, a fresh R_{min} sample should be prepared. It is also possible to extend the incubation time at this step to improve overall uniformity of the R_{min} sample.

14. Only the K_d for Oregon Green BAPTA-1 is used for $\Delta[Ca^{2+}]_c$ calculations (6.1) based on the assumption that the Oregon Green BAPTA-1 signal represents the larger delta, and the Fura Red delta fluorescence represents approximately 10% of Oregon Green BAPTA-1 signal.

15. Following calibration, baseline or resting platelet $\Delta[Ca^{2+}]_c$ values should fall into the range of 40–60 nM while maximal or peak $\Delta[Ca^{2+}]_c$ should fall within the range of 200–1,200 nM dependent on the nature of the adhesive surface or chemical stimulus.

16. Care should be taken to monitor overall platelet morphology and function throughout the assay. High magnitude stimuli that may affect platelet membrane integrity, such as procoagulant microparticle formation, may lead to excessive Ca^{2+} probe leakage and result in spurious $\Delta[Ca^{2+}]_c$ measurements.

17. In our experience, the use of this dual-dye ratiometric protocol is not suitable when monitoring platelet adhesion to immobilized type I collagen substrate, as the relative increase in platelet $\Delta[Ca^{2+}]_c$ significantly suppresses the Fura Red fluorescence output and leads to exaggerated and inaccurate $\Delta[Ca^{2+}]_c$ measurements. We generally recommend the use of a single ratiometric dye, such as Fura 2 for collagen-based flow experiments.

References

1. Arderiu, G., et al., External calcium facilitates signalling, contractile and secretory mechanisms induced after activation of platelets by collagen. Platelets, 2008. 19(3): p. 172–81.

2. Giuliano, S., et al., Bidirectional integrin alphaIIbbeta3 signalling regulating platelet adhesion under flow: contribution of protein kinase C. Biochem J, 2003. 372(Pt 1): p. 163–72.

3. Goncalves, I., et al., Integrin alpha IIb beta 3-dependent calcium signals regulate platelet-fibrinogen interactions under flow. Involvement of phospholipase C gamma 2. J Biol Chem, 2003. 278(37): p. 34812–22.

4. Goncalves, I., et al., Importance of temporal flow gradients and integrin alphaIIbbeta3 mechanotransduction for shear activation of platelets. J Biol Chem, 2005. 280(15): p. 15430–7.

5. Maxwell, M.J., et al., SHIP1 and Lyn Kinase Negatively Regulate Integrin alpha IIb beta 3 signaling in platelets. J Biol Chem, 2004. 279(31): p. 32196–204.

6. Mazzucato, M., et al., Distinct spatio-temporal Ca2+ signaling elicited by integrin alpha2beta1 and glycoprotein VI under flow. Blood, 2009. 114(13): p. 2793–801.

7. Mazzucato, M., et al., Sequential cytoplasmic calcium signals in a 2-stage platelet activation process induced by the glycoprotein Ibalpha mechanoreceptor. Blood, 2002. 100(8): p. 2793–800.

8. Nesbitt, W.S., et al., Intercellular calcium communication regulates platelet aggregation and thrombus growth. J Cell Biol, 2003. 160(7): p. 1151–61.

9. Nesbitt, W.S., et al., Distinct glycoprotein Ib/V/IX and integrin alpha IIbbeta 3-dependent calcium signals cooperatively regulate platelet adhesion under flow. J Biol Chem, 2002. 277(4): p. 2965–72.

10. Sun, D.S., et al., Calcium oscillation and phosphatidylinositol 3-kinase positively regulate integrin alpha(IIb)beta3-mediated outside-in signaling. J Biomed Sci, 2005. 12(2): p. 321–33.

11. van Gestel, M.A., et al., Real-time detection of activation patterns in individual platelets during thromboembolism in vivo: differences between thrombus growth and embolus formation. J Vasc Res, 2002. 39(6): p. 534–43.

12. Rink, T.J. and S.O. Sage, Calcium signaling in human platelets. Annu Rev Physiol, 1990. 52: p. 431–49.

13. Authi, K.S., TRP channels in platelet function. Handb Exp Pharmacol, 2007(179): p. 425–43.

14. Jardin, I., et al., Intracellular calcium release from human platelets: different messengers for multiple stores. Trends Cardiovasc Med, 2008. 18(2): p. 57–61.

15. Rosado, J.A., et al., Two pathways for store-mediated calcium entry differentially dependent on the actin cytoskeleton in human platelets. J Biol Chem, 2004. 279(28): p. 29231–5.

16. Rosado, J.A. and S.O. Sage, The actin cytoskeleton in store-mediated calcium entry. J Physiol, 2000. 526 Pt 2: p. 221–9.

17. Varga-Szabo, D., A. Braun, and B. Nieswandt, Calcium signaling in platelets. J Thromb Haemost, 2009. 7(7): p. 1057–66.

18. Suzuki-Inoue, K., et al., Murine GPVI stimulates weak integrin activation in PLCgamma2−/− platelets: involvement of PLCgamma1 and PI3-kinase. Blood, 2003. 102(4): p. 1367–73.

19. Berridge, M.J., M.D. Bootman, and H.L. Roderick, Calcium signalling: dynamics, homeostasis and remodelling. Nat Rev Mol Cell Biol, 2003. 4(7): p. 517–29.

20. Shattil, S.J. and L.F. Brass, Induction of the fibrinogen receptor on human platelets by intracellular mediators. J Biol Chem, 1987. 262(3): p. 992–1000.

21. Lian, L., et al., The relative role of PLCbeta and PI3Kgamma in platelet activation. Blood, 2005. 106(1): p. 110–7.

22. Quinton, T.M. and W.L. Dean, Multiple inositol 1,4,5-trisphosphate receptor isoforms are present in platelets. Biochem Biophys Res Commun, 1996. 224(3): p. 740–6.

23. El-Daher, S.S., et al., Distinct localization and function of (1,4,5)IP(3) receptor subtypes and the (1,3,4,5)IP(4) receptor GAP1(IP4BP) in highly purified human platelet membranes. Blood, 2000. 95(11): p. 3412–22.

24. Braun, A., et al., STIM1 is essential for Fcgamma receptor activation and autoimmune inflammation. Blood, 2009. 113(5): p. 1097–104.

25. Varga-Szabo, D., et al., The calcium sensor STIM1 is an essential mediator of arterial thrombosis and ischemic brain infarction. J Exp Med, 2008. 205(7): p. 1583–91.

26. Braun, A., et al., Orai1 (CRACM1) is the platelet SOC channel and essential for pathological thrombus formation. Blood, 2009. 113(9): p. 2056–63.

27. Vial, C., et al., A study of P2X1 receptor function in murine megakaryocytes and human platelets reveals synergy with P2Y receptors. Br J Pharmacol, 2002. 135(2): p. 363–72.

28. Hassock, S.R., et al., Expression and role of TRPC proteins in human platelets: evidence that TRPC6 forms the store-independent calcium entry channel. Blood, 2002. 100(8): p. 2801–11.

29. Sage, S.O., E.H. Yamoah, and J.W. Heemskerk, The roles of P(2X1) and P(2 T AC) receptors in ADP-evoked calcium signalling in human platelets. Cell Calcium, 2000. 28(2): p. 119–26.

30. Harper, A.G. and S.O. Sage, A key role for reverse Na+/Ca2+ exchange influenced by the actin cytoskeleton in store-operated Ca2+ entry in human platelets: evidence against the de novo conformational coupling hypothesis. Cell Calcium, 2007. 42(6): p. 606–17.

31. Tolhurst, G., et al., Interplay between P2Y(1), P2Y(12), and P2X(1) receptors in the activation of megakaryocyte cation influx currents by ADP: evidence that the primary megakaryocyte represents a fully functional model of platelet P2 receptor signaling. Blood, 2005. 106(5): p. 1644–51.

32. Nesbitt, W.S., et al., A shear gradient-dependent platelet aggregation mechanism drives thrombus formation. Nature Medicine, 2009. 15(6): p. 665–73.

33. Kulkarni, S., et al., Techniques to examine platelet adhesive interactions under flow. Methods Mol Biol, 2004. 272: p. 165–86.

34. Nesbitt, W.S., et al., Distinct glycoprotein Ib/V/IX and integrin alpha IIbbeta 3-dependent calcium signals cooperatively regulate platelet adhesion under flow. J Biol Chem, 2002. 277(4): p. 2965–72.

Chapter 7

Platelet Shape Change and Spreading

Joseph E. Aslan, Asako Itakura, Jacqueline M. Gertz, and Owen J.T. McCarty

Abstract

Hemostasis is dependent upon the successful recruitment and activation of blood platelets to the site of a breach in the vasculature. Platelet activation stimulates the rapid reorganization of the cortical actin cytoskeleton, resulting in the transformation of platelets from biconcave disks to fully spread cells. During this process, platelets extend filopodia and generate lamellipodia, resulting in a dramatic increase in the platelet surface area. Kohler-illuminated Nomarski Differential Interference Contrast microscopy has proved an effective tool to characterize platelet morphological changes in real time, and provides a useful tool to identify genetic and pharmacological regulators of platelet function.

Key words: Lamellipodia, Filopodia, Kohler-illuminated Nomarski DIC microscopy

1. Introduction

Platelets initiate the formation of a thrombus through their attachment, spreading, and aggregation at the vessel surface (1–3). Upon vessel injury, platelets activate through receptor-mediated cascades, inducing a rapid reorganization of the cortical actin cytoskeleton, a rounding of the cell, and attachment to an immobilized surface. This change in morphology increases platelet surface area as platelets extend filopodia and form lamellipodia to strengthen contact with the surface as well as to other platelets. Platelet spreading and aggregation are not uniquely important to acute vascular injury; platelet adhesion and thrombus formation at atherosclerotic plaques or vascular lesions are important causative factors in the development of thrombotic disorders, such as myocardial infarction and stroke.

Jonathan M. Gibbins and Martyn P. Mahaut-Smith (eds.), *Platelets and Megakaryocytes: Volume 3, Additional Protocols and Perspectives*, Methods in Molecular Biology, vol. 788, DOI 10.1007/978-1-61779-307-3_7, © Springer Science+Business Media, LLC 2012

In this chapter, we present an optimized protocol to characterize platelet shape changes during platelet adhesion and spreading on immobilized surfaces. This includes a procedure for the isolation and purification of platelets, the preparation of surface-immobilized adhesive proteins, and an outline of steps to image platelet adhesion and spreading in real time using Kohler-illuminated Nomarski Differential Interference Contrast (DIC) microscopy. Using this protocol, we and others have defined the molecular mechanisms that mediate platelet adhesion to extracellular matrix proteins (4–6), identified essential regulators of lamellipodia formation (7–9), described novel cytological features of platelets (10, 11), and identified receptors that mediate platelet-coagulation factor interactions (12–14). These methods can be readily incorporated into future studies of platelets from knockout mice or platelets treated with pharmacological agents to provide new insights into platelet physiology, thrombus formation, and disease.

2. Materials

2.1. Platelet Purification

1. Acid-citrate-dextrose (ACD): 85 mM sodium citrate, 110 mM glucose, 71 mM citric acid.

2. Sodium citrate is dissolved at 3.8% (w/v) in sterile water.

3. Modified Tyrode buffer: 129 mM NaCl, 0.34 mM Na_2HPO_4, 2.9 mM KCl, 12 mM $NaHCO_3$, 20 mM HEPES, 1 mM $MgCl_2$, 7.3 pH. Supplement with glucose (5 mM) prior to use.

4. Prostacyclin (PGI_2, Cayman Chemical) is dissolved at 1 mg/mL in 50 mM Tris buffer, pH 9.1.

5. A wide range of systems and approaches are available that may be used to count platelets. For instance, a 1:1,000 dilution of platelets can be counted using a Bright-Line Hemacytometer (Hausser Scientific, Horsham, PA). For more detailed information on platelet counting, please refer to Vol. 1 in this series, Chapter 3.

2.2. Differential Interference Contrast Imaging

1. Glass coverslips (12 or 25-mm diameter, number 1.5 thickness) from Fisher Scientific.

2. Phosphate-buffered saline (PBS): 138 mM NaCl, 2.7 mM KCl in sterile water, pH 7.4.

3. Bovine serum albumin (essentially fatty acid-free BSA, Sigma): Prepare a 5 mg/mL BSA solution in PBS fresh for each experiment. Place vessel in boiling water for 10 min, and then cool on ice. Filter through a 0.45-μm filter and store on ice until use.

4. Paraformaldehyde: Prepare a fresh 3.7% paraformaldehyde solution in PBS.

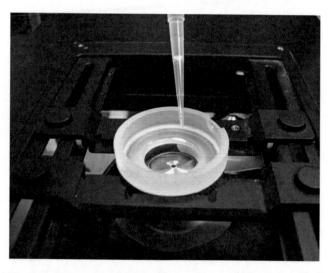

Fig. 1. High-profile, open-bath imaging chamber mounted on the stage of a Zeiss Axiovert 200 M microscope equipped with a Zeiss 63× oil-immersion 1.40 numerical aperture (NA) plan-apochromat lens.

5. Mounting media: Fluoromount-G (SouthernBiotech, Birmingham, AL).

6. Imaging chamber: High Profile Open Bath Chamber (Warner Instruments, LLC, Hamden, CT; Fig. 1).

7. Images used in this chapter were acquired using a Zeiss 63× oil immersion 1.40 numerical aperture (NA) plan-apochromat lens on a Zeiss Axiovert 200 M microscope equipped with a temperature-controlled XL-3 incubator on a Vistek Vibration Isolation Platform (Carl Zeiss Inc., Thornwood, NY). Time-lapse events were captured using a Zeiss Axiocam MRm CCD camera using Slidebook 5.0 (Intelligent Imaging Innovations, Inc., Denver, CO, USA).

3. Methods

Platelets undergo dramatic changes in morphology following adhesion and activation. The amount of actin assembled in platelets approximately doubles in response to activation as quiescent, discoid platelets undergo a series of morphological changes that include rounding, generation of filopodia and lamellipodia, and formation of stress fibers (8). Visualization of these morphological changes is increasingly been utilized as a technique to characterize the pathways that mediate platelet activation. This can be accomplished through the use of Kohler-illuminated Nomarski DIC microscopy. Shown is an example of the use of DIC microscopy to characterize platelet spreading in the presence of pharmacological

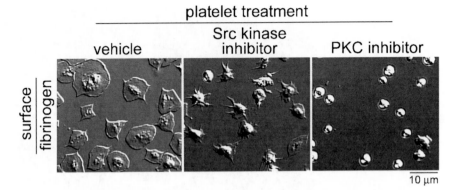

Fig. 2. Platelet adhesion to immobilized fibrinogen. Human-washed platelets (2×10^7/mL) were placed on fibrinogen-coated coverslips for 45 min at 37°C and imaged using DIC microscopy. In selected experiments, platelets were pretreated for 10 min with vehicle (0.1% (v/v) DMSO), Src kinase inhibitor PP2 (10 μM), or protein kinase C (PKC) inhibitor RO-318220 (10 μM). Full platelet spreading is observed on a surface of fibrinogen. Platelets only partially spread on fibrinogen in the presence of the Src kinase inhibitor PP2, whereas platelet spreading is abrogated by the PKC inhibitor RO-318220.

inhibitors of the Src kinase and protein kinase C-signaling pathways (see data in Fig. 2).

3.1. Preparation of Human Platelets

1. These steps outline the procedure for isolation and purification of human platelets (see Note 1). Obtain whole blood by venipuncture from human volunteers into polypropylene syringes containing sodium citrate at a ratio of 1 part 3.8% sodium citrate to 9 parts blood (0.38% sodium citrate final concentration). Add warmed ACD (30°C) to the blood at a ratio of 1 part ACD to 9 parts sodium citrate anticoagulated whole blood.

2. Transfer anticoagulated whole blood to a series of 5-mL polypropylene centrifuge tubes and centrifuge at $200 \times g$ for 20 min. Harvest the supernatant platelet-rich plasma (PRP) layer into a 50-mL centrifuge tube with a transfer pipette (see Note 2).

3. Add 0.1 μg/mL PGI_2 to the PRP prior to further centrifugation at $1,000 \times g$ for 10 min. Remove and discard the supernatant platelet-poor plasma (PPP) layer.

4. Gently resuspend the platelet pellet in 1 mL of modified Tyrode buffer containing 150 μL ACD. Take care to be sure that the platelet pellet is thoroughly resuspended. Add an additional 25 mL of modified Tyrode buffer and 3 mL of ACD.

5. Add 0.1 μg/mL PGI_2 to the platelet solution and wash by centrifugation at $1,000 \times g$ for 10 min. Resuspend the platelet pellet with modified Tyrode buffer, count platelets using a Bright-Line Hemacytometer, and adjust the final platelet concentration to 2×10^8/mL. Purified platelets should be kept at room temperature for no longer than 4 h before use in adhesion assays.

3.2. Static Adhesion Assay

Steps 1–3 outline the process to surface immobilize adhesive proteins on glass coverslips for the end-point analysis of platelet adhesion and spreading.

1. Place 12-mm glass coverslips into desired wells of a 24-well plate (Falcon).

2. Make stock solutions of soluble adhesive proteins. Pipette 50 µL of adhesive protein solution onto the center region of each coverslip and incubate for 1 h at room temperature or overnight at 4°C. Typically, ≥50 µg/mL solution is sufficient to provide maximum coating efficiency. Take care to minimize air bubbles in the coating solution to ensure an even coating.

3. After washing coverslips three times with PBS, pipette 300 µL of BSA blocking buffer, being sure that the entire coverslip is covered. Treating an uncoated coverslip with BSA provides a negative control. Incubate for 1 h at room temperature. Wash an additional three times with PBS.

4. Treat platelets with selected agonists or vehicle. Typically, treating ≥350 µL of platelets (2×10^7/mL) per agonist or antagonist for 10 min in a microcentrifuge tube is sufficient.

5. Gently pipette 300 µL of treated platelets (2×10^7/mL) into selected wells of the 24-well plate. Incubate for 45 min at 37°C.

6. Wash coverslips three times with PBS, being sure not to let the wells dry out between washes. Care should be taken to ensure that washing is sufficient to remove nonadherent platelets. Pipette 300 µL of 3.7% paraformaldehyde into each well. Incubate for 10 min at room temperature.

7. Wash coverslips three times with PBS, leaving PBS in the wells during the final wash. Carefully remove coverslips (a pipette tip may aid in removal) and invert into a drop of mounting medium on a microscope slide. Care should be taken to avoid breaking the fragile coverslips. After the mounting medium is allowed to set overnight at 4°C, use nail varnish to seal the sample (see Note 3).

8. The slides can now be viewed under DIC microscopy. See Note 4 for a brief tutorial on imaging with DIC optics. An example micrograph is shown in Fig. 2. Software can be used to quantify the degree of adhesion and surface area of adherent platelets (see Note 5).

3.3. Real-Time Spreading Assay

Steps 1–4 outline the process to image platelet adhesion and spreading in real time using Köhler-illuminated Nomarski DIC microscopy.

1. Coat the center of 25-mm glass coverslips with 150 µL of adhesive protein solution for 1 h at room temperature or overnight at 4°C. Wash coverslips three times with PBS and block

with 150 μL of BSA blocking buffer for 1 h at room temperature. Wash an additional three times with PBS, being sure to leave PBS on the coverslip after the last wash (see Note 6).

2. Load coverslip into imaging chamber. Pipette 250 μL of modified Tyrode buffer onto coverslip. Adjust DIC optics (see Note 4).

3. Treat platelets with selected agonists or vehicle and warm to 37°C. Gently pipette 250 μL of platelets (4×10^7/mL) into the imaging chamber (final concentration, 2×10^7/mL).

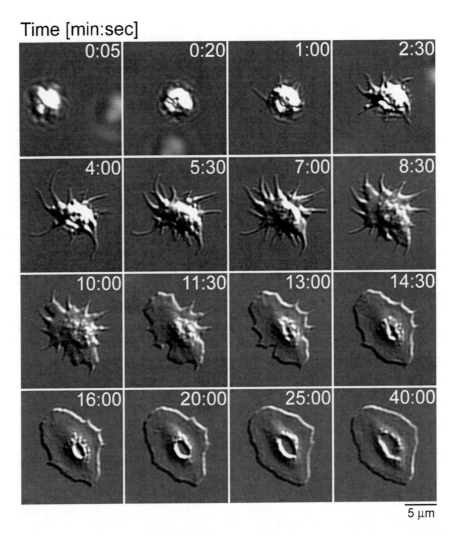

Fig. 3. Real-time imaging of platelet spreading on immobilized fibrinogen. Human-washed platelets (2×10^7/mL) were exposed to fibrinogen-coated coverslips and imaged in real time using Nomarski differential interference contrast microscopy. Upon initial contact with the fibrinogen-coated surface, platelets undergo rounding before generating short, dynamic filopodial protrusions (see the 4-min time point). Sheet-like lamellipodial membrane then proceeds to fill in the gaps between the filopodia, resulting in a fully spread platelet (see the 25-min time point).

4. Record images of platelet adhesion and spreading as platelets settle out of solution (see Note 7). Typically, recording an image every 5 s for at least 30 min is sufficient to capture a complete series of single-cell morphological changes. An example time course of platelet spreading is shown in Fig. 3. Software can be used to quantify the platelet surface area as a function of time (see Note 5).

4. Notes

1. This protocol can be adapted for the purification of mouse platelets (7–9).

2. Low-speed centrifugation results in the separation of platelets (top layer) from larger and more dense cells, such as leukocytes and erythrocytes (bottom layer). For certain rare platelet disorders which are characterized by larger than normal platelets, such as Bernard-Soulier Syndrome (2), PRP may be prepared by allowing whole blood to gravity separate for 2 h post venipuncture.

3. Care should be taken to minimize the presence of undesirable air bubbles in the mounting medium. Use of a thin layer of quick-drying, clear nail polish around the edges of the coverslip is preferable.

4. The proper alignment and use of DIC optics produces a distinct, relief-like, shadow-cast appearance that gives the illusion of three dimensionality. This is due to the ability of a system of dual-beam interference optics to transform local gradients in optical path lengths in a specimen into regions of contrast in an image. The DIC system consists of the following four optical components: (1) polarizer – converts collimated light emanating from a halogen lamp into plane-polarized light; (2) condenser prism – beam splitter that converts individual waves of polarized light into two closely spaced waves of light that are focused on the specimen by the condenser lens (for low-NA air objectives, a DIC II prism is used while for high-NA lens oil-immersion objectives, a DIC III prism is used); (3) objective prism – recombines the two waves of light produced by the condenser prism after traveling through the specimen and objective; (4) analyzer – converts the elliptically polarized light emanating from the objective prism into plane-polarized light. Precise alignment of all four components is required to produce an image with optimum contrast. Proper adjustment of the lateral position of the objective prism can be accomplished by turning the positioning screw located on the objective prism holder. An example of how objective prism adjustment can affect the DIC image is shown in Fig. 4.

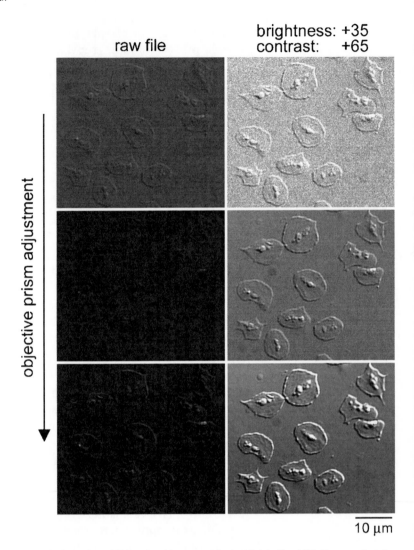

Fig. 4. Optimization of DIC optics. Human-washed platelets (2×10^7/mL) were placed on fibrinogen-coated coverslips for 45 min at 37°C and imaged using DIC microscopy. The objective prism is adjusted to optimize the distinct, relief-like, shadow-cast appearance that gives the illusion of three dimensionality. Brightness and contrast are adjusted offline to optimize image quality.

5. To compute the surface area of platelets, adherent platelets are manually outlined and the number of pixels within each outline computed using a Java plugin for Image J software (NIH, Bethesda, MD). Imaging of a graticule under the same conditions allows the conversion of pixel size to microns (13, 15).

6. Care should be taken to maintain a good meniscus on top of the coverslip during the coating and washing steps to avoid getting liquid under the coverslip. A razor blade may be required to carefully remove coverslips in the advent of spillage. Coverslips should be kept covered during incubations. We

routinely coat three 25-mm coverslips in one 100-mm tissue culture dish (Falcon).

7. High-powered objectives, such as 63× objectives, tend to drift out of focus, and therefore must be continually monitored and adjusted in order to achieve a crisp set of images. We have found that locating and focusing on one of the initial platelets that settles onto the coverslip allows the user to maintain the image in focus throughout the time course.

Acknowledgments

We would like to thank Tara White-Adams and Michelle Berny-Lang for helpful input and discussions. This work was supported by an American Heart Association Grant-in-Aid, 09GRNT2150003, and the National Institute of Health (1U54CA143906-01).

References

1. Watson, S.P. (2009) Platelet activation by extracellular matrix proteins in haemostasis and thrombosis. Curr Pharm Des 15: 1358–1372.

2. Wei, A.H., Schoenwaelder, S.M., Andrews, R.K. and Jackson, S.P. (2009) New insights into the haemostatic function of platelets. Br J Haematol 147: 415–430.

3. Watson, S.P., Auger, J.M., McCarty, O.J. and Pearce, A.C. (2005) GPVI and integrin alphaIIbbeta3 signaling in platelets. J Thromb Haemost 3: 1752–1762.

4. McCarty, O.J., Zhao, Y., Andrew, N., Machesky, L.M., Staunton, D., Frampton, J. and Watson, S.P. (2004) Evaluation of the role of platelet integrins in fibronectin-dependent spreading and adhesion. J Thromb Haemost 2: 1823–1833.

5. White, T.C., Berny, M.A., Robinson, D.K., Yin, H., DeGrado, W.F., Hanson, S.R. and McCarty, O.J. (2007) The leech product saratin is a potent inhibitor of platelet integrin alpha2beta1 and von Willebrand factor binding to collagen. Febs J 274: 1481–1491.

6. Eshel-Green, T., Berny, M.A., Conley, R.B. and McCarty, O.J. (2009) Effect of sex difference on platelet adhesion, spreading and aggregate formation under flow. Thromb Haemost 102: 958–965.

7. Calaminus, S.D., McCarty, O.J., Auger, J.M., Pearce, A.C., Insall, R.H., Watson, S.P. and Machesky, L.M. (2007) A major role for Scar/WAVE-1 downstream of GPVI in platelets. J Thromb Haemost 5: 535–541.

8. McCarty, O.J., Larson, M.K., Auger, J.M., Kalia, N., Atkinson, B.T., Pearce, A.C., Ruf, S., Henderson, R.H., Tybulewicz, V.L., Machesky, L.M. and Watson, S.P. (2005) Rac1 is essential for platelet lamellipodia formation and aggregate stability under flow. J Biol Chem 280: 39474–39484.

9. Pearce, A.C., McCarty, O.J., Calaminus, S.D., Vigorito, E., Turner, M. and Watson, S.P. (2007) Vav family proteins are required for optimal regulation of PLCgamma2 by integrin alphaIIbbeta3. Biochem J 401: 753–761.

10. Calaminus, S.D., Auger, J.M., McCarty, O.J., Wakelam, M.J., Machesky, L.M. and Watson, S.P. (2007) MyosinIIa contractility is required for maintenance of platelet structure during spreading on collagen and contributes to thrombus stability. J Thromb Haemost 5: 2136–2145.

11. Calaminus, S.D., Thomas, S., McCarty, O.J., Machesky, L.M. and Watson, S.P. (2008) Identification of a novel, actin-rich structure, the actin nodule, in the early stages of platelet spreading. J Thromb Haemost 6: 1944–1952.

12. Berny, M.A., White, T.C., Tucker, E.I., Bush-Pelc, L.A., Di Cera, E., Gruber, A. and McCarty, O.J. (2008) Thrombin mutant W215A/E217A acts as a platelet GPIb antagonist. Arterioscler Thromb Vasc Biol 28: 329–334.

13. White, T.C., Berny, M.A., Tucker, E.I., Urbanus, R.T., De Groot, P.G., Fernandez, J.A., Griffin, J.H., Gruber, A. and McCarty,

O.J. (2008) Protein C supports platelet binding and activation under flow: role of glycoprotein Ib and apolipoprotein E receptor 2. J Thromb Haemost 6: 995–1002.

14. White-Adams, T.C., Berny, M.A., Tucker, E.I., Gertz, J.M., Gailani, D., Urbanus, R.T., de Groot, P.G., Gruber, A. and McCarty, O.J. (2009) Identification of coagulation factor XI as a ligand for platelet apolipoprotein E receptor 2 (ApoER2). Arterioscler Thromb Vasc Biol 29: 1602–1607.

15. McCarty, O.J., Abulencia, J.P., Mousa, S.A. and Konstantopoulos, K. (2004) Evaluation of platelet antagonists in in vitro flow models of thrombosis. Methods Mol Med 93: 21–34.

Chapter 8

Clot Retraction

Katherine L. Tucker, Tanya Sage, and Jonathan M. Gibbins

Abstract

The study of clot retraction in vitro has been adopted as a simple and reproducible approach to assess platelet function. Plasma clots should retract away from the sides of a glass tube within a few hours allowing the rapid characterization of outside-in signaling through platelet integrin $\alpha_{IIb}\beta_3$. In this chapter, we describe the role of platelets in fibrin clot retraction and provide a detailed description of the methods used to assess this process.

Key words: Clot retraction, Platelets, Integrin $\alpha_{IIb}\beta_3$, Outside-in signaling

Abbreviations

AMPK α2   Adenosine monophosphate (AMP)-activated protein kinase α2
PLCγ2   Phospholipase Cγ2
PP2   4-Amino-5-(4-chlorophenyl)-7-(t-butyl)pyrazolo-D-3,4-pyrimidine
PRP   Platelet-rich plasma
RBC   Red blood cell

1. Introduction

The conversion of fibrinogen to fibrin at sites of injury results in the formation of threads of fibrin which form a clot trapping platelets and red blood cells (RBCs) to form a plug at the site of injury, and in the arterial circulation consolidating the developing platelet thrombus. Clot retraction is a process driven by outside-in signaling by platelet integrin $\alpha_{IIb}\beta_3$ that results in the contraction of the fibrin mesh. The contraction of the fibrin clot results in the blood clot becoming smaller and excess fluid is extruded. This draws the edges of damaged tissue together and forms a mechanically stable clot.

Jonathan M. Gibbins and Martyn P. Mahaut-Smith (eds.), *Platelets and Megakaryocytes: Volume 3, Additional Protocols and Perspectives*, Methods in Molecular Biology, vol. 788, DOI 10.1007/978-1-61779-307-3_8, © Springer Science+Business Media, LLC 2012

The study of the retraction of plasma clots formed in vitro has been adopted as a simple and reproducible approach to characterize outside-in signaling through platelet integrin $\alpha_{IIb}\beta_3$.

1.1. Clot Formation, Stabilization, and Retraction

Exposure of subendothelial collagen upon vessel injury activates platelets (1–3) and also initiates coagulation resulting in the production of thrombin, a potent platelet agonist (4). Platelet activation results in spreading, the secretion of granules containing pro-thrombotic factors, and the affinity of the fibrinogen receptor, integrin $\alpha_{IIb}\beta_3$ for its ligand, is increased (5). These "inside-out" signals result in platelet aggregation via bivalent binding to fibrinogen. As a consequence of fibrinogen binding, the cytoplasmic tail of the β_3 integrin subunit becomes phosphorylated triggering an "outside-in" signal that provides a second wave of activatory signaling that is necessary for irreversible platelet thrombus formation. Such "outside-in" signaling through integrin $\alpha_{IIb}\beta_3$ plays a key role in clot retraction (6).

The role of platelets in fibrin clot formation is long established. At sites of injury the activation of the tissue factor- (extrinsic) and contact- (intrinsic) dependent coagulation pathways leads to the activation of factor X, and through subsequent processing of prothrombin to thrombin, leads to the conversion of plasma fibrinogen to fibrin (Fig. 1). Activated platelets provide phospholipids released in the form of microparticles that act along with calcium as cofactors for the actions of factor X. The exposure of amino phospholipids such as phosphatidylserine on the surface of activated platelets provides a surface for the assembly of the prothrombinase complex, and thereby thrombin may be produced in the vicinity of a platelet thrombus as it forms, ensuring the incorporation of fibrin for thrombus stabilization. Through the activation of protease-activated receptors on the platelet surface, thrombin also acts as a powerful platelet agonist.

Platelets generate force to contract the fibrin matrix and draw the edges of the wound together. The retraction is driven by the interaction between the fibrin outside the cells and the actin–myosin cytoskeleton of the platelets, which is mediated by integrin $\alpha_{IIb}\beta_3$. Myosin binding occurs upon phosphorylation of the β_3 subunit of the integrin (7) and the ANK domain-containing Bcl-3 colocalizes with the cytoskeleton, which have been shown, along with Src-family kinases to play key regulatory roles in this process (8–10). Phospholipase Cγ2 (PLCγ2) is also implicated, and following fibrinogen–integrin $\alpha_{IIb}\beta_3$ binding, becomes tyrosine phosphorylated by a Src family kinase-dependent mechanism (10). Evidence suggests that tyrosine kinases regulate the attachment of the cytoskeleton to integrin $\alpha_{IIb}\beta_3$ and are essential for the transition of cellular contractile forces to the fibrin polymers (11). The process of retraction is energy demanding and the metabolite sensing kinase adenosine monophosphate (AMP)-activated protein kinase

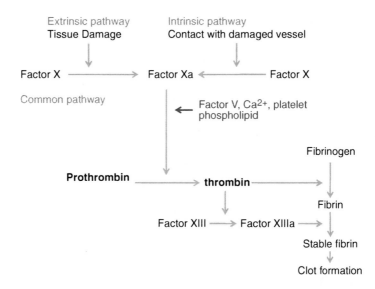

Fig. 1. The regulation of fibrin clot formation. Diagram of the final stages of the coagulation cascade where the intrinsic and extrinsic pathways meet and combine in the final common pathway, that leads to the conversion of fibrinogen to fibrin.

α2 (AMPK α2) affects the phosphorylation of the Src-family kinase Fyn, which in turn affects integrin $\alpha_{IIb}\beta_3$ signaling and thus clot retraction (12).

1.2. Clot Retraction as an Assay of Integrin $\alpha_{IIb}\beta_3$ Outside-in Signaling

The ability of platelets to drive clot retraction has become a surrogate measure of outside-in signaling following engagement of the integrin $\alpha_{IIb}\beta_3$ that may be readily measured with little required equipment. In this chapter, we describe a simple in vitro technique that can be used to assess platelet function in human or mouse blood.

2. Materials

1. Modified Tyrode's–Hepes buffer (134 mM NaCl, 0.34 mM Na_2HPO_4, 2.9 mM KCl, 12 mM $NaHCO_3$, 20 mM HEPES, 5 mM glucose, 1 mM $MgCl_2$, pH 7.3).

2. 4% (w/v) sodium citrate (see Note 1).

3. Thrombin (Sigma, Poole, UK), make a stock of 20 units/ml in Tyrode's–Hepes buffer.

4. Syringes, sealed glass pipettes (see Note 2), reusable tack adhesive (such as Blu-Tack), glass tubes (10 mm diameter tubes with a total capacity of 3 ml for mouse clot retraction assays, and 10 or 12 mm diameter tubes for assays with human blood) and test tube rack.

5. Camera and microbalance.

3. Methods

In this section, we describe the measurement of clot retraction in vitro using platelet-rich plasma (PRP), and note considerations for performing and designing experiments.

3.1. Preparation of Human Blood

1. Human blood is drawn into a syringe containing 4% (w/v) sodium citrate and PRP obtained by centrifugation at $100 \times g$ for 15–20 min at room temperature in the presence of prostacyclin (0.1 μg/ml) (see Note 3).

2. Remove the PRP into a clean tube and keep back some RBCs (see Note 4).

3.2. Preparation of Mouse Blood

1. Mouse blood (1 ml) is drawn into a syringe containing 100 μl of 4% (w/v) sodium citrate as an anticoagulant. PRP is prepared from whole blood by centrifugation at $200 \times g$ for 8 min at room temperature in the presence of prostacyclin (0.1 μg/ml).

3.3. Clot Retraction Assay

1. Rest the platelets in a warm water bath (30°C) for 30 min.

2. Count platelet numbers and adjust if necessary using platelet poor plasma (PPP) to ensure consistency (see Note 5).

3. Fill glass tubes with 745 μl of warm (30°C) Tyrode's–Hepes buffer and 5 μl RBC (used to color the clot (see Note 4)). Place all the tubes in the rack on a plain piece of paper or clean bench where it can be undisturbed for several hours.

4. Add 200 μl PRP (this can be reduced to 50 μl for mouse platelets) to each tube.

5. At this point any test compounds that may affect clot retraction or platelet function should be added and incubated if required (see Note 6).

6. Add 50 μl thrombin 20 unit/ml (final concentration of 120 unit/ml) and mix by flicking the tube quickly.

7. Finally and immediately after adding the thrombin add a sealed glass pipette to the center of the tube and secure with sticky tack (see Note 7).

3.3.1. Summary of Assay Setup

Tyrode's–Hepes buffer	745 μl
PRP	200 μl
Thrombin (20 units/ml stock)	50 μl
RBC	5 μl

Amounts can be reduced for the study of mouse platelets or where the volumes of blood samples are limited, using narrow glass tubes (10 mm diameter or less).

Tyrode's–Hepes buffer	175 µl
PRP	50 µl
Thrombin (20 units/ml stock)	20 µl
RBC	5 µl

3.3.2. Assessment of Results

Images of the experiment enable the kinetics of clot retraction to be observed, and photographs should be taken at time 0 and then every 30 min (or more frequently if desired). Images may also be analyzed to provide numerical data, such as apparent thrombus area, using image analysis software.

Results from clot retraction are assessed numerically by the weight of the clot or weight/volume of extruded serum. Typically an experiment is ended after 120 min (although platelet concentration, temperature, and type of experiment will affect this). A point to end the assay should be chosen before all samples reach completion since differences may be less apparent after several hours since all samples may eventually completely contract (some treatments may result in delay in contraction, rather than inhibition of contraction). This will require optimization for particular experimental outcomes (see Notes 8 and 9).

An example of clot retraction using this procedure is shown in Fig. 2, where the Src family kinase inhibitor PP2 was used to reduce or inhibit clot retraction.

3.3.3. Alternative Methods

One limitation of the standard method is that it may prove difficult to differentiate between coagulation and platelet defects. Plasma-free clot retraction in the presence of purified fibrinogen can be used to investigate the function of integrin $\alpha_{IIb}\beta_3$, without the complication of the effects of the coagulation cascade (10, 13, 14).

4. Notes

1. Sodium citrate is used as an anticoagulant as this reduces the availability of ionized calcium and magnesium without reducing the pH, which would reduce platelet function.

2. Glass pipettes can easily be sealed by holding the tip over a Bunsen flame for a few seconds.

3. Do not add ACD during platelet preparation, since this will lower pH and chelate ions preventing clot retraction from occurring.

4. RBCs are used to color the clot for easy visualization. Use the RBCs that have been pelleted in the preparation of PRP as these are more concentrated than whole blood and will provide more defined color.

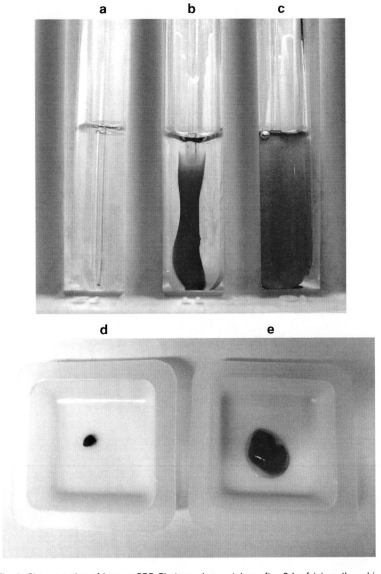

Fig. 2. Clot retraction of human PRP. Photograph was taken after 2 h of (**a**) no thrombin (no red blood cells added to this sample), (**b**) 1 unit/ml thrombin (and DMSO solvent control 0.1%), (**c**) with 1 unit/ml thrombin PP2 (25 μM) Src family kinase inhibitor (dissolved in 0.1% DMSO). Clots were weighed after 3 h (**d**) thrombin control and (**e**) thrombin and PP2 Src inhibitor.

5. To standardize platelet numbers, samples may be be diluted with PPP. With human work this will be more easily achieved as blood volumes are likely to be larger, however, with mouse experiments when comparing test and control animals this may involve taking blood from another animal (test animal for test samples and control for control samples) in order to prepare the necessary PPP. For this type of work, it is worth preparing samples from several animals at once to allow pairing up similar platelet concentrations, and use of PPP can be

shared. PPP may be prepared by centrifugation of PRP ($1,400 \times g$ for 10 min at room temperature) to remove platelets and can be stored at $-20°C$ for future use.

6. When testing compounds that may influence clot retraction it is important to incorporate a vehicle control.

7. Developing clots may collapse if disturbed. The sticky tack is used to ensure that the glass pipette does not move.

8. Setting up experiments in duplicate will allow clots to be weighed to be taken at a point where differences are pronounced while keeping another set to examine if the same end point is eventually reached.

9. Temperature can affect the speed of clot retraction. If the laboratory temperature is variable, assays should be conducted within an incubator at $20°C$.

References

1. Farndale, R. W., Sixma, J. J., Barnes, M. J., and de Groot, P. G. (2004) The role of collagen in thrombosis and hemostasis. *J Thromb Haemost* 2, 561–73.

2. Jackson, S. P., Nesbitt, W. S., and Kulkarni, S. (2003) Signaling events underlying thrombus formation. *J Thromb Haemost 1*, 1602–12.

3. Gibbins, J., Asselin, J., Farndale, R., Barnes, M., Law, C. L., and Watson, S. P. (1996) Tyrosine phosphorylation of the Fc receptor gamma-chain in collagen-stimulated platelets. *J Biol Chem 271*, 18095–9.

4. Coughlin, S. R. (2000) Thrombin signalling and protease-activated receptors. *Nature 407*, 258–64.

5. Shattil, S. J., and Newman, P. J. (2004) Integrins: dynamic scaffolds for adhesion and signaling in platelets. *Blood 104*, 1606–15.

6. Law, D. A., DeGuzman, F. R., Heiser, P., Ministri-Madrid, K., Killeen, N., and Phillips, D. R. (1999) Integrin cytoplasmic tyrosine motif is required for outside-in alphaIIbbeta3 signalling and platelet function. *Nature 401*, 808–11.

7. Shattil, S. J., Kashiwagi, H., and Pampori, N. (1998) Integrin signaling: the platelet paradigm. *Blood 91*, 2645–57.

8. Weyrich, A. S., Denis, M. M., Schwertz, H., Tolley, N. D., Foulks, J., Spencer, E., Kraiss, L. W., Albertine, K. H., McIntyre, T. M., and Zimmerman, G. A. (2007) mTOR-dependent synthesis of Bcl-3 controls the retraction of fibrin clots by activated human platelets. *Blood 109*, 1975–83.

9. Berndt, M. C., and Andrews, R. K. (2007) Full clot retraction: running on mTOR. *Blood 109*, 1791–2.

10. Suzuki-Inoue, K., Hughes, C. E., Inoue, O., Kaneko, M., Cuyun-Lira, O., Takafuta, T., Watson, S. P., and Ozaki, Y. (2007) Involvement of Src kinases and PLCgamma2 in clot retraction. *Thromb Res 120*, 251–8.

11. Schoenwaelder, S. M., Jackson, S. P., Yuan, Y., Teasdale, M. S., Salem, H. H., and Mitchell, C. A. (1994) Tyrosine kinases regulate the cytoskeletal attachment of integrin alpha IIb beta 3 (platelet glycoprotein IIb/IIIa) and the cellular retraction of fibrin polymers. *J Biol Chem 269*, 32479–87.

12. Randriamboavonjy, V., Isaak, J., Fromel, T., Viollet, B., Fisslthaler, B., Preissner, K. T., and Fleming, I. AMPK alpha2 subunit is involved in platelet signaling, clot retraction, and thrombus stability. *Blood 116*, 2134–40.

13. Osdoit, S., and Rosa, J. P. (2001) Fibrin clot retraction by human platelets correlates with alpha(IIb)beta(3) integrin-dependent protein tyrosine dephosphorylation. *J Biol Chem 276*, 6703–10.

14. Seiffert, D., Pedicord, D. L., Kieras, C. J., He, B., Stern, A. M., and Billheimer, J. T. (2002) Regulation of clot retraction by glycoprotein IIb/IIIa antagonists. *Thromb Res 108*, 181–9.

Chapter 9

Visualization and Manipulation of the Platelet and Megakaryocyte Cytoskeleton

Jonathan N. Thon and Joseph E. Italiano

Abstract

Driven by the application of immunofluorescence (IF) microscopy and modern molecular biology approaches to cytoskeletal manipulation, the last 5 years have yielded considerable progress to our understanding of the molecular mechanisms governing megakaryocyte development and platelet biogenesis. Such studies have visualized endomitotic spindle dynamics, characterized the maturation of the demarcation membrane system, delineated the mechanics of organelle transport and microtubule assembly in living megakaryocytes, described the process of platelet production in vivo, and revealed factors contributing to and the mechanisms driving proplatelet production and platelet release. Here, we describe methods to (1) culture megakaryocytes from murine fetal livers, (2) manipulate the tubulin and actin cytoskeleton of both platelets and cultured megakaryocytes, and (3) examine these by live-cell microscopy and fixed-cell immunofluorescence microscopy.

Key words: Platelet, Megakaryocyte, Cell culture, Whole blood, Platelet-rich plasma, Cytoskeleton, Immunofluorescence microscopy, Tubulin, Actin, Nuclear localization

1. Introduction

Platelet production represents the final stage of megakaryocyte development. This process, during which a giant endomitotic cell converts into thousands of individual platelets, is highly specialized and has great clinical significance. Megakaryocytes generate platelets by remodeling their cytoplasm into long proplatelet extensions, which progressively shorten as platelets release from the ends. Proplatelets have been observed extending from megakaryocytes in the bone marrow through junctions in the lining of blood sinuses; it is thought that their release into the circulation precedes their further fragmentation into platelets (1–5). Mice lacking distinct hematopoietic transcription factors have severe

Jonathan M. Gibbins and Martyn P. Mahaut-Smith (eds.), *Platelets and Megakaryocytes:*
Volume 3, Additional Protocols and Perspectives, Methods in Molecular Biology, vol. 788,
DOI 10.1007/978-1-61779-307-3_9, © Springer Science+Business Media, LLC 2012

thrombocytopenia and fail to produce proplatelets in culture, underscoring the correlation to platelet biogenesis in vivo (6–9). Because of the dramatic morphological changes that occur during proplatelet production, the cytoskeletal mechanics that drive platelet production have been the focus of many studies.

Proplatelet (and platelet) formation is highly dependent upon a complex network of protein filaments that represent the molecular struts and girders of the cell. Actin and tubulin are major components of this cytoskeletal network. Actin filaments and microtubules are formed from thousands of molecules that assemble into linear polymers, and can be regulated by a number of poisons which stabilize or disrupt their assembly (10, 11). During proplatelet formation, repeated actin-dependent bending and branching bifurcate the proplatelet shaft and serve to increase the number of free proplatelet ends (12). Branching occurs when a portion of the proplatelet shaft becomes bent, from which some of the microtubules within the loop separate from the bundle to form a new bulge in the shaft. Elongation/sliding of these microtubules then generate a new daughter proplatelet process. The microtubule marginal band is also essential for maintaining resting platelet discoid shape, and platelets in $\beta 1$-tubulin-nullizygous mice show reduced reactivity to thrombin, defective coil contraction preceding granule concentration, and altered granule release upon activation (13). Visualization and manipulation of actin and tubulin dynamics in megakaryocyte progenitors and released platelets, thus, offer a unique insight into the mechanics of megakaryocyte maturation, proplatelet branching, elongation, platelet activation, and granule release.

2. Materials

2.1. Megakaryocyte Cell Culture (Murine)

1. CD-1 strain pregnant mouse at day 13.5 of gestation (Charles River, Wilmington, MA) (see Note 1).

2. Dulbecco's Modified Eagle's Medium (DMEM) (Gibco/BRL, Bethesda, MD) supplemented with 10% fetal bovine serum (FBS), 50 U/mL penicillin, and 50 μg/mL streptomycin (Invitrogen Corporation, Carlsbad, CA) stored at 4°C.

3. Purified recombinant mouse c-Mpl ligand (Thrombopoietin – TPO) (Prospec-Tany TechnoGene Ltd., Rehovot, Israel) (see Note 3) stored in a 1-mL aliquot at –80°C until use and maintained at 4°C thereafter (14). Working solution of DMEM/TPO prepared at a final concentration of 0.5–1% (0.1 μg/mL) in DMEM supplemented with FBS, penicillin, and streptomycin, as previously described, stored at 37°C.

4. BD Falcon standard tissue culture dishes, 6-well and 24-well tissue culture plates (Fisher Scientific).

5. Millipore Steriflip sterile, disposable, vacuum, 0.22 μm filter units (Fisher Scientific).

6. BD Falcon 15 mL and 50 mL conical centrifuge tubes (Fisher Scientific).

7. Solution of 70% (v/v) histological grade ethanol (Fisher Scientific).

8. Fisherbrand standard dissecting scissors and jewelers straight microforceps (Fisher Scientific).

9. BD PrecisionGlide 22 and 25 gauge sterile disposable needles (Fisher Scientific).

10. BD Falcon 40 μm cell strainer (Fisher Scientific).

2.2. Purifying Megakaryocytes from Cell Culture (Day 4)

1. Solutions of 3 and 1.5% (w/v) protease-free (0.22-μm filtered), fatty acid-free bovine serum albumin (BSA) fraction V (Roche Diagnostics GmbH, Mannheim, Germany) dissolved in sterile 1× PBS (Mediatech Inc., Manassas, VA), pH 7.4, stored at 4°C.

2.3. Cytoskeletal Manipulation

1. Paclitaxel (Taxol, Sigma–Aldrich): Microtubule-stabilizing agent.

2. Nocodazole (Sigma–Aldrich): Microtubule-depolymerizing agent.

3. Cytochalasin D (Sigma–Aldrich): Actin-depolymerizing agent.

2.4. Visualization of Cytoskeletal Dynamics in Living Megakaryocytes

1. Semliki forest virus expressing EB3-GFP was generated in the laboratory of Dr. Neils Galjart.

2. GFP-beta-1 tubulin pWZL retroviral vector construct was generated in the laboratory of Dr. Ramesh Shivdasani.

3. Glass bottom, 35 mm culture dish (MatTek Corporation, Ashland, MA).

4. Solution of 65% (v/v) phenol-red-free DMEM complete (Gibco/BRL) and 35% (v/v) MethoCult M-3230 (Stemcell Technologies, Vancouver, Canada), stored at 4°C.

5. Oxyfluor (Oxyrase Inc., Mansfield, OH) prepared at a 1:100 dilution in 65% (v/v) phenol-red DMEM complete and 35% (v/v) MethoCult, stored at 4°C.

6. A range of microscopy systems could be used for this type of analysis. Our system comprises Zeiss Axiovert 200 microscope equipped with a 63× (numerical aperture (NA) = 1.4) oil-immersion objective, a 100 W mercury lamp and an XL-3 incubation chamber (Carl Zeiss, Germany), an Orca II charged coupled device (CCD) camera (Hamamatsu Photonics, Japan), an anti-vibration table (Technical Manufacturing Corporation, Peabody, MA), and running of the Metamorph image analysis software (Universal Imaging Corporation, Molecular Devices, Downingtown, PA).

2.5. Sample Collection

1. Fisherbrand 12 mm circle grade No. 1.5 microscope cover glass (Fischer Scientific) boiled in Milli-Q water (Millipore), then rinsed three times, and stored at RT in 70% ethanol (see Note 18).

2. Poly-L-lysine hydrobromide (Sigma–Aldrich, St. Louis, MO) dissolved at 10 mg/mL in Milli-Q water (Millipore) and stored in single-use, 100-µL aliquots at –20°C. Working solution of poly-L-lysine hydrobromide prepared at a final concentration of 1 mg/mL in Milli-Q water (Millipore).

3. Solution of 4% histology grade formaldehyde (EMD Chemicals Inc., Darmstadt, Germany) dissolved in sterile 1× Hanks Balanced Salt Solution (HBSS, 0.22 µm filtered) (Mediatech Inc.), stored at RT.

4. Solution of 0.5% (v/v) Triton-X-100 dissolved in sterile 1× HBSS (0.22 µm filtered) (Mediatech Inc.), stored at RT.

5. 50 mL immunofluorescence (IF) blocking buffer (0.22 µm filtered): 0.5 g BSA, 0.25 mL 10% sodium azide, 5 mL FCS, dissolved in sterile 1× PBS (Mediatech Inc.), pH 7.4, stored at 4°C.

2.6. Fluorescence Labeling

1. Primary antibody: Rabbit polyclonal antibody against mouse tubulin (Abcam Inc., Cambridge, MA) (see Note 23) diluted 1:500 in IF blocking buffer, stored at 4°C.

2. Wash buffer (0.22 µm filtered): Sterile 1× PBS (Mediatech Inc.) stored at RT.

3. Secondary antibody: Goat anti-rabbit antibody conjugated to an Alexa 488 fluor (Invitrogen Corporation) diluted 1:500 in IF blocking buffer, stored at 4°C (see Note 25).

4. Filamentous actin stain: 1 mg/mL phalloidin conjugated to Alexa 568 fluor (Invitrogen) stored at 4°C. Working solution of phalloidin prepared at a final concentration of 1 µg/mL in IF blocking buffer stored at 4°C (see Note 26).

5. Nuclear stain: 25 µg/mL 4′,6-diamidino-2-phenylindole (DAPI) (Sigma–Aldrich) dissolved in Milli-Q water (Millipore) stored in 20-µL aliquots at –20°C. Working solution of DAPI prepared at a final concentration of 1 µg/mL in IF blocking buffer stored at 4°C.

6. Fisherfinest premium plain 3″×1″×1 mm microscope slides (Fischer Scientific).

7. Aqua Poly/Mount mounting media (Polysciences Inc., Warrington, PA) stored at 4°C (see Note 27).

8. Whatman filter papers (GE Healthcare Biosciences, Pittsburgh, PA).

3. Methods

Primary megakaryocytes can be obtained from fetal livers recovered aseptically from mice between embryonic days 13–15, with optimal megakaryocyte purity on day 13.5 (15). It is imperative that as much connective tissue as possible be removed from livers during dissection to limit the number of contaminating cells in the culture. Hematopoietic stem cells and progenitors are suspended in a medium supplemented with FBS and TPO, and develop into large polyploid megakaryocytes relatively quickly. These begin to dominate the culture by day 3, and can be observed producing proplatelets with yields averaging roughly 60% on day 5 of culture. Gradient sedimentation of the culture on day 4 enriches for round megakaryocytes by virtue of their size, but should not be permitted to go for longer than an hour as smaller (contaminating) cells inevitably sediment as well.

Manipulation of the microtubule and actin cytoskeleton in both megakaryocytes and platelets can be achieved by incubating these cells with a host of cytoskeletal poisons. Taxol/nocodazole stabilize/disrupt microtubule assembly, whereas cytochalasin D inhibits actin polymerization. Visualization of the microtubule and actin cytoskeleton is achieved by spinning the cells down onto poly-L-lysine hydrobromide-coated cover glass (to prevent glass-induced contact activation), fixing, lightly permeabilizing, and then probing for each. The machinery that drives proplatelet elongation includes the highly divergent β1-tubulin isoform that is expressed exclusively and in its predominant isoform in megakaryocytes and platelets (16). While the functional significance of this restricted expression is unclear, β1-tubulin specificity may be used to distinguish megakaryocytes/platelets from nonhematopoietic (contaminating) cells. Less is known about the role of actin in platelet biogenesis; however, repeated actin-dependent bending and branching bifurcate the proplatelet shaft and serve to increase the number of free proplatelet ends from which platelets are thought to release (12). DAPI, a fluorescent nuclear stain that binds DNA, can be used to label both live and fixed cells, and is useful in distinguishing round and proplatelet-producing megakaryocytes (multilobed nuclei) from released proplatelets and individual platelets (both anuclear) in culture.

3.1. Megakaryocyte Cell Culture (Murine)

Flow diagram of megakaryocyte cell culture and gradient sedimentation protocols are shown in Fig. 1.

1. Ensure that female mouse is pregnant and at day 13.5 of gestation (see Note 1).

2. Prepare collection media and dissection dishes ahead of time by transferring 40 mL DMEM (+10% (v/v) FCS and 1% (w/v)

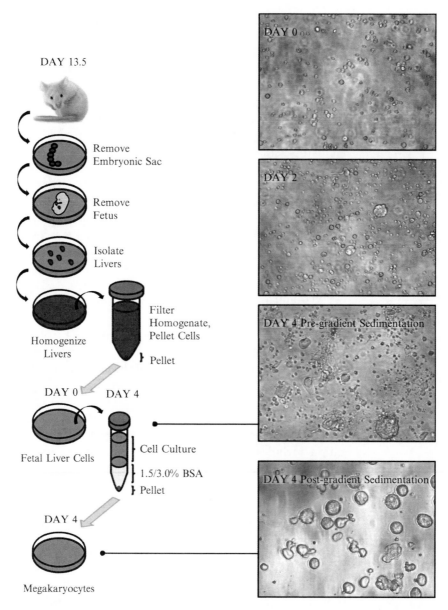

Fig. 1. Fetal liver cell culture and megakayocyte isolation. This figure illustrates a work flow diagram for fetal liver cell culture and megakaryocyte isolation. Phase-contrast pictures were captured using a 20× objective lens and illustrate megakaryocyte maturation and proplatelet production over a period of 4 days. Gradient sedimentation enables isolation of round megakaryocytes from fetal liver cell culture. These begin extending proplatelets and releasing platelets within 24 h of culture.

Pen/Strep) to a 50-mL conical centrifuge tube and filtering the solution through a Steriflip sterile, disposable, vacuum filter unit (see Note 2).

3. Pipette exactly 10 mL of the filtered DMEM solution into one standard tissue culture dish and about 5 mL each into another two standard tissue culture dishes, and place dishes in a 37°C incubator set to 5% CO_2 until needed.

4. Transfer ~12 mL of the filtered DMEM solution into a second 50-mL conical centrifuge tube (mouse fetal liver collection tube) and place in a 37°C water bath. Allow sufficient time for media to warm.

5. Sterilize dissection tools.

6. Gather dissection tools, 15 mL 70% (v/v) ethanol in 50-mL conical centrifuge tube (to soak scissors and forceps during dissection), and mouse fetal collection tube with ~12 mL filtered DMEM solution (warmed). Prepare materials for dissection.

7. Euthanize pregnant mouse by CO_2 asphyxiation and cervical dislocation, then place pregnant mouse belly up on a paper towel, spray abdomen with ethanol, and cut through abdominal skin (careful not to damage the internal organs). Once the initial incision is made, cut two diagonal lines upward from the abdomen to either side of the chest exposing the abdominal cavity.

8. Carefully remove the embryonic sac, which should appear as a series of beads, cutting any connective tissue to free the string. Count fetuses and immediately put in mouse fetal collection tube. Record the number of fetuses (necessary for a later step).

9. Clean tools, discard ethanol container, and dispose of female mouse.

10. Remove all three standard tissue culture dishes from 37°C incubator and transfer fetuses into the first 5-mL culture dish (see Note 2).

11. Use forceps to remove fetuses from the embryonic sac and move them into the second 5-mL culture dish. Soak forceps in ethanol after all fetuses have been transferred.

12. Use forceps to dissect fetal livers and transfer them to the 10-mL standard tissue culture dish (see Note 4).

13. Use syringe fitted with a 22-gauge needle to draw up fetal livers with the media twice, and then use the 25-gauge needle to draw up homogenized livers twice.

14. Transfer homogenate to a 50-mL conical centrifuge tube through 40-μm cell strainer and centrifuge at $650 \times g$ for 5 min to collect hematopoietic stem cells. During centrifugation, prepare DMEM + TPO at a volume equivalent to 2× the number of fetal livers collected + 5 mL (e.g., for ten fetal livers, prepare 25 mL DMEM + TPO) and filter the solution through a Steriflip sterile, disposable, vacuum filter unit.

15. Discard supernatant and resuspend pellet in 3–4 mL DMEM + TPO. Prepare an appropriate number of tissue culture dishes for the calculated volume (step 14) such that each dish contains ~8–10 mL DMEM + TPO. Aliquot the resuspended cell culture evenly between them and place dishes in a 37°C incubator with 5% CO_2 (see Note 5). Allow cells to culture for 96 h (day 4) (see Note 6).

3.2. Purifying Megakaryocytes from Cell Culture (Day 4)

1. Warm 1.5 and 3% BSA (w/v) solutions (filtered) to 37°C in water bath.

2. Add 1.5 mL of 3% BSA to a 15-mL conical centrifuge tube and carefully layer 1.5 mL of the 1.5% BSA solution onto this volume (see Note 7). Repeat this for each megakaryocyte culture dish.

3. Layer ~10 mL of megakaryocyte culture onto the BSA gradient (see Note 8) and transfer gradient to 37°C incubator with 5% CO_2 for 1 h (see Note 9).

4. Prepare 4 mL DMEM + TPO per gradient (filtered) in a 50-mL conical centrifuge tube. Place the 50-mL conical tube in a water bath and allow solutions to warm to 37°C during above incubation.

5. Following gradient sedimentation (1 h), discard supernatant and resuspend megakaryocyte pellet in 4 mL of DMEM + TPO. Transfer megakaryocyte cultures to six-well tissue culture plate and culture at 37°C with 5% CO_2.

6. Released proplatelets and proplatelet-producing megakaryocytes can be further isolated from round megakaryocytes by performing a second gradient sedimentation on culture day 5 (see Subheadings 3.2, step 3–5). Remove the proplatelet/platelet-rich supernatant and centrifuge ($200 \times g$, 5 min) to separate released proplatelet (pellet) and platelet (supernatant) fractions. The proplatelet pellet should be resuspended in 2.5 mL of DMEM + TPO. The BSA fraction is enriched in proplatelet-producing megakaryocytes, and can be cultured separately. The megakaryocyte pellet should be resuspended in 4 mL of DMEM + TPO. Transfer fractions to a six-well tissue culture plate and culture at 37°C with 5% CO_2.

3.3. Cytoskeletal Manipulation

This protocol can be applied to washed platelets.

1. Incubate cells with 5 μM taxol (microtubule-stabilizing agent), 5 μM nocodazole (microtubule-depolymerizing agent), or 5 μM cytochalasin D (actin-depolymerizing agent) for 1 h at 37°C with 5% CO_2 (see Note 10). These can be added to the culture dish during live-cell microscopy to visualize an effect on megakaryocyte/platelet cytoskeleton in real time (see Note 11).

3.4. Visualization of Cytoskeletal Dynamics by Live Cell Microscopy

To visualize cytoskeletal dynamics in living megakaryocytes and platelets, one can retrovirally direct megakaryocytes to express GFP-labeled proteins, and track the movements of these proteins by fluorescence microscopy as megakaryocytes transition into platelets. Alternatively, megakaryocytes can be cultured from mice genetically engineered to express GFP-labeled proteins. This chapter describes two systems we have used to observe microtubule assembly and organization in cultured megakaryocytes in the process of proplatelet formation: the semliki forest virus (SFV) (17) and a pWZL retroviral system (18). Examples of phase-contrast and fluorescence live-cell microscopy images of cultured murine megakaryocytes are shown in Fig. 2.

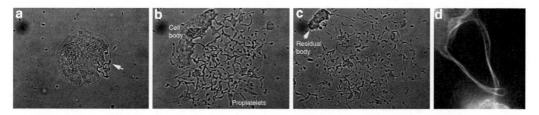

Fig. 2. Formation of proplatelets by a mouse megakaryocyte. Time-lapse sequence of a maturing megakaryocyte, showing the events that lead to elaboration of proplatelets in vitro. (**a**) Platelet production commences when the megakaryocyte cytoplasm starts to erode at one pole. (**b**) The bulk of the megakaryocyte cytoplasm has been converted into multiple proplatelet processes that continue to lengthen and form swellings along their length. These processes are highly dynamic and undergo bending and branching. (**c**) Once the bulk of the megakaryocyte cytoplasm has been converted into proplatelets, the entire process ends in a rapid retraction that separates the released proplatelets from the residual cell body. (**d**) Fluorescence image of a proplatelet-producing megakaryocyte expressing GFP-beta-tubulin. Microtubule bundles line the proplatelet and form loops at the proplatelet ends.

3.4.1. Expression of Constructs Using the Semliki Forest Virus

To visualize microtubule assembly in living megakaryocytes and platelets, SFV-mediated gene delivery was used to express EB3-GFP in cultured megakaryocytes. EB3-GFP labels the growing (+) ends of microtubules and appears as a comet or rocket when visualized using live-cell fluorescence microscopy (*method adopted from the laboratory of Dr. Neils Galjart*) (19–21).

1. Infect cultured megakaryocytes by adding 1 μL of SFV infectious replicons to 400 μL megakaryocyte culture.

2. Visualize EB3-GFP movements by fluorescence microscopy 8–48 h post infection, as outlined below (see Subheading 3.4.3). To visualize microtubule dynamics in platelets, infected megakaryocytes should be cultured for a period of 72–96 h post infection (to allow platelets to be generated) and analyzed for the presence of released platelets (identified by size and presence of a circumferential microtubule band in GFP-expressing cells).

3.4.2. Expression of Constructs Using the pWZL Retroviral Vector

To visualize microtubule organization in living megakaryocytes and platelets, the pWZL retroviral system was used to express GFP-beta-1 tubulin in cultured megakaryocytes (*GFP-beta-1 was generated by, and the megakaryocyte transduction protocol was adopted from, the lab of Dr. Ramesh Shivdasani*) (18).

1. Megakaryocytes isolated from fetal liver cultures should be resuspended in DMEM containing FBS, 5 μg/mL polybrene, and the retroviral supernatant on day 2.5 of culture and incubated at 37°C and 5% CO_2 for 24 h (see Note 12).

2. Replace medium with DMEM complete, and culture megakaryocytes for an additional 48 h prior to analysis. Fluorescence microscopy should be used to identify infected megakaryocytes based on the expression of GFP, as outlined below (see Subheading 3.4.3).

3.4.3. Video Microscopy of Living Megakaryocytes

Coat glass bottom of 35-mm culture dish with 3% (w/v) BSA for 3 h (see Note 13).

1. Replace 3% (w/v) BSA solution with 2 mL semisolid medium containing 65% (v/v) DMEM complete and 35% (v/v) MethoCult M-3230 that has been warmed to 37°C (see Note 14).
   For visualization of EB3-GFP, add 1:100 dilution of Oxyfluor to culture medium (see Note 15).

2. Add 400 μL (resuspended) megakaryocyte culture dropwise into the central recession of the glass-bottom, 35-mm culture dish, and allow megakaryocytes to settle at RT for 15 min.

3. Examine cells on a microscope equipped with a 63× objective (NA 1.4) and 1.6× optivar. Cells should be maintained at 37°C with constant 5% CO_2 gas infusion using an incubation chamber. An antivibration table should be used to minimize vibrations (see Note 16).

4. Frames should be captured at 60-s, 5-, or 10-min intervals, and illumination should be shuttered between exposures (see Note 17).
   For visualization of EB3-GFP, frames should be captured every 2–5 s, with an average image capture time of 100–500 ms.

5. A stage micrometer should be used for measurement of the rate of proplatelet elongation, and length measurements should be made using image analysis software.
   For visualization of EB3-GFP, comet velocity should be determined by dividing the distance travelled by the time elapsed. Include only comets that can be followed for a minimum of 15 s.

3.5. Sample Collection

This protocol can be applied to washed platelets with the following changes, highlighted in italics.

1. Place washed cover glass in a 24-well tissue culture plate and coat with 300–500 μL poly-L-lysine hydrobromide (1 mg/mL) for 15–20 min (see Note 19).

2. Wash cover glass with Milli-Q water and add 500 μL platelet buffer (filtered, warmed to 37°C).
   For washed platelets, substitute platelet buffer with 4% (w/v) formaldehyde (filtered, warmed to 37°C) (see Note 20).

3. Add 250 μL (resuspended) cell culture to each well and centrifuge at 1,850×*g* for 5 min (see Notes 21 and 22).
   For washed platelets, proceed directly to step 5.

4. Discard supernatant and fix megakaryocytes with 300–500 μL of 4% (w/v) formaldehyde solution (filtered, warmed to 37°C) for 15 min.

5. Wash cover glass with 1× HBSS (filtered).

6. Permeabilize megakaryocytes with 0.5% (v/v) Triton-X-100 for 5 min.

7. Wash cover glass with 1× PBS (filtered).

8. Add 500 μL of IF blocking buffer (filtered) to each well and store overnight at 4°C.

3.6. Fluorescence Labeling

1. Discard blocking buffer, add 300–500 μL of primary antibody (Tubulin, see Notes 23 and 24), and allow samples to incubate for 1 h at RT.

2. Wash wells 3× with 1× PBS (filtered), allowing 3 min per wash.

3. Add 300–500 μL of secondary antibody (see Notes 24–26) and incubate for 1 h in the dark at RT.

4. Wash wells 3× with 1× PBS (filtered), allowing 3 min per wash.

5. Add 300–500 μL fluorescent phalloidin (1 μg/mL, see Notes 24–27) and incubate for 20 min in the dark at RT.

6. Wash wells 3× with 1× PBS (filtered), allowing 3 min per wash.

7. Add 300–500 μL DAPI (1 μg/mL, see Notes 24–26) and incubate for 30 min in the dark at RT.

8. Wash wells 3× with 1× PBS (filtered), allowing 3 min per wash. Leave 300 μL 1× PBS in well with cover glass.

9. Carefully remove cover glass from well and invert into a drop of mounting medium on a microscope slide (see Notes 27–30).

10. Allow samples to dry overnight in the dark at RT.

3.7. Visualization

1. Microscope slides should be viewed under phase-contrast microscopy (to locate the cells and identify the focal plane) and under IF microscopy. Excitation at 488 nm induces Alexa Fluor fluorescence (green emission) for the tubulin, excitation at 568 nm induces Alexa Fluor fluorescence (red emission) for actin, and excitation at 364 nm induces DAPI fluorescence (blue emission) (see Notes 31–43). Software can be used to overlay the phase-contrast and fluorescence images. Examples of phase-contrast and fluorescence images of tubulin-, actin-, and DAPI-stained murine cultured megakaryocytes and washed platelets are shown in Fig. 3.

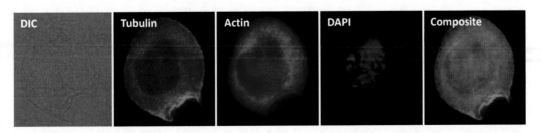

Fig. 3. Phase-contrast and immunofluorescence microscopy pictures of isolated megakaryocytes. Megakaryocytes were probed with a rabbit polyclonal antibody against detyrosinated tubulin (*green*). Phalloidin was used to label actin (*red*), and DAPI was used to label the megakaryocyte nucleus (*blue*). Megakaryocytes were analyzed by differential interference contrast (DIC) and immunofluorescence microscopy, and a composite image was generated. Pictures were taken with a 63× objective lens.

4. Notes

Megakaryocyte Cell Culture (Murine) (~2 h)

1. Use a large litter strain, such as the CD-1 mouse. Primary megakaryocytes are obtained from fetal livers recovered aseptically from mice between embryonic days 13–15, and cultured in the presence of TPO. This represents the most desirable culture method for studying proplatelet production and platelet release in vitro due to a rapid megakaryocyte maturation period (4 days) and high proplatelet yields (~60%).

2. All work should be conducted in a biological safety cabinet with proper sterile technique to minimize the risk of culture contamination.

3. Purified recombinant mouse c-Mpl ligand can be prohibitively expensive. Fetal liver cultures can instead be supplemented with 1% tissue culture supernatant from a fibroblast cell line engineered to secrete recombinant human TPO. This cell line was produced in Paris in the labs of William Vainchenker and Francoise Wendling, and is available for general use and distribution (14).

4. Livers look like a dark red spot in the middle of the fetus. Try to remove as much of the connective tissue (pink) from the liver as possible. Be patient!

5. Cells should be resuspended in a volume equivalent to 2× the number of fetal livers and split into an appropriate number of standard tissue culture dishes (8–12 mL each) to ensure suitable concentrations for culture.

6. By convention, the initial day of culture is considered day 0. Day 4 refers to 96 h post culture.

Purifying Megakaryocytes from Cell Culture (Day 4) (~1.5 h, with Incubation)

7. Use a serological pipette for continuous flow and tilt the 15-mL conical during transfer to minimize disruption of the BSA layers.

8. Each megakaryocyte culture dish should yield one gradient. Ensure that cultures are resuspended well before loading gradients as cells will have likely settled to the bottom of the dish.

9. Do not allow cultures to sediment on gradient for longer than 1 h as this significantly affects the purity of your megakaryocyte isolation.

Cytoskeletal Manipulation (~1 h)

10. Taxol stabilizes the microtubule structure by binding to the β subunit of tubulin, inhibiting the normal function of microtubule breakdown while permitting further microtubule assembly. Conversely, nocodazole functions by interfering with the polymerization of tubulin monomers, thus promoting microtubule disassembly. Vinblastine can serve as an alternative microtubule-depolymerizing agent and bind to tubulin, thereby inhibiting the assembly of microtubules. Cooling the cell culture to 4°C for 1 h has a similar (albeit reversible) effect, driving microtubule disassembly; tubulin reassembly can be induced by returning the culture to 37°C. Latrunculin and cytochalasin D are both potent inhibitors of actin polymerization and promote actin disassembly.

11. Taxol, nocodazole, vinblastine, latrunculin, and cytochalasin should be dissolved in DMSO, as per the manufacturer's instructions, and stored at –20°C. Compounds are typically used at a final concentration range of 0.1–10 μM and cell cultures should not exceed 0.5% DMSO (v/v).

Visualization of Cytoskeletal Dynamics in Living Megakaryocytes (Variable)

12. The retroviral supernatant should be varied to establish the best concentration for expression.

13. Three percent BSA serves to inhibit contact activation of megakaryocytes with the cover glass.

14. The methylcellulose in the MethoCult reduces the viscosity of the medium and stabilizes the proplatelets.

15. This step is necessary to quench free radical formation.

16. If your microscope system does not contain a fully enclosed incubation chamber, an alternative method can be used. Fill video dish with 65% Leibowitz L-15 medium and 35% DMEM complete with no phenol red (Gibco/BRL). Add cells, as described, and completely cover the dish with topical light USP mineral oil (Roxane Laboratories, Columbus, Ohio) to prevent evaporation. Cells should be maintained at 37°C with a bipolar temperature controller (Medical Systems Corporation, Greenville, NY).

17. Constant exposure to the halogen bulb causes megakaryocytes to retract proplatelets and round up.

Sample Collection (~1 h)

18. High-resolution microscope objectives are designed to be used with a 170-μm-thick cover glass, which corresponds to grade No. 1.5. Microscope objectives perform poorly when imaging through thinner or thicker glass. Using the wrong thickness of

cover glass introduces spherical aberrations in the image, which results in dimmer fluorescence.

19. Poly-L-lysine hydrobromide serves to adhere megakaryocytes/platelets to the cover glass. Ensure that cover glass is uniformly coated and not floating above solution. Add more solution or press cover glass to the bottom of well if necessary.

20. Washed platelets may activate during centrifugation and upon contact with glass. Fixation at this step ensures that platelets remain inactive during adhesion to cover glass. Conversely, fixation of megakaryocytes at this step dramatically reduces their capacity to adhere to cover glass and should be avoided.

21. Mechanical disruptions of cells should be kept to a minimum in order to avoid compromising the quality of the image. For cells on coverslips, aspirate solutions gently from the side of the dish with one hand and add new solutions gently to the other side with your other hand. Do not drop solutions directly onto the cells and do not allow cells to dry out.

22. Use microscope (20× magnification for megakaryocytes, 40× magnification for platelets) to ensure that cells have adhered onto coverslips. If need be, spin again.

Fluorescence Labeling (~3.5 h, with Incubation)

23. A noncommercial alternative is the SuperGlu rabbit polyclonal antibody against detyrosinated tubulin produced by Dr. Chloe Bulinski, Columbia University, NY.

24. In general, the best fluorescence is obtained by sequentially incubating in individual antibodies (primary, secondary, primary, secondary). Primary/secondary antibodies and all stains should be titrated to minimize nonspecific labeling, as any bleed through between fluorescence channels during observation makes it almost impossible to assess colocalization. Bleed through can be minimized with the appropriate choice of band-pass excitation and emission filters. A filter that blocks the first color can also be inserted into the light path when viewing the second color. Single-label controls should be initially included to confirm the general localization of test antigens.

25. Choose a bright (high quantum yield, high extinction coefficient) and photostable fluorophore.

26. If you notice high background, filter antibody through a 0.2-μm syringe filter or spin in a microcentrifuge.

27. Do not incubate for longer than 20 min; highly fluorescent molecules, such as fluorescent phalloidin, are frequently sticky and increase background staining with longer incubations.

28. Ensure that mounting media is of high glycerol content (enhances fluorescence) and contains polyvinylalcohol-vinylacetate

(prevents photobleaching) to retain fluorescent stains. A higher glycerol concentration improves the fluorescence image at the expense of the differential interference contrast (DIC, Nomarski) image. For IF microscopy, 90% glycerol provides one with the brightest and highest resolution image. If a DIC image is required as well, one should use a mounting media comprising 50% glycerol.

29. Use ONLY about 6–8 μL of mounting media per 18-mm cover glass. The solution should slowly spread to the edges after you place the coverslip onto the media. Whatman filter paper should be used to wick the excess solution that leaks out the sides.

30. Use one set of microforceps to create a small hook on one tip of a second (dedicated) pair of microforceps. This hook significantly improves your ability to remove the cover glass from the well without breaking the cover glass.

Visualization

The following methods dramatically improve the quality of the generated IF images. Adapted from Waters, J.C. 2009 (22).

31. Use a high numerical aperture objective lens with lowest acceptable magnification (63×).

32. Use a cooled CCD camera with at least 60% quantum efficiency.

33. Turn off the room lights.

34. Use software to monitor intensity values in the image to choose the best acquisition settings. Avoid high camera gain and saturating pixels in the image.

35. Use the full dynamic range of the camera for fixed specimens.

36. For live-cell imaging, mount specimen in minimally fluorescent medium (e.g., without phenol red).

37. For live-cell work, it is often necessary to sacrifice signal-to-noise ratio (SNR) to minimize specimen exposure to light and maintain cell health and viability.

38. *Always* save the raw images (use either no compression or lossless compression, if necessary).

39. Use flat-field correction to correct for uneven illumination during image processing.

40. Be certain that image processing software that is used prior to quantification preserves relative intensity values.

41. Subtract local background value from intensity measurements, and use only uncompressed images for analysis.

42. Calculate and report the error in your measurements.

43. The following information *must* be documented and should be included with the Materials and Methods or Figure Legend.

Adapted from Waters, J.C. 2009 (22).

1. Manufacturer and model of microscope.

2. Objective lens magnification, NA, and correction for aberration (e.g., 60×1.4 NA Plan-Apochromat).

3. Fluorescence filter set manufacturer and part number and/or the transmission and bandwidth (e.g., 490/30).

4. Illumination light source (including wavelength for laser illumination).

5. Camera manufacturer and model.

6. Software program(s) and version.

7. Manufacturer and model of other acquisition hardware, including confocal, filter wheels, focus motors, motorized stage, shutters, etc.

8. Image acquisition settings, including exposure times, gain, and binning.

9. Other acquisition parameters, including focus step size (for z-series), time between images (for time lapse), etc.

10. Description of image-processing routine used to create figures.

11. Description of segmentation and image analysis routine, and method of validation.

Acknowledgments

This work was supported in part by the National Institutes of Health Grant HL68130 (J.E.I.). J.E.I. is an American Society of Hematology Junior Faculty Scholar.

References

1. Choi ES, Nichol JL, Hokom MM, Hornkohl AC, Hunt P. Platelets generated in vitro from proplatelet-displaying human megakaryocytes are functional. Blood. 1995;85:402–413.

2. Becker RP, DeBruyn PP. The transmural passage of blood cells into myeloid sinusoids and the entry of platelets into the sinusoidal circulation; a scanning electron microscopic investigation. Am J Anat. 1976;145:1046–1052.

3. Behnke O. An electron microscope study of the rat megakaryocyte. II. Some aspects of platelet release and microtubules. J Ultrastruct Res. 1969;26:111–129.

4. Radley JM, Scurfield G. The mechanism of platelet release. Blood. 1980;56:996–999.

5. Tavassoli M, Aoki M. Migration of entire megakaryocytes through the marrow-blood barrier. British Journal of Hematology. 1981;48:25–29.

6. Lecine P, Villeval J, Vyas P, Swencki B, Yuhui X, Shivdasani RA. Mice lacking transcription factor NF-E2 provide in vivo validation of the proplatelet model of thrombocytopoiesis and show a platelet production defect that is intrinsic to megakaryocytes. Blood. 1998;92:1608–1616.

7. Shivdasani R, Orkin S. The transcriptional control of hematopoiesis. Blood. 1996;87:4025–4039.

8. Shivdasani RA. Molecular and transcriptional regulation of megakaryocyte differentiation. Stem Cells. 2001;19:397–407.

9. Shivdasani RA, Rosenblatt MF, Zucker-Franklin D, et al. Transcription factor NF-E2 is required for platelet formation independent of the actions of thrombopoietin/MGDF in megakaryocyte development. Cell. 1995;81:695–704

10. Tablin F, Castro M, Leven RM. Blood platelet formation in vitro. The role of the cytoskeleton in megakaryocyte fragmentation. J Cell Sci. 1990;97 (Pt 1):59–70.

11. Handagama PJ, Feldman BF, Jain NC, Farver TB, Kono CS. In vitro platelet release by rat megakaryocytes: effect of metabolic inhibitors and cytoskeletal disrupting agents. Am J Vet Res. 1987;48:1142–1146.

12. Hartwig JH, Italiano JE, Jr. Cytoskeletal mechanisms for platelet production. Blood Cells Mol Dis. 2006;36:99–103.

13. White JG, Krivit W. An ultrastructural basis for the shape changes induced in platelets by chilling. Blood. 1967;30:625–635.

14. Villeval JL, Cohen-Solal K, Tulliez M, et al. High thrombopoietin production by hematopoietic cells induces a fatal myeloproliferative syndrome in mice. Blood. 1997;90:4369–4383.

15. Italiano JE, Jr., Lecine P, Shivdasani RA, Hartwig JH. Blood platelets are assembled principally at the ends of proplatelet processes produced by differentiated megakaryocytes. J Cell Biol. 1999;147:1299–1312

16. Lecine P, Italiano JE, Jr., Kim SW, Villeval JL, Shivdasani RA. Hematopoietic-specific beta 1 tubulin participates in a pathway of platelet biogenesis dependent on the transcription factor NF-E2. Blood. 2000;96:1366–1373.

17. Patel SR, Richardson JL, Schulze H, et al. Differential roles of microtubule assembly and sliding in proplatelet formation by megakaryocytes. Blood. 2005;106:4076–4085.

18. Schulze H, Korpal M, Bergmeier W, Italiano JE, Jr., Wahl SM, Shivdasani RA. Interactions between the megakaryocyte/platelet-specific beta1 tubulin and the secretory leukocyte protease inhibitor SLPI suggest a role for regulated proteolysis in platelet functions. Blood. 2004;104:3949–3957.

19. Stepanova T, Slemmer J, Hoogenraad CC, et al. Visualization of microtubule growth in cultured neurons via the use of EB3-GFP (end-binding protein 3-green fluorescent protein). J Neurosci. 2003;23:2655–2664.

20. Ehrengruber MU, Lundstrom K, Schweitzer C, Heuss C, Schlesinger S, Gahwiler BH. Recombinant Semliki Forest virus and Sindbis virus efficiently infect neurons in hippocampal slice cultures. Proc Natl Acad Sci USA. 1999;96:7041–7046.

21. Lundstrom K, Schweitzer C, Rotmann D, Hermann D, Schneider EM, Ehrengruber MU. Semliki Forest virus vectors: efficient vehicles for in vitro and in vivo gene delivery. FEBS Lett. 2001;504:99–103.

22. Waters JC. Accuracy and precision in quantitative fluorescence microscopy. J Cell Biol. 2009;185:1135–1148.

Chapter 10

Measurement of Platelet Microparticles

Jeffrey I. Zwicker, Romaric Lacroix, Françoise Dignat-George, Barbara C. Furie, and Bruce Furie

Abstract

Platelet microparticles are submicron vesicles that can support thrombin generation on externalized negatively charged phospholipids. Increased numbers of circulating platelet microparticles have been investigated as the basis of hypercoagulability in a variety of prothrombotic conditions. Measurement of platelet microparticles is not standardized and a number of preanalytic considerations can influence accurate analysis. We describe methodology for light scatter-based flow cytometry as well as impedance-based flow cytometry for the enumeration and characterization of platelet microparticles.

Key words: Platelet microparticles, Microparticles, Impedance-based flow cytometry

1. Introduction

Originally labeled as "platelet dust," platelet microparticles promote coagulation activation and thrombin generation following exposure of negatively charged phospholipids (i.e., phosphatidylserine) on the external membrane leaflet (1, 2). The physiologic relevance of platelet microparticles continues to be explored as pathologic increases in numbers of circulating platelet microparticles are implicated in a variety of prothrombotic disorders such as heparin-induced thrombocytopenia (3), sickle cell crisis (4), inflammation (5–7), and myocardial infarction (8). The plasma concentration of platelet-derived microparticles reported in the literature varies considerably both in healthy individuals and in disease, this is likely due to heterogeneous methodologies for detection as well as considerable variation in preanalytic conditions (9, 10).

Platelet microparticles are generally defined as vesicular structures derived from platelets or megakaryocytes that measure < 1 μm

Jonathan M. Gibbins and Martyn P. Mahaut-Smith (eds.), *Platelets and Megakaryocytes:*
Volume 3, Additional Protocols and Perspectives, Methods in Molecular Biology, vol. 788,
DOI 10.1007/978-1-61779-307-3_10, © Springer Science+Business Media, LLC 2012

in diameter. In contradistinction, platelet exosomes are generally considered smaller (<0.1 μm), more homogenous, and lack phosphatidylserine on the outer membrane leaflet (11). Platelet microparticles may be generated in vitro using both physiologic (e.g., thrombin, collagen, or adenosine diphosphate) (12–14) or nonphysiologic (e.g., calcium ionophore) (13) agonists. Stimuli for the production of platelet-derived microparticles in vivo may also include shear stress (15, 16), oxidative species (17), complement proteins C5b-9 (18), or CD40 ligand (19). Circulating platelet (CD41+) microparticles were previously assumed to be exclusively derived from the mature platelet following activation. However, murine data suggest that megakaryocytes are the predominant source of CD41+ microparticles under physiologic conditions (20).

Microparticles are commonly characterized by the loss of membrane asymmetry. Phospholipid transporters regulate the asymmetry of membrane phospholipids with phosphatidylserine and phosphatidylethanolamine concentrated on the inner leaflet. Platelet activation is followed by an influx of cytosolic calcium that modulates phospholipid transport activity (i.e., flippase, floppase, scramblase). The exposure of phosphatidylserine on the outer membrane is followed by the formation of platelet microparticles. The relevance of phospholipid transport activity and platelet microparticle generation is exemplified by Scott syndrome, a rare bleeding disorder characterized by deficient scramblase activity, deficient phosphatidylserine exposure, and decreased microparticle generation (21). However, the precise role of phospholipid transport proteins in the generation of platelet microparticles is uncertain as mice deficient in scramblase (PSLCR1 or PLSCR3) do not exhibit loss of phosphatidylserine asymmetry (22, 23), platelet-derived microparticle formation occurs in the absence of phosphatidylserine exposure (24), and a large percentage of circulating platelet microparticles lack externalized phosphatidylserine (25). Other mechanisms that regulate platelet microparticle formation continue to be explored, including apoptotic pathways (i.e., caspase-3) and calpain-mediated phosphorylation of cytoskeletal proteins (26, 27).

1.1. Measurement of Platelet Microparticles by Flow Cytometry

Flow cytometry is the standard technique used to quantify and size platelet-derived microparticles in plasma. It is widely available and provides a reliable means to characterize cells by size and antigen expression thus providing an attractive method to characterize microparticles. However, there are notable limitations in the application of flow cytometry based on light scattering for the analysis of microparticles. Forward scatter is dependent on variables that are independent of size including wavelength of incident light, particle shape, absorptive material, and relative refractive indices of particles and suspension medium (28). One key limitation of the use of light scattering for sizing of biologic microparticles is that polystyrene beads have a much greater refractive index than the microparticles. The latest generation flow cytometers appear to have improved sensitivity for the detection of microparticles although smaller particles are likely below the lower limits of detection (29). Alternative methods for microparticle detection are currently being evaluated, including atomic force microscopy (30), dynamic light scatter (31), capture or activity assays (10, 32), and impedance-based flow cytometry (33).

We currently utilize an impedance-based flow cytometer for microparticle sizing, characterization, and concentration determination (33). This instrument (Cell Lab Quanta SC, Beckman Coulter, Miami, FL) is based on the Coulter principle whereby a particle suspended in an electrolyte solution enters an aperture and displaces an equal volume of electrolyte solute. The displaced solute increases the impedance across the circuit generating a voltage spike that is proportional to the volume of the microparticle. Simultaneous measurement of fluorescence using a 488-nm laser permits characterization of surface antigens. The sensitivity of instrument into the submicron range largely is dependent on the aperture size. Refinement of Coulter-type instruments in the 1970s enabled the characterization of submicron structures including bacteria and viruses (34, 35). The standard diameter aperture for the Cell Lab Quanta SC is 125 μm with a resolution of approximately 2.5 μm. We have modified this instrument using smaller (40 or 25 μm diameter) flow cells to improve submicron resolution. Although this methodology offers an improved sensitivity compared with standard flow cytometers based on light scattering, very small particles cannot be resolved under the current configuration.

2. Materials

2.1. Impedance Analysis

1. Beckman Coulter Cell Lab Quanta SC flow cytometer (Beckman Coulter, Miami, FL) has been refitted with a smaller aperture flow cell. The standard instrument is equipped with a 125-μm diameter flow cell which is not adequate for microparticle analysis. In general, the lower limit of microparticle sizing by impedance is 2% of the aperture diameter. Our current system is configured with a 40-μm diameter aperture for microparticle analysis.

2. Fluorescent polystyrene microspheres (Dragon Green 0.78 μm fluorescent microspheres from Bangs Laboratories, Fishers, IN).

3. Sheath fluid (Iso-Diluent, Cat#629967, Beckman Coulter, Miami, FL).

4. Antibodies.

   (a) For platelet microparticle analysis we utilize mouse anti-human CD41a-FITC and mouse irrelevant isotype-matched IgG1-FITC (BD Diagnostics, Franklin Lakes, NJ). A number of platelet antigens can be considered for platelet microparticle analysis including antibodies to platelet glycoproteins IIb (CD41), IX (CD42a), Ibα (CD42b), IIIa (CD61), or P selectin (CD62b) (36). The expression of these antigens can be influenced by platelet activation or mechanism of microparticle formation (37, 38). All the concentrations of antibodies must be optimized. The stock concentration of our CD41a-FITC from BD Diagnostics is 6.25 µg/ml while that of isotype-matched control immunoglobulin IgG1-FITC is 50 µg/ml and thus the isotype control antibody is diluted 1:8 in filtered buffer.

2.2. Standard Flow Cytometry Analysis

1. FC500 Cytomics flow cytometer (Beckman Coulter, Miami, FL) with optimal laser alignment and clean flow cell.

2. Megamix Beads (BioCytex, Marseille, France): a blend of monodisperse fluorescent beads with three diameters: 0.5, 0.9, and 3 µm. The specifications of these beads indicate an accurate numerical ratio of 2:1 between the 0.5- and 0.9-µm beads.

3. Sheath fluid (Iso-Flow, Beckman Coulter) passed through a 0.22-µm filter.

4. Labeling Reagents: For platelet microparticle analysis, we use

   (a) Annexin V labeled with FITC (Annexin V FITC/7AAD, Beckman Coulter ref PN IM3614).

   (b) Mouse antihuman CD41-PE antibody (clone PL2-49, BioCytex catalogue number 5112-PE100T).

   (c) Mouse antihuman IgG1-PE isotype-matched control antibody (clone 2DNP-2H11/2H12 BioCytex, catalogue number 5108-PE100T)
       All antibodies are centrifuged for 2 min at 13,000×g before use.

5. Counting beads: Flow count® fluorospheres, Beckman Coulter ref 7547053.

3. Methods

A number of strategies have been utilized to evaluate microparticle populations in platelet-free plasma (PFP). Herein is a description of the methodology for impedance-based and light-scatter flow cytometry. However, there are a number of preanalytic considerations that can influence platelet microparticle measurements which need to be standardized for accurate analysis (see Notes 1–7).

3.1. Enumeration and Sizing of Polystyrene Beads by Impedance-Based Flow Cytometry

1. Prior to running biologic samples, polystyrene beads are evaluated on a daily basis to ensure proper calibration of size and counting (Fig. 1). For these purposes, we utilize fluorescent 0.78-µm beads from Bangs Laboratories. These have a known concentration and are accurately sized by ±0.5%. Based on a stock concentration of $4.2 \times 10^6/\mu l$, beads are diluted 1:10,000 in filtered sheath solution.

2. Aliquot 1.5 ml of vortexed bead dilution into sample cup

3. Set fluorescence photomultiplier tube voltage (PMT) such that beads are detected between the 10^2 and 10^3 FL1 units on the x-axis. Note that fluorescent beads typically require lower PMT voltage than biologic specimens labeled with fluorescent antibodies. The fluorescent threshold for detection of events is set above 10^1. Events are triggered based on fluorescence rather than electronic volume.

4. Analyze size and concentration of polystyrene beads to ensure accuracy of the instrument.

3.2. Measuring Microparticles by Impedance-Based Flow Cytometry

1. Each sample will be analyzed in the presence of irrelevant isotype immunoglobulins and again in the presence of antigen-specific antibodies. Antibody concentrations need to be optimized. For platelet microparticle analysis using BD

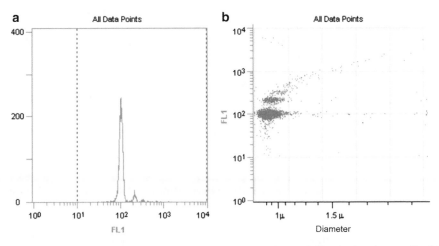

Fig. 1. Calibration of the Cell Lab Quanta SC using polystyrene microspheres. Calibration beads are utilized to validate microparticle measurements. Analysis is performed using 0.78-µm fluorescent beads diluted 1:10,000 in filtered sheath fluid. (a) Histogram of fluorescence events recorded. (b) The same fluorescence events are shown relative to microsphere diameter. The predominance of events are shown at 10^2 (FL1 arbitrary units) corresponding to a diameter of 0.78 µm. Populations at increased fluorescence and size represent microsphere doublets.

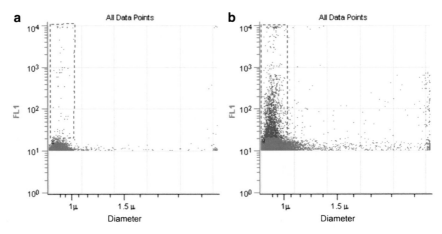

Fig. 2. Measurement of plasma microparticles by impedance-based flow cytometry. The presence of different populations of microparticles can be determined using fluorescent antibodies in platelet poor plasma with a Cell Lab Quanta SC. (**a**) Irrelevant isotype-matched antibodies are used to establish background fluorescence PMT voltage. The *rectangular* region defining microparticles is based on size (diameter <1 µm) and fluorescence (FL1). (**b**) Antigen-positive microparticles are quantified in the *rectangular* region above baseline fluorescence (*x*-axis, diameter; *y*-axis, fluorescence).

Diagnostics CD41a antibodies as above the antibody (20 µl) is added to 40 µl of plasma sample (Fig. 2).

2. Incubate 30 min in the dark.

3. Perform a 1:50 dilution of sample-antibody solution sheath solution passed through a 0.22-µm filter prior to use in the assay (e.g., 10 µl with 490 µl of sheath solution).

4. Analyze the microparticles treated with isotype-matched control antibody and set fluorescence PMT voltage such that background fluorescence is just above lower discriminator (10^1 units on FL1). This is illustrated in Fig. 2, where microparticle events are captured based on a diameter set between 0 and 1.0 µm and fluorescence (FL1) above the background observed for fluorescently labeled isotype-matched antibodies.

5. Setup software analysis for two samples (one treated with isotype-matched control antibody and one treated with specific antibody) to be run in triplicate.

6. Analyze labeled samples.

7. The concentration of platelet microparticles is calculated by subtracting the background number of microparticles determined using the isotype-matched control from the number of microparticles determined using the antigen-specific antibody.

3.3. Measuring Microparticles by Light Scattering-Based Flow Cytometry

Light scattering-based flow cytometry is the method most commonly used for analysis of microparticle populations. Forward light scatter is the most appropriate parameter to analyze microparticle size. The limit of resolution on forward scatter between instrument noise and microparticles depends, in part, on fine optical adjustments, fluidics, and clean optics. These variables are prone to change over time and can vary between platforms. Therefore, in order to obtain reliable and reproducible results on standard flow cytometers, a bead-based strategy is recommended to identify the lower sensitivity of size-related forward scatter (39). These latex beads are not intended to give an accurate sizing of the microparticle population but to standardize analysis. It should be noted that because manufacturers of flow cytometers do not use the same optical design for forward scatter measurements results may vary for different instruments (29, 40). However, with adequate optical centering the relative position between the same reference beads and microparticle populations does not appear to vary significantly for specific cytometry instruments.

3.3.1. Standardization Protocol

This protocol establishes methods for reproducible measurement of platelet microparticles using a standardized window with Megamix beads (Fig. 3). Megamix beads are run every day to check the stability of the instrument. The 0.9-μm beads identify the upper limit of the microparticle region; the 0.5-μm beads identify the forward scatter threshold and lower microparticle limit at a standardized level corresponding to its median value. The analysis is performed using the intrinsic 2:1 ratio of 0.5 μm beads to 0.9 μm beads, always including the same percentage of the 0.5-μm beads in the analysis. The following steps describe how to create the standardized window for microparticle analysis.

Step 1

1. Run Megamix beads with discriminator (threshold) on FL1 parameter.
2. Gate each individual bead subset on SS (Side Scatter) × FL1 cytogram.
3. Build forward scatter histogram for 0.5 and 0.9 μm beads.
4. At this step, all the 0.5 and 0.9 μm beads are included in the analysis; so, percentages of 66 and 33% should be found for 0.5 and 0.9 μm beads, respectively.

Step 2

1. Run Megamix beads with discriminator on forward scatter parameter.
2. Monitor percentage of 0.5 μm beads (left peak) and optimize forward scatter settings to approach 50%. This threshold level is selected as the lower limit for microparticle analysis.
3. Visualize side scatter/forward scatter cytogram. Create an elliptical autogate on the 0.9 μm beads. Set up the microparticle gate to tangent the autogate. This defines the upper limit of the microparticle window.

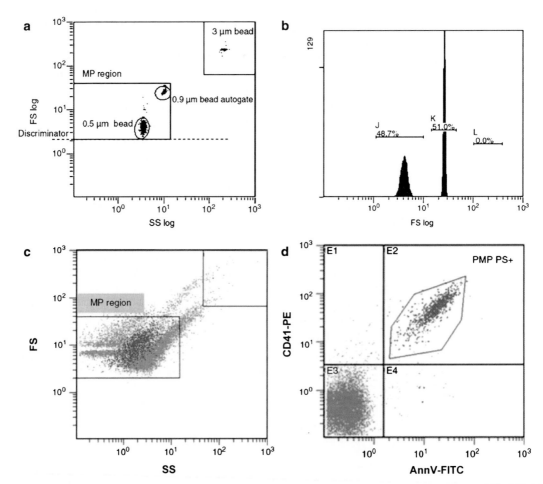

Fig. 3. Measurement of plasma microparticles by scatter-based flow cytometry using a standardized protocol. Platelet microparticles can be reproducibly measured on a standardized window by Megamix beads using a Cytomics FC500. *Upper panel*: Construction of the microparticle (MP) region in the microparticle protocol. (**a**) On a FS (forward scatter) log SS (side scatter) log cytogram, the microparticle region is defined by the lower part of the FS cut-off (discriminator, shown here as a *dotted line*), set up using the 0.5-μm bead percentage, and the upper part of the 0.9-μm bead autogate. (**b**) FS distribution of the Megamix submicrometer beads. In this bimodal histogram gated on the union of singlets from the 0.5 to 0.9 μm beads, the percentage of 0.5 μm beads (region J) varies with FS settings. The value of 48.7% shown here corresponds to a cut-off that rejects about half of the 0.5-μm beads. *Lower panel*: Platelet microparticle (PMP) analysis in the microparticle (MP) protocol. (**c**) Phosphatidylserine (PS+) PMP forward scatter. PS + PMPs are seen as *red dots* among background (*gray dots*). (**d**) Dual fluorescence analysis of a platelet-free plasma stained with annexin V (AnnV)–fluorescein isothiocyanate (FITC) and anti-CD41–phycoerythrin (PE). PS + PMPs are represented as *red dots*, and background noise or other MPs as *gray dots* (this figure is kindly provided by S. Robert).

3.4. Measuring Platelet Microparticles by Forward-Based Scatter Flow Cytometry

1. Mix 30 μl of PFP with 10 μl CD41-PE + 10 μl AnnV-FITC (Tube 3). For the negative control, the same volume of PFP is mixed with 10 μl CD41-PE + 10 μl AnnV-FITC (Tube 1, which will be diluted in a buffer without calcium) and 10 μl IgG1-PE + 10 μl AnnV-FITC (Tube 2). Vortex.

2. Importantly, antibody concentrations were adapted by BioCytex so that 10 μl CD41-PE matched (12.5 μg/ml) with 10 μl IgG1-PE (25 μg/ml).

3. Incubate 30 min at room temperature with samples protected from light.

4. Add 500 μl of binding buffer (2.5 mM $CaCl_2$ in HEPES buffer) to tubes 2 and 3 and 500 μl of PBS (phosphate-buffered saline) to tube 1.

5. Vortex Flow-Count® fluorospheres for 10 s and add 30 μl to each vial.

6. Keep stained samples at room temperature and immediately proceed to flow cytometry analysis.

7. Run each labeled sample using the analysis protocol described above (time stop set to 1 min; low flow-rate), with a maximum delay of 30 min after the end of labeling.

8. Analyze platelet microparticles in a dual fluorescence cytogram (FL1 × FL2) gated on the microparticle scatter gate established above.

9. Set FL1 and FL2 compensations using the samples from tubes 1 and 2.

10. Platelet microparticles are defined as AnnV+/CD41+ ± AnnV–/ CD41+ events. Absolute values are determined using counting beads (Flow-Count, Beckman-Coulter) as follows: platelet microparticles (events/μl) = (platelet microparticle events × beads concentration)/beads events.

This standardized strategy provides reliable results for platelet microparticle enumeration. However due to limitations in resolution, very small platelet microparticles are not detected under these conditions. Recent technological improvements of different cytometry instruments, such as Gallios (Beckman Coulter) and BD influx (Becton Dickinson) have partially overcome these limitations opening the avenue for improved detection of microparticle populations of smaller sizes (29).

3.5. Immunoassays

Enzyme linked immunosorbent assays (ELISA) can also be utilized for the analysis of platelet microparticles (32). These assays have certain potential advantages including increased sensitivity for the evaluation of microparticles below the resolution of flow cytometry or identification of particles with weakly expressed antigens. For instance, 96-well plates can be coated with antiplatelet antibodies (e.g., antiglycoprotein IX) to bind platelet microparticles. The number of platelet microparticles can be quantified using biotinylated antiglycoprotein Ib using a peroxidase substrate (32). Michelsen et al. described a method that quantifies platelet microparticles by an immunofluorometric assay using lanthanide (europium chelate) as time-delayed fluorescent reporter (41). Hybrid assays such as Zymuphen-MP® technology (Hyphen Biomed, Andresy, F) combine solid-phase capture of MP on platelet antibodies and determine prothrombinase activity (42, 43).

4. Notes

Accurate measurement of platelet microparticles in plasma requires the standardization and optimization of a number of preanalytic variables. Manipulations from venipuncture to thawing a specimen can influence the number of platelet microparticles measured in a specimen. The optimal methods to control for such variables have not been established but may include the following:

1. Venipuncture

   Platelet activation may occur during phlebotomy. Perform venipuncture with a 20- or 21-gauge needle applied to the antecubital vein following the application of a light tourniquet. The first 2–3 ml of blood should be discarded (38, 44).

2. Collection tubes

   Activation of platelets prior to centrifugation can generate platelet microparticles. Sodium citrated Vacutainer® tubes (BD Diagnostics, Franklin Lakes, NJ) are most commonly utilized to prevent platelet activation. The addition of platelet inhibitors as contained in CTAD tubes (0.109 M buffered sodium citrate, 15 mM theophylline, 3.7 mM adenosine, and 0.198 mM dipyridamole) may be more effective in inhibiting in vitro platelet microparticle formation (38, 45–47). EDTA leads to dissociation of platelet glycoprotein IIb/IIIa heterodimers (48, 49) and platelet activation thus should be avoided (36).

3. Transportation

   Transporting blood in upright rather than horizontal position may limit the number of platelet microparticles formed in vitro (50).

4. Time to centrifugation

   Platelet-derived microparticles increase over time following blood collection (51). Microparticle numbers appear to be stable up to 2 h in citrated or CTAD tubes (45, 47).

5. Centrifugation

   The centrifugation speeds and time vary widely among studies (44). Typically, the initial spin to generate cell-free plasma is $1,500$–$2,500 \times g$ for 15–20 min. However, platelets may persist after a single centrifugation step (36) and an additional centrifugation step of $13,000 \times g$ for 2 min ensures PFP (39, 52). Following centrifugation, the plasma should be carefully aspirated leaving the bottom 1 cm undisturbed (9).

6. Storage

   If samples must be stored prior to analysis, snap freezing of PFP at $-80°C$ can be considered (39).

7. Thawing

Different methods have been described for thawing microparticle samples including slow thaw on ice (9). Thawing for several minutes at 37°C is preferred by some investigators (45, 53, 54). A single freeze–thaw does not increase platelet microparticle numbers appreciably but repeated freeze–thaw cycles may significantly alter the number of platelet microparticles (4, 51, 55). Stability of platelet microparticles should be confirmed for specific applications.

Acknowledgment

We thank S. Robert for his input on forward scatter-based flow cytometry.

References

1. Berckmans, R. J., Neiuwland, R., Boing, A. N., Romijn, F. P., Hack, C. E., and Sturk, A. (2001) Cell-derived microparticles circulate in healthy humans and support low grade thrombin generation, *Thromb Haemost 85*, 639–646.

2. Wolf, P. (1967) The nature and significance of platelet products in human plasma, *Br J Haematol 13*, 269–288.

3. Hughes, M., Hayward, C. P., Warkentin, T. E., Horsewood, P., Chorneyko, K. A., and Kelton, J. G. (2000) Morphological analysis of microparticle generation in heparin-induced thrombocytopenia, *Blood 96*, 188–194.

4. Shet, A. S., Aras, O., Gupta, K., Hass, M. J., Rausch, D. J., Saba, N., Koopmeiners, L., Key, N. S., and Hebbel, R. P. (2003) Sickle blood contains tissue factor-positive microparticles derived from endothelial cells and monocytes, *Blood 102*, 2678–2683.

5. Daniel, L., Fakhouri, F., Joly, D., Mouthon, L., Nusbaum, P., Grunfeld, J. P., Schifferli, J., Guillevin, L., Lesavre, P., and Halbwachs-Mecarelli, L. (2006) Increase of circulating neutrophil and platelet microparticles during acute vasculitis and hemodialysis, *Kidney Int 69*, 1416–1423.

6. Pereira, J., Alfaro, G., Goycoolea, M., Quiroga, T., Ocqueteau, M., Massardo, L., Perez, C., Saez, C., Panes, O., Matus, V., and Mezzano, D. (2006) Circulating platelet-derived microparticles in systemic lupus erythematosus. Association with increased thrombin generation and procoagulant state, *Thromb Haemost 95*, 94–99.

7. Pamuk, G. E., Vural, O., Turgut, B., Demir, M., Umit, H., and Tezel, A. (2006) Increased circulating platelet-neutrophil, platelet-monocyte complexes, and platelet activation in patients with ulcerative colitis: a comparative study, *Am J Hematol 81*, 753–759.

8. van der Zee, P. M., Biro, E., Ko, Y., de Winter, R. J., Hack, C. E., Sturk, A., and Nieuwland, R. (2006) P-selectin- and CD63-exposing platelet microparticles reflect platelet activation in peripheral arterial disease and myocardial infarction, *Clin Chem 52*, 657–664.

9. Biro, E., Nieuwland, R., and Sturk, A. (2004) Measuring circulating cell-derived microparticles, *J Thromb Haemost 2*, 1843–1844.

10. Jy, W., Horstman, L. L., Jimenez, J. J., Ahn, Y. S., Biro, E., Nieuwland, R., Sturk, A., Dignat-George, F., Sabatier, F., Camoin-Jau, L., Sampol, J., Hugel, B., Zobairi, F., Freyssinet, J. M., Nomura, S., Shet, A. S., Key, N. S., and Hebbel, R. P. (2004) Measuring circulating cell-derived microparticles, *J Thromb Haemost 2*, 1842–1851.

11. Heijnen, H. F., Schiel, A. E., Fijnheer, R., Geuze, H. J., and Sixma, J. J. (1999) Activated platelets release two types of membrane vesicles: microvesicles by surface shedding and exosomes derived from exocytosis of multivesicular bodies and alpha-granules, *Blood 94*, 3791–3799.

12. Takano, K., Asazuma, N., Satoh, K., Yatomi, Y., and Ozaki, Y. (2004) Collagen-induced generation of platelet-derived microparticles in whole blood is dependent on ADP released from red

blood cells and calcium ions, *Platelets 15*, 223–229.

13. Tans, G., Rosing, J., Thomassen, M. C., Heeb, M. J., Zwaal, R. F., and Griffin, J. H. (1991) Comparison of anticoagulant and procoagulant activities of stimulated platelets and platelet-derived microparticles, *Blood 77*, 2641–2648.

14. Siljander, P., Carpen, O., and Lassila, R. (1996) Platelet-derived microparticles associate with fibrin during thrombosis, *Blood 87*, 4651–4663.

15. Holme, P. A., Orvim, U., Hamers, M. J., Solum, N. O., Brosstad, F. R., Barstad, R. M., and Sakariassen, K. S. (1997) Shear-induced platelet activation and platelet microparticle formation at blood flow conditions as in arteries with a severe stenosis, *Arterioscler Thromb Vasc Biol 17*, 646–653.

16. Miyazaki, Y., Nomura, S., Miyake, T., Kagawa, H., Kitada, C., Taniguchi, H., Komiyama, Y., Fujimura, Y., Ikeda, Y., and Fukuhara, S. (1996) High shear stress can initiate both platelet aggregation and shedding of procoagulant containing microparticles, *Blood 88*, 3456–3464.

17. Nardi, M. A., Gor, Y., Feinmark, S. J., Xu, F., and Karpatkin, S. (2007) Platelet particle formation by anti GPIIIa49-66 Ab, Ca2+ ionophore A23187, and phorbol myristate acetate is induced by reactive oxygen species and inhibited by dexamethasone blockade of platelet phospholipase A2, 12-lipoxygenase, and NADPH oxidase, *Blood 110*, 1989–1996.

18. Sims, P. J., Faioni, E. M., Wiedmer, T., and Shattil, S. J. (1988) Complement proteins C5b-9 cause release of membrane vesicles from the platelet surface that are enriched in the membrane receptor for coagulation factor Va and express prothrombinase activity, *J Biol Chem 263*, 18205–18212.

19. Prasad, K. S., Andre, P., He, M., Bao, M., Manganello, J., and Phillips, D. R. (2003) Soluble CD40 ligand induces beta3 integrin tyrosine phosphorylation and triggers platelet activation by outside-in signaling, *Proc Natl Acad Sci USA 100*, 12367–12371.

20. Flaumenhaft, R., Dilks, J. R., Richardson, J., Alden, E., Patel-Hett, S. R., Battinelli, E., Klement, G. L., Sola-Visner, M., and Italiano, J. E., Jr. (2009) Megakaryocyte-derived microparticles: direct visualization and distinction from platelet-derived microparticles, *Blood 113*, 1112–1121.

21. Zwaal, R. F., Comfurius, P., and Bevers, E. M. (2005) Surface exposure of phosphatidylserine in pathological cells, *Cell Mol Life Sci 62*, 971–988.

22. Wiedmer, T., Zhao, J., Li, L., Zhou, Q., Hevener, A., Olefsky, J. M., Curtiss, L. K., and Sims, P. J. (2004) Adiposity, dyslipidemia, and insulin resistance in mice with targeted deletion of phospholipid scramblase 3 (PLSCR3), *Proc Natl Acad Sci USA 101*, 13296–13301.

23. Zhou, Q., Zhao, J., Wiedmer, T., and Sims, P. J. (2002) Normal hemostasis but defective hematopoietic response to growth factors in mice deficient in phospholipid scramblase 1, *Blood 99*, 4030–4038.

24. Razmara, M., Hu, H., Masquelier, M., and Li, N. (2007) Glycoprotein IIb/IIIa blockade inhibits platelet aminophospholipid exposure by potentiating translocase and attenuating scramblase activity, *Cell Mol Life Sci 64*, 999–1008.

25. Connor, D. E., Exner, T., Ma, D. D., and Joseph, J. E. The majority of circulating platelet-derived microparticles fail to bind annexin V, lack phospholipid-dependent procoagulant activity and demonstrate greater expression of glycoprotein Ib, *Thromb Haemost 103*, 1044–1052.

26. Shcherbina, A., and Remold-O'Donnell, E. (1999) Role of caspase in a subset of human platelet activation responses, *Blood 93*, 4222–4231.

27. Fox, J. E., Austin, C. D., Reynolds, C. C., and Steffen, P. K. (1991) Evidence that agonist-induced activation of calpain causes the shedding of procoagulant-containing microvesicles from the membrane of aggregating platelets, *J Biol Chem 266*, 13289–13295.

28. Shapiro, H. M. (2003) *Practical Flow Cytometry*, 4th ed., Wiley-Liss, New York.

29. Lacroix, R., Robert, S., Poncelet, P., and Dignat-George, F. (2010) Overcoming limitations of microparticle measurement by flow cytometry, *Semin Thromb Hemost 36*, 807–818.

30. Yuana, Y., Oosterkamp, T. H., Bahatyrova, S., Ashcroft, B., Garcia Rodriguez, P., Bertina, R. M., and Osanto, S. (2010) Atomic force microscopy: a novel approach to the detection of nanosized blood microparticles, *J Thromb Haemost 8*, 315–323.

31. Lawrie, A. S., Albanyan, A., Cardigan, R. A., Mackie, I. J., and Harrison, P. (2009) Microparticle sizing by dynamic light scattering in fresh-frozen plasma, *Vox Sang 96*, 206–212.

32. Nomura, S., Inami, N., Shouzu, A., Omoto, S., Kimura, Y., Takahashi, N., Tanaka, A., Urase, F., Maeda, Y., Ohtani, H., and Iwasaka, T. (2009) The effects of pitavastatin, eicosapentaenoic acid and combined therapy on platelet-derived microparticles and adiponectin in hyperlipidemic, diabetic patients, *Platelets 20*, 16–22.

33. Zwicker, J. I., Liebman, H. A., Neuberg, D., Lacroix, R., Bauer, K. A., Furie, B. C., and Furie, B. (2009) Tumor-derived tissue factor-bearing microparticles are associated with

venous thromboembolic events in malignancy, *Clin Cancer Res 15*, 6830–6840.

34. DeBlois, R. W., and Wesley, R. K. (1977) Sizes and concentrations of several type C Oncorna-viruses and Bacteriophage T2 by the resistive-pulse technique, *J Virol 23*, 227–233.

35. Feuer, B. I., Uzgiris, E. E., Deblois, R. W., Cluxton, D. H., and Lenard, J. (1978) Length of glycoprotein spikes of vesicular stomatitis virus and Sindbis virus, measured in situ using quasi elastic light scattering and a resistive-pulse technique, *Virology 90*, 156–161.

36. Enjeti, A. K., Lincz, L. F., and Seldon, M. (2007) Detection and measurement of microparticles: an evolving research tool for vascular biology, *Semin Thromb Hemost 33*, 771–779.

37. Perez-Pujol, S., Marker, P. H., and Key, N. S. (2007) Platelet microparticles are heteroge-neous and highly dependent on the activation mechanism: studies using a new digital flow cytometer, *Cytometry A 71*, 38–45.

38. Kim, H. K., Song, K. S., Lee, E. S., Lee, Y. J., Park, Y. S., Lee, K. R., and Lee, S. N. (2002) Optimized flow cytometric assay for the mea-surement of platelet microparticles in plasma: pre-analytic and analytic considerations, *Blood Coagul Fibrinolysis 13*, 393–397.

39. Robert, S., Poncelet, P., Lacroix, R., Arnaud, L., Giraudo, L., Hauchard, A., Sampol, J., and Dignat-George, F. (2009) Standardization of platelet-derived microparticle counting using calibrated beads and a Cytomics FC500 routine flow cytometer: a first step towards multicenter studies?, *J Thromb Haemost 7*, 190–197.

40. Nebe-von-Caron, G. (2009) Standardization in microbial cytometry, *Cytometry A 75*, 86–89.

41. Michelsen, A. E., Wergeland, R., Stokke, O., and Brosstad, F. (2006) Development of a time-resolved immunofluorometric assay for quantifying platelet-derived microparticles in human plasma, *Thromb Res 117*, 705–711.

42. Mallat, Z., Benamer, H., Hugel, B., Benessiano, J., Steg, P. G., Freyssinet, J. M., and Tedgui, A. (2000) Elevated levels of shed membrane microparticles with procoagulant potential in the peripheral circulating blood of patients with acute coronary syndromes, *Circulation 101*, 841–843.

43. Hugel, B., Socie, G., Vu, T., Toti, F., Gluckman, E., Freyssinet, J. M., and Scrobohaci, M. L. (1999) Elevated levels of circulating procoagu-lant microparticles in patients with paroxysmal nocturnal hemoglobinuria and aplastic anemia, *Blood 93*, 3451–3456.

44. Piccin, A., Murphy, W. G., and Smith, O. P. (2007) Circulating microparticles: pathophysi-ology and clinical implications, *Blood Rev 21*, 157–171.

45. Mody, M., Lazarus, A. H., Semple, J. W., and Freedman, J. (1999) Preanalytical requirements for flow cytometric evaluation of platelet activa-tion: choice of anticoagulant, *Transfus Med 9*, 147–154.

46. Bode, A. P., Orton, S. M., Frye, M. J., and Udis, B. J. (1991) Vesiculation of platelets dur-ing in vitro aging, *Blood 77*, 887–895.

47. Pearson, L., Thom, J., Adams, M., Oostryck, R., Krueger, R., Yong, G., and Baker, R. (2009) A rapid flow cytometric technique for the detec-tion of platelet-monocyte complexes, activated platelets and platelet-derived microparticles, *Int J Lab Hematol 31*, 430–439.

48. Ma, Y., and Wong, K. (2007) Reassociation and translocation of glycoprotein IIB-IIIA in EDTA-treated human platelets, *Platelets 18*, 451–459.

49. Nomura, S., Suzuki, M., Kido, H., Yamaguchi, K., Fukuroi, T., Yanabu, M., Soga, T., Nagata, H., Kokawa, T., and Yasunaga, K. (1992) Differences between platelet and microparticle glycoprotein IIb/IIIa, *Cytometry 13*, 621–629.

50. Dignat-George, F. (2010) Standardization and pre-analytical variables, in *Micro & Nanovesicles in Health and Disease*, Oxford, England.

51. Keuren, J. F., Magdeleyns, E. J., Govers-Riemslag, J. W., Lindhout, T., and Curvers, J. (2006) Effects of storage-induced platelet microparticles on the initiation and propaga-tion phase of blood coagulation, *Br J Haematol 134*, 307–313.

52. Dignat-George, F., Sabatier, F., Camoin, L., and Sampol, J. (2004) Numeration of circulating microparticles of various cellular origin by flow cytometry, *J Thromb Haemost 2*, 1844–1845.

53. Simak, J., and Gelderman, M. P. (2006) Cell membrane microparticles in blood and blood products: potentially pathogenic agents and diag-nostic markers, *Transfus Med Rev 20*, 1–26.

54. Trummer, A., De Rop, C., Tiede, A., Ganser, A., and Eisert, R. (2009) Recovery and compo-sition of microparticles after snap-freezing depends on thawing temperature, *Blood Coagul Fibrinolysis 20*, 52–56.

55. Ahmad, S., Amirkhosravi, A., Langer, F., Desai, H., Amaya, M., and Francis, J. L. (2005) Importance of pre-analytical variables in the measurement of platelet derived microparticles, *J Thromb Haemost 3*, OR372.

Chapter 11

Assessing Protein Synthesis by Platelets

Hansjörg Schwertz, Jesse W. Rowley, Neal D. Tolley,
Robert A. Campbell, and Andrew S. Weyrich

Abstract

Platelets are anucleate cytoplasts that circulate in the bloodstream for approximately 9–11 days. Because they lack nuclei, platelets were considered incapable of protein synthesis. However, studies over the last decade have revealed that platelets use a variety of translational control pathways to synthesize proteins.

A variety of protocols can be employed to assess protein synthesis by platelets. These protocols are scattered throughout the literature and, more often than not, lack critical details. In this chapter, we thoroughly outline methods used in our laboratory to assess protein synthesis by platelets.

Key words: Platelet protein synthesis, Pre-mRNA processing, MicroRNA processing, Protein Translation, Spliceosome factors

1. Introduction

For years, platelets were considered "sacs of glue," whose primary and only function is to halt blood flow. This simplistic caricature, however, has lost momentum as we moved into the twenty-first century. The last decade has revealed that platelets regulate a variety of complex biologic and pathologic responses, such as innate and acquired immunity, angiogenesis, and cell proliferation (1, 2). Regulation occurs at several checkpoints of control, including the release of stored products and de novo synthesis of new proteins (1).

Protein synthesis by platelets was discovered over 60 years ago (reviewed in Weyrich et al. (3)). Steady progress has occurred since that time with significant advances in the last 10 years. It is now clear that anucleate platelets use "nuclear-like" pathways to synthesize proteins. These include pre-mRNA processing (4–7), microRNA processing (8), and translation of mRNA via the mammalian target of

Jonathan M. Gibbins and Martyn P. Mahaut-Smith (eds.), *Platelets and Megakaryocytes:*
Volume 3, Additional Protocols and Perspectives, Methods in Molecular Biology, vol. 788,
DOI 10.1007/978-1-61779-307-3_11, © Springer Science+Business Media, LLC 2012

rapamycin (mTOR) (9–11). Our current understanding of protein synthesis in platelets has been the focus of recent reviews (3, 12, 13). Here, we briefly describe studies that point toward the biologic significance of protein synthesis by platelets. Then we describe standard protocols used in our laboratory to study this process in platelets.

1.1. The Biologic Significance of Protein Synthesis by Platelets

Our group began studying protein synthesis in the late 1990s after we serendipitously discovered that activated platelets accumulate protein for B-cell lymphoma 3 (Bcl-3) (11). Since that time, Bcl-3 has served as a standard for identifying and dissecting signaling pathways that control translation in platelets (9–11, 14). In addition, it is now known that Bcl-3 regulates platelet cytoskeletal events providing initial insights into the biological function of protein synthesis by platelets. Specifically, after it accumulates in platelets, Bcl-3 binds the SH3 domain of Fyn (11). Interactions with Fyn facilitate the retraction of extracellular fibrin networks, a response that is impaired in Bcl-3 deficient murine platelets (10, 11).

Synthesis of Bcl-3 by activated platelets is controlled at multiple levels, including outside-in signaling through integrin $\alpha_{IIb}\beta_3$ (9, 14). Bcl-3 synthesis is markedly reduced in platelets from Glanzmann thrombasthenic patients while adherence to fibrinogen, the primary ligand for integrin $\alpha_{IIb}\beta_3$, increases Bcl-3 synthesis in platelets isolated from healthy subjects (14). Recent studies by Yang and colleagues (15) demonstrate that fibrinogen also controls the expression of P-selectin in murine and human platelets. The authors found that P-selectin expression was significantly decreased in platelets from fibrinogen-deficient $(Fg^{-/-})$ mice or hypofibrinogenemic patients. Transfusion of fibrinogen, rescued platelet P-selectin levels in the $(Fg^{-/-})$ mice, a response that required interactions with β_3 integrins. P-selectin levels also accumulated after platelets were cultured ex vivo in the presence of fibrinogen (15).

Several other proteins are known to be synthesized by platelets, including interleukin-1β (4, 7, 16). Synthesis of IL-1β is regulated by pre-mRNA splicing in the cytoplasm, a novel mechanism of posttranscriptional control (4, 7). Requisite spliceosome factors and IL-1β pre-mRNA get transferred from megakaryocytes to platelets (4). After splicing occurs, the mature transcript is subsequently translated into IL-1β protein over several hours (4). Newly synthesized IL-1β is primarily sequestered in platelets or packaged into microparticles that increase the adhesiveness of endothelial cells for polymorphonuclear leukocytes (16). A recent development indicates that platelet-derived IL-1 amplifies inflammation in arthritis (17, 18). Using in vivo murine and human models of arthritis, Boilard and colleagues (17) found that platelet-derived microparticles were pro-inflammatory, eliciting cytokine responses from synovial fibroblasts via IL-1α and β. The authors demonstrated that collagen, which is present in the inflamed synovium, served as the key trigger for generating platelet microparticles. IL-1 bearing microparticles are generated over hours providing compelling evidence that alternative functions of platelets, such as protein synthesis, may contribute to prolonged inflammatory responses.

Protein synthesis may also be important for new functions of platelets that have recently been identified. In this regard, we found that human platelets develop new cell bodies when they are cultured ex vivo in growth medium or whole blood (19). The new cell bodies are packed with α-granules, respiring mitochondria, and they express requisite adhesion molecules. Progeny formation occurs over several hours and is associated with striking shifts in protein expression patterns. In this regard, numerous proteins are upregulated or downregulated in cultured platelets indicating that the platelet proteome is very dynamic. We found that cultured platelets constitutively synthesize protein (19) and others have demonstrated that platelets have a functional proteasome pathway (20, 21). This suggests that much like nucleated cells, platelets are continually synthesizing and degrading proteins as they circulate in the bloodstream.

Our understanding of protein synthesis by platelets has increased rapidly over the last decade, in the large part, because we have access to more sophisticated technologies and high-end reagents. As studies emerge, the contributions of protein synthesis to traditional and new functions of platelets will become clearer. In this chapter, we outline current strategies employed by our laboratory to study protein synthesis in anucleate platelets keeping in mind that we are refining and updating our approaches on a continual basis.

1.2. Studying Protein Synthetic Pathways in Platelets

Our group commonly employs several methods to prove that platelets synthesize proteins. These include initial detection of specific mRNAs in platelets by semiquantitative and real-time PCR, in situ detection of the mRNA, and rigorous assessment of protein accumulation in the presence or absence of translational inhibitors. The types of protein assays we use depend on the candidate protein but usually involve a combination of well-established techniques such as flow cytometry, Western analyses, ELISA, and immunocytochemistry. Ultimately, we determine if the protein of interest incorporates amino acids. The methods we use for amino acid incorporation are described below.

2. Materials

Most of the materials used for protein synthesis studies in platelets are widely available. Specific recommendations and comments for critical materials are provided below.

2.1. Assessing Protein Synthesis by Platelets

1. Labeled amino acids: L-[^{35}S] methionine, L-[^{35}S] cysteine (11.0 mCi/mL) from vendor of choice.

2. Culture Medium: DMEM without L-glutamine, L-methionine, L-cysteine with glucose. When using this medium, it is important to add back 4.5 g/L of L-glutamine to the medium. Complete DMEM (all amino acids) with glucose 4.5 g/L is also needed for steps that involve washing with complete medium. M199 without L-methionine and L-cysteine can also be used but needs to be specially ordered.

3. Prostaglandin E$_1$ (PGE$_1$): Analytical grade suspended as a 3-mM stock solution in DMSO.

4. Platelet enrichment: CD45 MicroBeads (Miltenyi Biotec) and either an autoMACS Separator cell sorter or columns and magnetic stand.

5. Translation inhibitors: 2–50 μg cycloheximide or 0.1–1.0 mM puromycin.

2.2. Measuring Global Protein Synthesis in Platelet Lysates and Supernatants

1. Cell lysis buffer: 1× RIPA (1× phosphate-buffered saline (PBS), 1% NP-40, 0.5% sodium deoxycholate, and 0.1% SDS in water).

2. Protease inhibitors: A combination of 0.1 mg/mL PMSF, 0.3 TIU/mL aprotinin, and 1 mM sodium orthovanadate. Many vendors supply premixed cocktails and tablets that also work well.

3. Protein precipitation and capture: 10% trichloroacetic acid (TCA) in water; 2.5-cm glass microfiber filter discs (GF/C, Whatman); filter tray (such as a disposable bottle top filter, a Hirsch or a Büchner funnel); vacuum; and 99.7% acetone (Fisher Scientific, A949-4).

4. Counting radiolabel incorporation: 20-mL scintillation vials (Beckman Poly-Q Vials, 566350), and scintillation fluid (Optifluor, PerkinElmer, 6013199).

5. 1× Laemmli buffer: 62.5 mM Tris–HCl pH 6.8, 10% glycerol, 2% SDS, 5% β-mercaptoethanol, and 0.001% bromophenol blue.

2.3. Characterizing Differential Synthetic Patterns by Platelets

1. 2-Dimensional gel electrophoresis apparatus: 2D electrophoresis is the preferred method for the separation of labeled proteins for identification. Both an IEF (strip or tube) and slab gel apparatus will be needed. For simple demonstration of protein synthesis only a slab, or 1D gel, apparatus is required.

2. Sample buffers: 2D lysis buffer [7 M urea, 2 M thio urea, 4% CHAPS, 40 mM TRIS buffer, 1 tablet Complete Mini Protease Inhibitor (Roche) per 10 mL solution]; rehydration buffer [7 M urea, 2 M thio Urea, 2% CHAPS, 0.5% IPG buffer,

0.001% bromophenol blue (water soluble) 18.2 mM DTT]; equilibration buffer [50 mM Tris–HCl, pH 8.8, 6 M urea, 30% (v/v) glycerol, 2% (w/v) SDS, 0.001% bromophenol blue, 65 mM DTT].

3. IEF strips: Several suppliers and pH ranges are available. When beginning, start with a wide pH range and as proteins of interest are identified move to narrower pH ranges.

4. Agarose: 0.5% agarose and 0.001% bromophenol blue in SDS-PAGE electrophoresis buffer, to overlay the IEF strip on SDS-PAGE gel.

5. SDS-PAGE gel: A gradient gel ranging from 4 to 20% or a 12% continuous gel is good for viewing most proteins, however, other percentage gels may be used for looking at proteins of specific molecular weights. A standard 12% SDS-PAGE gel contains, 12% acrylamide mix (National Diagnostics EC-890), 0.378 M Tris–HCl (pH 8.8), 0.1% SDS, 0.1% ammonium persulfate (APS in water), and 0.04% TEMED made up in double distilled water.

6. Fluorographic reagent (GE Healthcare): This is not required but may decrease the time needed for exposure of dried gel to film.

7. Detection of radio-labeled proteins: Two systems are common for detection of radiolabeled proteins on gel. The first is to dry the gel on a gel dryer and expose the gel to film. A newer technique is to expose the gel (wet or dry depending on the system) to a radiosensitive screen which can be placed in the requisite scanner and the image is captured digitally. The required materials will depend on the system used.

2.4. Confirming the Synthesis of a Specific Protein

1. Concentrating samples: Molecular weight cutoff filters such as Millipore Microcon Filters of the correct size to retain protein of interest.

2. Capture protein of interest: ELISA plate coated with antibody specific to protein of interest.

2.5. Assessing Global Protein Synthesis in Platelets by Confocal Microscopy

1. Click-iT L-azidohomoalanine (AHA) (Invitrogen C10102).

2. Paraformaldehyde (PFA): 4% PFA in 1× PBS pH 7.2 and 0.2 μm filtered.

3. PBS: 0.01 M phosphate buffer, 0.0027 M potassium chloride, and 0.137 M sodium chloride, pH 7.4, at 25°C.

4. Bovine serum albumin (BSA): 3% (w/v) in PBS. Several suppliers available.

5. Confocal microscope: Olympus FV1000 Spectral confocal, Olympus FV300 confocal, BD Pathway Confocal Bioimager, and a custom 2-Photon confocal microsope have all been used successfully.

3. Methods

3.1. Assessing Protein Synthesis by Platelets

In the following sections, we describe classic methods used to study protein synthesis in platelets. The techniques described below can be used to quantitate global protein synthesis (Subheading 3.2), determine differential synthetic expression patterns (Subheading 3.3), confirm that specific proteins are being synthesized by platelets (Subheading 3.4), and visualize global protein synthesis in platelets by microscopy (Subheading 3.5).

In regards to platelet preparation, we strongly encourage investigators to deplete contaminating leukocytes from washed platelet preparations so they can be confident that the synthesized proteins are derived from platelets. We briefly outline our strategy for removing leukocytes from human platelet isolates (Subheadings 3.1.1–3.1.2), which we employ on a daily basis for mRNA profiling and protein expression/synthesis studies. A similar bead-depletion protocol is used to isolate leukocyte-reduced murine platelets (not shown).

3.1.1. Isolation of Leukocyte-Reduced Platelets

Investigators use a variety of strategies to reduce the number of leukocytes present in platelet-rich plasma (PRP), such as increased speed and/or duration of the centrifugation step to sediment greater numbers of large cells (i.e., leukocytes and large platelets) from 2 to 3 μm platelets. Although these types of strategies typically reduce the number of contaminating leukocytes in platelet preparations, additional enrichment steps, such as positive selection of leukocytes and/or passing platelets through leukoreduction filters, ensures greater purity (see Note 1).

3.1.2. Preparing Leukocyte-Reduced Human Platelets for Protein Synthesis Studies

1. Isolate PRP using standard procedures (further details available in Platelets and Megakaryocytes, Vol. 1, Functional Assays, Chapter 2).

2. Add fresh PGE_1 (300 nM) to the PRP, centrifuge at $500 \times g$, remove the plasma, and resuspend the platelet pellet in PIPES saline glucose (PSG) containing fresh PGE_1.

3. Add 3 μL of CD45 MicroBeads (Miltenyi Biotec) per mL of original PRP to the cells (i.e., if you have started with 10 mL PRP add 30 μL MACS CD45 beads) (see Note 2).

4. Incubate platelets with microbeads for 20 min at room temperature (RT), mixing periodically.

5. After this incubation period, add fresh PGE_1 and negatively select platelets by bead sorting using standard push columns or a sorting machine (see Note 3).

6. Add fresh PGE_1 to the negatively sorted platelets, centrifuge, and resuspend the platelets in small volumes of warm (37°C)

amino acid-deficient culture medium (see below), count the platelets (detailed counting methodologies can be found in Platelets and Megakaryocytes Vol. 1, Functional Assays, Chapter 3), and further dilute the suspensions to desired cell numbers. Typical platelet concentrations for these studies are between 1×10^7 and 1×10^9 cells/mL.

7. The platelets should be cultured in the presence or absence of translational inhibitors (i.e., 2–50 μg cycloheximide or 0.1–1.0 mM puromycin, for 1×10^8–1×10^9 platelets total) for 1–2 h.

8. After this preincubation period, radiolabeled amino acids (i.e., [^{35}S]methionine plus or minus [^{35}S]cysteine, [^{3}H]-leucine, etc.) are added to the platelets in the presence or absence of stimuli (i.e., soluble agonists, adherence to matrix, high/low glucose, etc.).

9. Incorporation of radiolabeled amino acids into proteins at select time periods (usually 2–24 h) is subsequently measured as described in Subheading 3.3–3.5.

3.2. Measuring Global Protein Synthesis in Platelet Lysates and Supernatants

If an investigative group is interested in determining the magnitude of platelet protein synthesis under various experimental conditions, they may choose to first monitor the extent of radiolabeled incorporation into total cellular proteins. These types of studies are typically achieved by precipitation of total intracellular and/or secreted proteins (see Note 4).

3.2.1. Methods to Determine Radiolabeled Amino Acid Incorporation into Proteins

1. At the end of each experimental timepoint, centrifuge radiolabeled samples to pellet platelets. Remove the supernatant and place on ice for subsequent analysis (step 7). Wash the platelet pellet three times with culture medium containing all amino acids.

2. Place the pellet in ice-cold RIPA buffer containing standard protease inhibitors. Resuspend the pellet thoroughly and place the lysate on ice for 30 min.

3. After 30 min, either remove the insoluble cellular material by high-speed centrifugation (i.e., retain the cell-free supernatant after centrifugation) or place the crude lysate in an equal of volume ice-cold 10% TCA. When using the crude lysate, be sure to vortex the lysate vigorously and place on ice for 30 min.

4. Filter the lysate onto 2.5-cm glass microfiber filter discs (GF/C, Whatman) in a filtration apparatus under vacuum.

5. Wash the discs three times with 5 mL of ice-cold 10% TCA solution and twice with ice-cold ethanol. Let the sample air-dry for 30 min.

6. Transfer the discs from step 5 to 20-mL scintillation vials, add 5 mL scintillation fluid, and measure the radioactivity in a scintillation counter. You should observe increased counts into platelet proteins over time.

7. If interested in determining protein synthesis in supernatants, add 99.7% acetone to the supernatants collected in step 1 (2:1 acetone:supernatants).

8. Mix and vortex the solution and freeze at −70°C for 2–24 h.

9. Thaw the acetone/supernatant mixture, centrifuge the microfuge tube, and wash the protein-rich pellet three times with culture medium containing all amino acids.

10. Resuspend the pellet in 200 μL 1× Laemmli buffer and boil the samples for 5 min, and immediately centrifuge 5 min at $20,000 \times g$ to pellet nonsoluble components (i.e., salts, etc.). Remove the supernatant, transfer to microfiber filter discs, and process as described above (see Notes 3–5).

3.2.2. Things to Consider When Assessing Global Protein Synthesis by Platelets

When working with radioactive materials, the investigator needs to take appropriate precautions to avoid radioactive contamination of the investigator and surrounding areas. TCA is also extremely caustic, so investigators should protect eyes and avoid contact with skin when preparing and handling TCA solutions.

3.3. Characterizing Differential Synthetic Patterns by Platelets

It is often important to visualize synthesized proteins based on their relative amount, molecular weight, or pH. For these types of studies, we separate platelet proteins by two-dimensional gel electrophoresis. Proteins can also be separated by one-dimensional electrophoresis if investigators are only interested in the molecular weight of synthesized proteins.

3.3.1. Separation of Radiolabeled Proteins Using Two-Dimensional Electrophoresis

1. At the end of each experimental timepoint, pellet platelets and collect supernatants as described in Subheading 3.2.1. Soluble proteins from acetone-precipitated supernatants should be processed as outlined in step 7 of Subheading 3.2.1.

2. Wash the platelet pellets (i.e., intracellular proteins) or acetone-precipitated soluble proteins three times with culture medium containing all amino acids.

3. Resuspend the pellets in standard urea-based two-dimensional lysis buffer. We recommend separating proteins immediately by two-dimensional gel electrophoresis to avoid decay of the radioisotope. However, pellets can be frozen (−70°C) for a short period of time until the investigator is ready to load the first dimension strips.

4. Load the samples together with standard rehydration buffer on IEF strips (i.e., Biorad, GE Healthcare) covering the appropriate

pH range. Strips need to be overlayed with mineral oil to prevent evaporation and protein oxidation.

A standard running protocol for the first dimension could look like the following:

Rehydration active	12 h	50 V
Linear gradient	1 h	500 V
Linear gradient	1 h	1,000 V
Linear gradient	2 h	6,000 V
Quick ramp and hold	25 min	6,000 V

5. After running the first dimension, the IEF strips need to be equilibrated to match the SDS conditions of the second dimension. Load the strips on the second dimension SDS gel without trapping air bubbles between the IEF strips and the second dimension gel and then fix in position by using warm agarose.

6. After the second dimension electrophoresis is completed, we recommend staining the proteins with either silver or coomassie blue using standard protocols (see Note 5).

7. After staining the proteins, amplify the radioactive signal by soaking the gels in fluorographic reagent (GE Healthcare) for 15–30 min.

8. Dry the gels in a vacuum gel dryer for a minimum of 2 h at 72°C and then expose the gels to film. The duration of exposure should be empirically tested but, in part, will depend on the intensity of the radioactive amino acids used in the study.

3.4. Confirming the Synthesis of a Specific Protein

The protocols outlined above will inform investigators if the experimental milieu alters global protein synthesis (Subheading 3.2) or expression patterns of synthesized proteins (Subheading 3.3). These strategies, however, do not elucidate the identity of synthesized proteins. Thus, we normally capture synthesized proteins by immunoprecipitation (see Note 6).

3.4.1. Immunoprecipitation of Radiolabeled Proteins

1. At the end of the experimental period, centrifuge platelets to obtain pellets and cell-free supernatants. Resuspend the pellets on ice (30 min) in appropriate lysis buffers (i.e., RIPA, etc.).

2. Centrifuge the lysates and collect the cell-free supernatants (discard the pellet).

3. Immunoprecipitate your target protein (intracellular or secreted) using standard protocols. Be sure to incubate lysates or releasates with appropriate controls such as isotype-matched antibodies, peptide quenching, etc.

4. The immunoprecipitated protein is resuspended in SDS-PAGE loading buffer and the protein is subsequently separated by one- or two-dimensional electrophoresis.

5. Gels are subsequently stained, dried, and exposed to film as described in Subheading 3.3.1.

3.4.2. Capture of Radiolabeled Protein by ELISA

1. After your experiment is completed, centrifuge platelets to obtain pellets and cell-free supernatants. Resuspend the pellets on ice (30 min) in appropriate lysis buffers (i.e., RIPA) and then clear the insoluble material.

2. Concentrate the supernatants using MW cutoff Millipore Microcon Filters according to the manufacturer's instructions. Microcon filters should be chosen based on the size of the candidate protein.

3. After reducing the volume of the samples, wash the column in cold culture medium containing all amino acids according to the manufacturer's specifications.

4. Transfer and invert spin column to clean tubes, centrifuge, and collect the concentrated sample.

5. Capture your candidate protein by loading ~50–100 μL of each sample (lysate or concentrated supernatant) to a standardized ELISA. It is essential that this ELISA has been characterized in detail in regards to its specificity for a specific protein. We recommend using commercial ELISAs that have been subjected to extensive quality controls.

6. Incubate for 2 h at room temperature.

7. Wash the wells extensively and then release the captured proteins by placing 200 μL of laemelli buffer in each well.

8. Remove the laemelli buffer and count in scintillation buffer as described above.

3.5. Assessing Global Protein Synthesis in Platelets by Confocal Microscopy

The studies described above use radioactive-based strategies to assess protein synthesis by platelets. We recently employed a non-radioactive method to show, by immunocytochemistry, that cultured platelets synthesize protein (19). This method is referred to as "Click-iT" (see Note 7).

3.5.1. In Situ Detection of Protein Synthesis by Platelets

1. As described above, incubate platelets for 2 h in methionine-free culture medium. After this preincubation period, add Click-iT L-AHA (Invitrogen C10102) to the cells and perform experiments as desired.

2. At the end of the experimental period, fix the cells in 4% PFA (20 min) by incubating 20 min at room temperature and prepare the cells for immunocytochemistry using standard procedures.

3. Wash the platelets three times (3 min for each wash) in PBS containing 3% (w/v) BSA.

4. Permeabilize the cells with 0.25% NP40 in PBS (15 min).

5. Wash the platelets three times (3 min for each wash) in PBS containing 3% (w/v) BSA.

6. Then, click stain synthesized proteins for 15 min at room temperature as described below:

   Prepare mixture immediately before use (e.g., 30–40 min ahead) using all the components of the Click-iT® Tetramethyl-rhodamine (TAMRA) Protein Analysis Detection Kit (Invitrogen C33370), except the 2× buffer and the TAMRA reagents.

   Solubilize Additive 1 and Additive 2 per instructions in the kit.

Per 1 mL of Click-iT® reaction mixture	Final concentration
100 µL of 1 M Tris pH 8.0	100 mM
5 µL of the 1 mM AF488 alkyne stock	5 µM
50 µL of 40 mM CuSO$_4$ (6.4 mg/mL in water)	2 mM
50 µL of 3 M NaCl (optional if staining fixed cells)	150 mM
50 µL of Additive 1 fill up to 1 mL with water vortex solution well and then add 20 µL of Additive 2 and vortex again	

7. Wash the platelets three times (3 min for each wash) in PBS containing 3% (w/v) BSA.

8. At this point the cells can be analyzed using a confocal laser scanning microscope or the cells can be costained using antibodies or other cell stains.

4. Notes

1. *Assessing Protein Synthesis by Platelets.* Removing leukocytes using protocols like those described in Subheading 3.1.2 has the distinct advantage of increasing the likelihood that synthesized proteins are derived from platelets. However, there are several limitations that need to be considered when employing these types of methods: (1) extended platelet preparation time; (2) leukoreduction can inadvertently activate platelets; (3) higher cost; (4) platelet yield is reduced, typically by 40–50%; and (5) trace numbers of leukocytes often bypass the leukocyte depletion step.

2. *Assessing Protein Synthesis by Platelets.* Multiple ligand binding beads can be added at this step to increase the efficiency of leukocyte depletion (i.e., CD14, CD15, etc.) (7) and/or remove contaminating red blood cells (i.e., Ter119).

3. *Measuring Global Protein Synthesis in Platelet Lysates and Supernatants.* We recommend that investigators check the purity of the negatively selected platelets by flow cytometry and/or immunocytochemistry.

4. *Measuring Global Protein Synthesis in Platelet Lysates and Supernatants.* It is important to note that with the exception of red blood cells, platelets typically synthesize less protein than other eukaryotic cells. In addition, platelets synthesize a number of cytoskeletal proteins as well as proteins that attach to the membrane skeleton. Thus, different types of lysis procedures are often needed to release synthesized proteins from insoluble platelet fractions. Lastly, translational inhibitors should always be used to confirm that radiolabeled amino acids incorporate specifically into protein.

5. *Measuring Global Protein Synthesis in Platelet Lysates and Supernatants.* This step has the twofold advantage of washing away nonspecific radiolabel and marking all of the resolved proteins on the gel for subsequent analyses.

6. *Confirming the Synthesis of a Specific Protein.* The strategies employed in Subheadings 3.4.1 and 3.4.2 should determine whether a candidate protein is synthesized by platelets. However, to ensure specificity of the signal the investigator should include proper controls that are commonly used for immunoprecipitation or ELISA. A limitation of these techniques is that they are not quantitative. Therefore, other strategies are required to estimate the amount of protein that is synthesized by platelets.

7. *Assessing Global Protein Synthesis in Platelets by Confocal Microscopy.* The recently developed AHA system allows investigators to determine that platelets synthesize protein. However, there are several limitations that need to be considered: (1) the method is semiquantitative; (2) the AHA system provides no information on the types of proteins synthesized by platelets; (3) this technique is extremely labor-intensive. Although unsuccessful in our hands, it should be noted that the Click-iT system comes with other labels that are intended to be used for the separation of synthesized proteins by gel electrophoresis.

References

1. Weyrich, A.S., S. Lindemann, and G.A. Zimmerman. (2003). The evolving role of platelets in inflammation. J Thromb Haemost, 1, 1897–905.

2. Semple, J.W. and J. Freedman. (2010). Platelets and innate immunity. Cell Mol Life Sci, 67, 499–511.

3. Weyrich, A.S., et al. (2009). Protein synthesis by platelets: historical and new perspectives. J Thromb Haemost, 7, 241–6.

4. Denis, M.M., et al. (2005). Escaping the nuclear confines: signal-dependent pre-mRNA splicing in anucleate platelets. Cell, 122, 379–91.

5. Gerrits, A.J., et al. (2010). Platelet tissue factor synthesis in type 2 diabetic patients is resistant to inhibition by insulin. Diabetes, 59, 1487–95.

6. Schwertz, H., et al. (2006). Signal-dependent splicing of tissue factor pre-mRNA modulates the thrombogenecity of human platelets. J Exp Med, 203, 2433–2440.

7. Shashkin, P.N., et al. (2008). Lipopolysaccharide is a direct agonist for platelet RNA splicing. J Immunol, 181, 3495–502.

8. Landry, P., et al. (2009). Existence of a microRNA pathway in anucleate platelets. Nat Struct Mol Biol, 16, 961–6.

9. Lindemann, S., et al. (2001). Integrins regulate the intracellular distribution of eukaryotic initiation factor 4E in platelets. A checkpoint for translational control. J Biol Chem, 276, 33947–51.

10. Weyrich, A.S., et al. (2007). mTOR-dependent synthesis of Bcl-3 controls the retraction of fibrin clots by activated human platelets. Blood, 109, 1975–1983.

11. Weyrich, A.S., et al. (1998). Signal-dependent translation of a regulatory protein, Bcl-3, in activated human platelets. Proc Natl Acad Sci USA, 95, 5556–61.

12. Weyrich, A.S., Lindemann, S., Tolley, N.D., Kraiss, L.W., Dixon, D.A., Mahoney, T.M., Prescott, S.M., McIntyre, T.M., Zimmerman, G.A. (2004). Change in protein phenotype without a nucleus: translational control in platelets. Seminars in Thrombosis and Hemostasis, 30, 493–500.

13. Zimmerman, G.A. and A.S. Weyrich. (2008). Signal-dependent protein synthesis by activated platelets: new pathways to altered phenotype and function. Arterioscler Thromb Vasc Biol, 28, s17-24.

14. Pabla, R., et al. (1999). Integrin-dependent control of translation: engagement of integrin alphaIIbbeta3 regulates synthesis of proteins in activated human platelets. J Cell Biol, 144, 175–84.

15. Yang, H., et al. (2009). Fibrinogen is required for maintenance of platelet intracellular and cell-surface P-selectin expression. Blood, 114, 425–36.

16. Lindemann, S., et al. (2001). Activated platelets mediate inflammatory signaling by regulated interleukin 1beta synthesis. J Cell Biol, 154, 485–90.

17. Boilard, E., et al. (2010). Platelets amplify inflammation in arthritis via collagen-dependent microparticle production. Science, 327, 580–3.

18. Zimmerman, G.A. and A.S. Weyrich. (2010). Immunology. Arsonists in rheumatoid arthritis. Science, 327, 528–9.

19. Schwertz, H., et al. (2010). Anucleate platelets generate progeny. Blood, 115, 3801–9.

20. Ostrowska, H., et al. (2003). Human platelet 20S proteasome: inhibition of its chymotrypsin-like activity and identification of the proteasome activator PA28. A preliminary report. Platelets, 14, 151–7.

21. Yukawa, M., et al. (1991). Proteasome and its novel endogenous activator in human platelets. Biochem Biophys Res Commun, 178, 256–62.

Chapter 12

A Rapid and Efficient Platelet Purification Protocol for Platelet Gene Expression Studies

Stefan Amisten

Abstract

Isolation of pure platelet samples from whole blood is crucial for the study of platelet gene expression. The main obstacles to overcome in order to successfully isolate platelets from whole blood include (1) platelet activation; (2) leukocyte and red blood cell contamination, and (3) time-dependent platelet mRNA degradation. This chapter describes a rapid and highly efficient method for isolating human circulating platelets from small volumes of whole blood based on efficient inhibition of platelet activation and leukocyte removal by filtration followed by magnetic bead-depletion of residual contaminating leukocytes and red blood cells. Also described are methods for RNA extraction, cDNA synthesis, and platelet gene expression studies using both quantitative real-time PCR and microarray.

Key words: Platelet purification, Platelet activation, Leukocyte-depletion, Gene expression, Real-time PCR, qPCR, Microarray, RNA extraction

1. Introduction

Microarray and real-time PCR have both been used to study the platelet transcriptome (1–5). However, successful platelet gene expression profiling is often hampered by the contamination of platelet preparations by other cell types, such as leukocytes.

Platelets are anuclear and, as a consequence, all nonmitochondrial platelet mRNA is derived from their precursor cell, the megakaryocyte. As a result, the amount of mRNA a platelet contains is dependent on the amount of mRNA inherited when it budded off from the megakaryocyte as well as the rate of platelet RNA degradation. Therefore, when studying platelet gene expression, it is advantageous to use a fast platelet purification method in order to minimize further degradation of mRNAs during the

Jonathan M. Gibbins and Martyn P. Mahaut-Smith (eds.), *Platelets and Megakaryocytes: Volume 3, Additional Protocols and Perspectives*, Methods in Molecular Biology, vol. 788, DOI 10.1007/978-1-61779-307-3_12, © Springer Science+Business Media, LLC 2012

purification process. Also, it is important to minimize the contamination of the platelet preparations by other cell types such as red blood cells and leukocytes. This is particularly important for leukocytes, which have been reported to contain as much as 10,000 times more RNA than platelets (6).

This chapter describes a fast, flexible and low-cost method for the generation of a highly purified population of human platelets from less than 100 mL of whole blood (1) (see Note 1). The protocol combines efficient inhibition of platelet activation with leukocyte removal using filters developed for the preparation of platelets for clinical use (7) and antibody-mediated magnetic depletion of residual leukocytes and erythrocytes. The purification can be performed in less than 2 h and without the need for expensive or bulky specialized equipment. The chapter also describes the use of real-time PCR to assess the expression of two platelet P2Y receptors (P2RY1 and P2RY12) and the level of residual contaminating leukocyte-specific messages (CD45).

2. Materials

2.1. Platelet Isolation

1. Dynabeads® Pan Mouse IgG (Invitrogen).
2. Mouse anti-human CD235a (red blood cell surface marker) (BD Inc, NJ, USA).
3. Mouse anti-human CD45 (leukocyte surface marker) (BD Inc, NJ, USA).
4. Dynabead-compatible magnet (DynaMag™-2 or DynaMag™-15) (Invitrogen) (see Note 2).
5. Dynabead wash buffer: PBS with 0.1% (w/v) BSA, pH 7.4.
6. Sterile anticoagulant: citrate-dextrose solution (ACD) (Sigma Aldrich).
7. Prostaglandin E1 (PGE1) (Sigma Aldrich): 1 mM stock in ethanol.
8. Acetylsalicylic acid (Sigma Aldrich): 30 mM stock in ethanol.
9. EDTA, 0.5 M (Sigma Aldrich).
10. 50 and 15 mL centrifuge tubes and 1.5 and 2 mL Eppendorf tubes (see Note 2).
11. Intravenous venflon cannula 18 gauge with port and flexible tubing (Becton Dickinson Infusion Therapy AB, Sweden).
12. Rotating and tilting test tube carousel.
13. AutoStop™ BC high efficiency filter for leukocyte removal (Pall Inc, Cat. No. ATSBC1E, see http://www.pall.com/medical_48187.asp).
14. TRIzol (Invitrogen).

2.2. RNA Purification and cDNA Generation

1. RNase-free filter tips: 10, 100 or 200 and 1,000 μL.
2. RNase-free tubes 1.5 and 0.6 mL (Axygen MCT-060-A and MCT-150-C).
3. Isopropanol, molecular biology grade (Sigma Aldrich).
4. Ethanol, molecular biology grade (Sigma Aldrich).
5. Water, molecular biology grade (RNase-free) (Sigma Aldrich).
6. Chloroform, molecular biology grade (Sigma Aldrich).
7. UltraPure™ Glycogen (Invitrogen, Cat. No. 10814–010).
8. PCR tubes, RNase free.
9. TaqMan reverse transcription kit (Applied Biosystems).
10. RNeasy MinElute Cleanup Kit (Qiagen).
11. Centrifuge with cooling for spinning at $12,000 \times g$.
12. NanoDrop (Thermo Scientific) or equivalent spectrophotometer for assaying nucleotide content in small sample volumes.
13. *Optional* – 2100 Bioanalyzer (Agilent Technologies).

2.3. Quantitative Real-Time PCR

1. QuantiFast SYBR Green PCR Kit (Qiagen).
2. QuantiTect Primer Assays (Qiagen) (see Note 18):
   (a) P2RY1 (QT00199983) – platelet ADP receptor used to demonstrate the qPCR protocol.
   (b) P2RY12 (QT01770111) – platelet ADP receptor used to demonstrate the qPCR protocol.
   (c) ACTB (QT01680476) – housekeeping gene.
   (d) P2RY11 (QT01150282) – GPCR expressed on leukocytes but not on platelets.
   (e) CD45 (PTPRC) (QT01869931) – abundant leukocyte protein, not present on platelets.
3. PCR tubes.
4. Filter tips, PCR grade.
5. Equipment for agarose gel electrophoresis of 50–200 bp qPCR products.

3. Methods

The platelet purification method has four main phases: blood collection, generation of platelet-rich plasma (PRP), PRP filtration and magnetic bead-mediated leukocyte and erythrocyte depletion. Appropriate ethical permission to take human blood should be obtained. To minimize platelet activation during blood collection, a 1.2-mm diameter (18 gauge) intravenous cannula is inserted into a suitable vein, and the blood is collected through the attached tubing via gravity flow into the platelet inhibition cocktail. In the author's experience, this method is superior to collecting blood using vacutainers, as it reduces mechanical platelet activation caused by the vacuum. Once purified, the platelet samples are dissolved in TRIzol and frozen until RNA extraction.

3.1. Preparation of Antibody-Conjugated Magnetic Beads

1. Mix 1 mL Dynabead wash buffer and 250 μL Dynabead slurry.
2. Place the tube in the Dynamag magnet for 1 min, remove the liquid, and repeat for a total of two washes.
3. Remove the washed beads from the magnet and resuspend the washed Dynabeads up to the original bead volume (i.e. 250 μL) using Dynabead wash buffer.
4. Add 15 μL of anti-CD235a antibody and 15 μL anti-CD45 antibody to the beads and mix.
5. Incubate the bead mix for at least 30 min with gentle tilting and rotation.
6. Place the tube in the magnet for 1 min and remove the supernatant. Resuspend the beads up to the original bead volume (i.e. 250 μL) using Dynabead wash buffer and store the antibody-conjugated magnetic beads at 4°C.

3.2. Preparation of Platelet Inhibition Cocktail

The volumes stated below apply to 100 mL whole blood collected into twelve 15 mL tubes.

1. Mix 18 mL ACD, 12 μL of 1 mM PGE$_1$, 120 μL of 30 mM acetylsalicylic acid, and 480 μL of 0.5 M EDTA.
2. Add 1.5 mL of the cocktail into each of the twelve 15 mL tubes.

3.3. Collection of Blood and Purification of Platelets

1. Weigh an empty 15 mL tube with cap and note down the tube weight.
2. Implant the intravenous cannula (1.2 mm diameter) and attach the flexible tubing. Collect 8.5 mL of blood by self-propagated flow through the flexible tubing into each tube containing 1.5 mL platelet inhibition cocktail. When the total volume reaches 10 mL, replace the cap and invert the tube gently to enable thorough mixing of the blood with the platelet inhibition cocktail. To fill 12 tubes you will need 102 mL of whole blood (see Notes 1, 3, and 4).
3. Place the tubes in a centrifuge and spin at room temperature ($200 \times g$, 20 min).
4. Carefully transfer the top 85% of the upper layer, the PRP, to new 15 mL tubes.

5. Spin at room temperature ($200 \times g$, 10 min) to further remove leukocytes and red blood cells.

6. Carefully transfer the top 85% of the PRP to a 50-mL tube.

7. With a sterile pair of scissors or a surgical blade, cut off excess tubing from the Pall Leukocyte removal filter pack and mount it on a rack.

8. Place the drainage tubing in a 50-mL collection tube and manually inject the PRP into the filter at a flow rate of 15 mL per minute. Collect the filtrate in a 50-mL tube.

9. Add all of the antibody-conjugated magnetic beads to the tube containing the filtered PRP and place the tube in a tilting and rotating carousel at room temperature for 45 min.

10. Transfer the PRP/bead mix into 2 mL tubes (see Note 2) and leave the tubes in the DynaMag™-2 for 2 min. The DynaMag™-2 can accommodate 16 tubes at a time, so steps 10–13 will have to be repeated if more than 32 mL of PRP is used.

11. With the tubes still held in the magnetic rack, transfer the supernatants into new magnet compatible tubes without disturbing the magnetic beads that are held by the magnetic field.

12. Put the tubes containing the depleted PRP in the magnet.

13. After 2 min, without disturbing the beads captured by the magnetic field, transfer the supernatant into new 15 mL tubes and harvest the platelets by centrifugation at $800 \times g$, 10 min at room temperature.

14. Discard the supernatants. Weigh the 15-mL tubes containing the platelet pellets and calculate the weight of each platelet pellet by subtracting the empty tube weight from step 1 above. Dissolve the platelet pellets using 1 mL TRIzol per 100 mg platelets (use a minimum of 1 mL TRIzol for platelet samples weighing less than 100 mg) and leave for 5 min at room temperature (see Note 5).

15. Transfer the dissolved platelets into 1.5 mL tubes (1 mL per tube) and store the samples at –80°C until RNA purification. RNA samples in TRIzol are stable for up to 1 month at –80°C or below.

3.4. Extraction of Platelet Total RNA

The following protocol for isolation of platelet total RNA is based on a modified version of the TRIzol method (1). An advantage of the TRIzol method over column-based RNA purification methods is that besides high-quality RNA, platelet protein may also be extracted from the same platelet TRIzol sample (see Notes 5–9). The RNA isolated using TRIzol is then further purified and concentrated using an RNeasy MinElute Cleanup Kit (Qiagen). The RNeasy MinElute Cleanup Kit columns remove contaminating residual salts and phenol as well as RNAs shorter than 200 bp, such as 5.8 S rRNA, 5 S rRNA, and tRNAs, which together comprise 15–20% of the total RNA, resulting in an enrichment of mRNAs (8).

1. Cool isopropanol to –20°C.

2. Cool centrifuge to 4°C.

3. Place the platelet samples dissolved in TRIzol in a fume hood, defrost, and leave for 5 min at room temperature.

4. Add 200 µL chloroform to each tube containing 1 mL TRIzol (see Notes 5 and 10).

5. Manually shake the tubes 15–30 times and leave for 2–3 min at room temperature.

6. Spin at $12,000 \times g$ for 15 min at 4°C.

7. Add 10 µg ultra-pure glycogen (0.5 µL of 20 µg/µL stock solution) to new RNase-free tubes (see Note 11).

8. Transfer the aqueous phase of each tube containing TRIzol/chloroform mix to tubes containing glycogen using a 1,000-µL filter tip. Leave about 3–4 mm of the aqueous phase above the interphase (Fig. 1) to minimize carryover of contaminating DNA.

9. Add 500 µL of cooled (–20°C) isopropanol to each tube of aqueous phase/glycogen mix.

10. Invert to mix and leave over night at –20°C or at –80 C for 1 h.

11. Spin the samples at $12,000 \times g$ for 15 min at 4°C.

12. Remove all the liquid and add 1 mL of 75% ethanol (freshly made up with molecular biology grade water). RNA pellets from the same donor that were split into separate 1 mL aliquots in TRIzol (steps 14 and 15, Subheading 3.3) may be combined by pooling the RNA pellets into one tube in a total volume of 1 mL 75% ethanol. RNA samples in 75% ethanol can be stored at –20°C for at least 1 year.

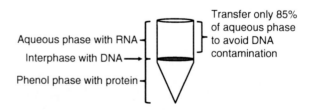

Fig. 1. Phase separation of TRIzol RNA samples after addition of chloroform. The clear aqueous phase (*top*) contains the RNA, the white interphase (*middle*) contains DNA, and the pink phenol phase (*bottom*) contains protein. To avoid contamination of the RNA by DNA from the interphase it is recommended that only 85% of the aqueous phase is transferred to a new tube.

13. Spin at $7500 \times g$ for 10 min at 4°C.

14. Remove all the liquid. It is important to remove ALL the liquid, otherwise it will take a very long time to dry the RNA pellet.

15. Dry the pellets by placing the tubes in a heating block at 35–37°C for 1–3 min with the lids open.

16. When the RNA pellet appears dry, dissolve the RNA in 101 μL water, vortex, and keep on ice.

17. Transfer 1 μL RNA into an RNase-free 1.5 mL tube containing 3 μL RNAse-free water (makes 4 μL of one in four diluted RNA). Freeze the remaining 100 μL RNA at –80 C until RNeasy MinElute Cleanup column purification.

18. Use 1 μL of the one in four diluted RNA to measure the total RNA concentration on a NanoDrop or equivalent spectrophotometer.

19. Calculate the total amount of RNA present in each sample and purify each RNA sample using the RNeasy MinElute Cleanup kit according to the manufacturer's instructions. Do not load more than 45 μg RNA dissolved in 100-μL RNase-free water on each spin column (8).

20. The total elution volume from the RNeasy MinElute column is 12 μL. Transfer 0.5 μL of the RNeasy MinElute kit purified RNA into a new RNase-free tube containing 1.5 μL molecular grade water (makes 2 μL of one in four diluted purified RNA). Store the remaining 11.5 μL of RNA at –20°C to –80°C until cDNA synthesis.

21. Use 1 μL of the one in four diluted RNA to measure the RNA concentration on a NanoDrop or equivalent. Calculate the concentration of the undiluted purified RNA sample. The purified RNA can be used for both cDNA synthesis and microarray hybridization. However, it is recommended that the RNA integrity of all samples intended for microarray hybridizations are verified using a 2100 Bioanalyzer (Agilent Technologies).

3.5. Generation of Platelet cDNA

Reverse transcription of RNA into cDNA can be achieved using a variety of kits. In our hands, the TaqMan reverse transcription kit (Applied Biosystems) has worked very well, and it will be used as an example below. The TaqMan reverse transcription kit is designed to transcribe 60–2,000 ng of total RNA into cDNA. If larger amounts of RNA are needed, it is recommended that several cDNA reactions not exceeding 2,000 ng RNA per reaction are performed, followed by pooling of all the cDNA reactions into one.

1. Program the thermal cycler for reverse transcription PCR (Table 1).

2. Prepare a reverse transcription master mix (Table 2) and distribute to RNase and DNase-free PCR tubes (see Note 12).

Table 1
Thermal cycling conditions for reverse transcription using the TaqMan reverse transcription kit

Temperature (°C)	Time (min)
25	10
48	30
95	5
4	Eternity

Table 2
Preparation of master mix for cDNA synthesis

Reaction volume	25 μL
Make for number of reactions	1×
10x TaqMan RT-buffer	2.50
25 mM MgCl$_2$	5.50
10 mM dNTP	5.00
Random hexamers	1.25
RNase inhibitor	0.50
MultiScribe Transcriptase	0.65
Master mix to each tube	15.40
RNA to each tube (60–2,000 ng)	9.60

All volumes are in μL

Add 60–2,000 ng RNA to each tube, vortex, and spin down. Always keep all the samples and reagents on ice. The total reaction volume can be scaled up or down according to the amount of cDNA needed.

3. Place PCR tubes containing reverse transcription mix and RNA in the thermal cycler. Ensure that the PCR tube caps are tightly closed.

4. Once the thermal cycling is completed, the cDNA can be stored short term at 4°C or long term at –20°C.

3.6. Relative Quantitative Real-Time PCR (qPCR) of Platelet Genes

To illustrate the use of quantitative real-time PCR (qPCR) the following protocol describes the quantification of the platelet ADP receptor genes P2RY1 and P2RY12 relative to the house keeping gene beta-actin (ACTB). The approach uses a Rotor-Gene 6000

real-time PCR cycler, the QuantiFast SYBR Green PCR Kit and
QuantiTect primer assays (all from Qiagen) (see Notes 13 and 14).
To evaluate the degree of residual leukocyte contamination, two leukocyte specific genes may also be quantified: *PTPRC*, which encodes
the leukocyte surface protein CD45, and *P2RY11*, which encodes a
G-protein coupled receptor that is expressed on leukocytes but not
on platelets. The degree of leukocyte depletion in the purified platelet samples can easily be measured by qPCR quantification of CD45
in cDNA from whole blood or PRP (from step 6 in Subheading 3.3)
compared to purified platelets (from step 13 in Subheading 3.3).

1. Program the Rotor-Gene 6000 thermal cycler (Table 3 and see
   Note 15).

2. Prepare a master mix for pilot qPCR quantification (Table 4 and
   see Note 16). Keep all the samples on ice and avoid exposing your
   mastermix to light, as the SYBR Green in the mastermix is light
   sensitive. Increased qPCR sensitivity and reduced reaction cost

Table 3
Rotor-Gene 6000 cycling conditions for qPCR using the QuantiFast SYBR Green PCR Kit

	Step	Temperature (°C)	Time (s)	Flourescence acquisition
1	Initial denaturation	95	300	None
2	Cycling denaturation	95	10	None
3	Cycling annealing/extension	60	30	Sybr/FAM
4	Repeat steps 2–3 40 cycles			
5	Melt curve	65–95[a]	5–30[a]	Sybr/FAM

[a]Rising by 1°C each step, wait for 30 s on first step, then wait for 5 s for each step afterward

Table 4
Preparation of master mix solution for pilot qPCR quantification using the QuantiFast SYBR Green PCR Kit

	$1 \times 10\ \mu L$	$3.5 \times 10\ \mu L$
Number of 10 μL assays		
Mastermix	5	17.5
Primer assay mix	0	0
Diluted cDNA	1	3.5
Volume mastermix added to each qPCR reaction	6	6
Volume QuantiTect Primer assay diluted one in four added to each qPCR reaction	4	4

All volumes are in μL

can be achieved by reducing the total reaction volume from 25 μL (i.e. the volume recommended by Qiagen) to 10 or 5 μL without reducing the total amount of cDNA template added to each qPCR reaction (see Note 16). For the pilot qPCR it is recommended to pool cDNA from each biological replicate and to dilute the pooled cDNA at least one in four, as this approach will use up less cDNA from each biological replicate.

> **Example**
>
> Platelet cDNA has been generated from three different platelet donors. For the initial pilot qPCR of the genes P2RY1, P2RY12, and ACTB, 1 μL of one in four diluted pooled cDNA will be used for each 10 μL reaction. It is very hard to predict the optimal cDNA dilution for each gene, since the expression of two different genes can vary by several orders of magnitude. Therefore, by running a pilot qPCR using a standard dilution it is possible to determine the optimal cDNA dilution factor required for the quantification of each gene of interest. In order to include cDNA from all the three platelet samples, 0.5 μL cDNA is pooled from each platelet sample, giving a total volume of 1.5 μL pooled platelet cDNA. For a one in four dilution of the cDNA, 4.5 μL water is added, resulting in 6 μL of one in four diluted pooled platelet cDNA. 3.5 μL of this cDNA is used for the pilot qPCR (see Table 4), and the remaining is frozen for future use.

3. Close the lids tightly, place the samples in the thermal cycler and run the program setup in step 1.

4. Perform a melt curve analysis for each primer assay to verify that only one amplification product has been produced in each qPCR sample. Please see the product documentation included with the QuantiFast SYBR Green PCR Kit and your qPCR thermal cycler for specific instructions on how to perform a melt curve analysis. If a gene of interest is not expressed in your sample, or if it is present at only very low levels, the QuantiTect primer assay may generate a nonspecific qPCR product (Fig. 2b). These nonspecific amplifications will be detected by the melt curve analysis and by agarose gel electrophoresis of the qPCR samples. Due to the small size of the qPCR products (50–200 bp), a 2% agarose gel is recommended for clear resolution of the individual bands. Only primer assays that generate a single melt curve peak (Fig. 2a) and a single qPCR product corresponding to the size given in the QuantiTect primer assay product literature should be used.

5. Calculate the Ct (Cycle threshold) values for all the samples in the qPCR run. Please see the product documentation included

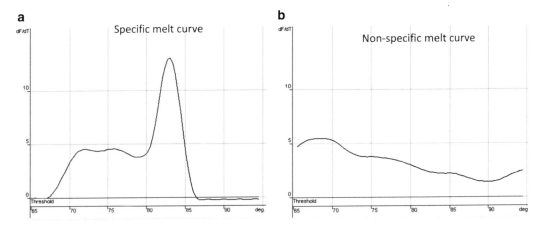

Fig. 2. Melt curve analysis following a real-time PCR run. A single peak (**a**) suggests that only one amplification product is generated whereas no single peak is present in the nonspecific melt curve (**b**).

with the QuantiFast SYBR Green PCR Kit and your qPCR thermal cycler for specific instructions on how to calculate the Ct values. In our hands, when using the Rotor-Gene 6000 and the RotorGene version 6.0 software, threshold values between 0.02 and 0.05 often give satisfactory Ct values.

6. Determine the optimal dilution of your cDNA samples for the quantification of your genes of interest. By diluting the cDNA, it is possible to save platelet cDNA without compromising the quality of the qPCR quantifications (see Note 17).

> **Example**
>
> The pilot quantification using pooled cDNA diluted one in four (see steps 2–5 above) returned the following Ct values: P2RY1: 25.91; P2RY12: 23.26; ACTB: 18.79. Since all Ct values are between 17 and 30 cycles, no further dilution of the cDNA is necessary, and a one in four dilution of the cDNA is sufficient for the quantification of all three genes. However, if there is a need to conserve cDNA, the cDNA used to quantify ACTB can be diluted up to 16 times more and P2RY12 twice more without compromising the qPCR quality.

7. An ideal primer pair has an amplification efficiency (E) of 2 (i.e. each PCR cycle results in a doubling of the number of template DNA molecules). However, sometimes the E value deviates slightly from the theoretical value. In order to determine the actual amplification efficiency for each primer assay, each gene needs to be quantified using serially diluted cDNA. The Ct values from the dilution series is then used to calculate the amplification efficiency for each gene:

$$E = 10^{(-1/k)}$$

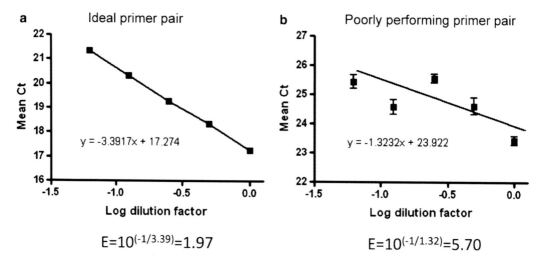

Fig. 3. Primer efficiency plots. Following quantification of serially diluted cDNA template (*see* Table 7), an optimal qPCR assay (**a**) will be fit by a line with a slope (*k*) close to −3.3, which will translate into a primer efficiency (*E*) of 2.0. (**b**) A nonfunctional assay will give a slope which can be either greater than, or less than the ideal slope of −3.3, resulting in a primer efficiency considerably different from 2.0.

where k is the slope of the linear function of the Ct values plotted against the log transformed cDNA dilution factors. Most functional primer assays will generate primer efficiencies (E) of 1.9–2.1 (Fig. 3a), whereas nonfunctional primer assays will not generate straight Ct vs. log dilution factor curves (Fig. 3b) or have primer efficiencies much greater or smaller than 2.

To determine the E value for a primer assay, prepare serially diluted cDNA (see Table 7) and quantify each gene in triplicate as described in steps 2–6 above. For each gene, prepare a master mix for 17×10 μL qPCR reactions (see Table 5).

Example

The primer efficiency (E) for the primer assays P2RY1, P2RY12, and ACTB needs to be determined experimentally using serially diluted platelet cDNA. The required volumes for the evaluation of three primer assays in triplicate can be calculated as shown in Tables 6 and 7.

The average Ct values for each dilution step are plotted in a graph against the log values of the dilution factors and the equation for the linear trendline is calculated (Table 8 and Fig. 3a). Finally, the amplification efficiency (E) for each primer assay is calculated according to the formula $E = 10^{(-1/k)}$. Since k (i.e., the slope of the trendline) for the ACTB primer assay dilution curve is −3.3917, the amplification efficiency (E) for the ACTB primer assay is

$$E = 10^{(-1/-3.3917)} = 1.97$$

Table 5
Preparation of master mix for determination of primer assay efficiency

Number of 10 µL assays	1 × 10 µL	17.5 × 10 µL
Mastermix	5	87.5
QuantiTect Primer assay	1	17.5
Volume mastermix added to each qPCR reaction	6	6
Volume diluted cDNA added to each qPCR reaction	4	4

All volumes are in µL

Table 6
Calculations of required amounts of cDNA for determination of primer amplification efficiency for the genes P2RY1, P2RY12, AND ACTB

Number of primer assays to test +1	4
Technical replicates	3
Diluted cDNA needed for each dilution factor step (number of assays × technical replicates)	12
Extra water to add so that 4 µL diluted cDNA can be added per assay[a]	36

[a]It is easier to accurately pipette 4 µL than 1 µL, so extra water is added to the diluted cDNA to allow 4 µL to be added to each reaction tube
All volumes are in µL

Table 7
Volumes required for serial dilutions of cDNA for primer amplification efficiency determination

Final cDNA dilution factor	Volume cDNA and water added to each dilution step
1	6 µL of undiluted cDNA stock + 90 µL water
2	48 µL water + 48 µL cDNA from the above dilution step
4	48 µL water + 48 µL cDNA from the above dilution step
8	48 µL water + 48 µL cDNA from the above dilution step
16	48 µL water + 48 µL cDNA from the above dilution step

Table 8
Calculation of average Ct and dilution factor log values for the dilution curve of ACTB

Dilution factor	Log dilution factor	Ct repl. 1	Ct repl. 2	Ct repl. 3	Average Ct
1	0.00	17.19	17.29	17.29	17.26
2	−0.30	18.38	18.42	18.20	18.33
4	−0.60	19.20	19.33	19.34	19.29
8	−0.90	20.30	20.37	20.35	20.34
16	−1.20	21.31	21.39	21.38	21.36

Table 9
Preparation of mastermix for the quantification of one gene in duplicate reactions in at least three biological replicates, resulting in a master mix for seven reactions ($3 \times 2 + 1 = 7$ reactions)

	$1 \times 10 \,\mu L$	$7 \times 10 \,\mu L$
Number of 10 μL assays		
Mastermix	5	35
Volume undiluted QuantiTect Primer assay	1	7
Volume mastermix added to each qPCR reaction	6	6
Volume diluted cDNA added to each qPCR reaction	4	4

All volumes are in μL

Using the same method, the E values for the P2RY1 and P2RY12 primer assays are found to be 2.01 and 1.98, respectively.

8. Make a new mastermix (Table 9) and quantify all genes of interest. It is recommended that at least three biological replicates are used and that each gene is quantified at least in duplicate reactions. In order to detect subtle changes in gene expression between samples and controls, it is recommended to use at least eight biological replicates in each group.

> **Example**
>
> A total of 18 reactions are run for the quantification in duplicates of P2RY1, P2RY12, and ACTB using platelet cDNA from three individuals, resulting in the following Ct values for platelet cDNA samples 1–3: P2RY1: 25.98, 25.32, 25.79; P2RY12: 22.7, 22.72, 22.92; ACTB: 18.58, 18.43, 18.6.

9. Calculate the expression of your genes of interest relative to your housekeeping gene using the formula:

$$\text{Relative expression } (R) = \frac{\left(E_{\text{gene of interest}}^{-Ct\text{ gene of interest}}\right)}{\left(E_{\text{housekeeping gene}}^{-Ct\text{ housekeeping gene}}\right)}.$$

Example

The expression of P2RY1 and P2RY12 relative to ACTB in platelet cDNA sample 1 is:

$$R_{\text{P2RY1}} = \left(2.01^{-25.98}\right)/\left(1.97^{-18.58}\right) = 0.0039$$
$$R_{\text{P2RY12}} = \left(1.98^{-22.7}\right)/\left(1.97^{-18.58}\right) = 0.054$$

3.7. Preparation of RNA Samples for Platelet Microarray Analysis

There are many different microarray platforms available, and some of the more commonly used ones are Illumina and Affymetrix. All microarray platforms have their pros and cons, and in most cases, the choice of microarray platform is based on cost and which microarrays are available locally. Many academic institutions now have core facilities specializing in microarray hybridizations and data analysis, and it is highly recommended to let your local core facility handle all microarray hybridizations.

If insufficient platelet RNA is available, it may also be necessary to either pool RNA from several different platelet preparations (9–12) or to perform a double amplification of the platelet RNA (13). Both methods have their pros and cons, and it is up to the researcher and the local microarray core facility to find an optimal solution suitable for the intended project and the local conditions.

In order to validate gene expression results obtained by microarray, it is recommended that 5–10% of the total amount of the RNA intended for microarray hybridization is set aside for cDNA synthesis and qPCR validation of the microarray results.

4. Notes

1. The expected platelet yield may vary between donors. In general, dehydrated donors will give a lower proportion of PRP compared to the red blood cell fraction than well-hydrated donors. To increase platelet yield, it is therefore recommended that all blood donors are well hydrated prior to blood donation. Various diseases and medications may also affect the platelet count, which will influence the final platelet yield.

2. Depending on the type of magnet that is used (DynaMag™-2 or DynaMag™-15), either 1.5 or 2 mL tubes (DynaMag™-2)

or 15 mL tubes (DynaMag™-15) are suitable for magnetic cell depletion. For the sake of consistency, 2 mL tubes are stated throughout this protocol.

3. Self-propagated blood flow is preferred in order to minimize platelet activation during the blood collection process.

4. Platelet yield and gene expression profiles may show natural variations, in addition to variations due to different disease states, medication, and pregnancy.

5. TRIzol and chloroform are hazardous chemicals and should only be handled in a well-ventilated fume hood. Please read appropriate safety information before use.

6. RNA is very sensitive to degradation. Never touch any tubes, tips, or surfaces that are to be used for RNA isolation with bare hands. Avoid unnecessary talking and breathing toward the samples while performing RNA extractions.

7. Always use filter tips during RNA isolation. Never reuse filter tips or use tips that have accidentally touched any contaminated surface.

8. All volumes in this protocol are based on tubes containing platelets dissolved in 1 mL TRIzol. If a platelet sample has been dissolved in more than 1 mL TRIzol, the sample will need to be subdivided into several tubes of 1 mL TRIzol each during the RNA purification. The resulting RNA pellets from the same donor can then be pooled into one tube during the RNA pellet ethanol wash step (Subheading 3.4, step 12).

9. In order to avoid accidental sample loss and contamination of the RNA samples, clear RNAse-free 1.5 mL tubes are preferred over large ultracentrifuge tubes for precipitation and washing of the RNA samples.

10. Chloroform easily drips out of 1,000 μL filter tips. To dispense chloroform, it is recommended to use either 100 or 200 μL tips.

11. A greater RNA yield can be achieved by precipitating the RNA at −20°C over night or at −80°C for 1 h in the presence of 10 μg glycogen.

12. In our hands, reverse transcription using random hexamer primers has resulted in better cDNA yields than cDNA generated using oligo-dT primers.

13. There are many different platforms for qPCR to choose from. We have had good success with Qiagen's QuantiFast SYBR Green kit and QuantiTect Primer assays using a Rotor-Gene 6000 PCR cycler. For an overview of other qPCR platforms such as TaqMan probes, please see http://en.wikipedia.org/wiki/Real-time_PCR.

14. Several different housekeeping genes are available for relative real-time-PCR, including GAPDH and ACTB that we have successfully used for platelet qPCR (1, 14).

15. For thermal cyclers other than Rotor-Gene 6000, see the QuantiFast SYBR Green PCR Kit manual for optimal cycling conditions.

16. Scaling down the total qPCR reaction volume from 25 to 10 µL without reducing the amount of cDNA added will result in a 60% reduction in QuantiFast SYBR Green cost and a 2.5 step lower Ct value for most qPCR reactions. Scaling down the reaction volume from 10 to 5 µL without reducing the amount of cDNA per reaction will further reduce the reagent cost but will provide slightly higher Ct values compared to the 10 µL reactions.

17. For most primer pairs, optimal linear quantification is achieved in the 17–30 Ct range (i.e., 17–30 PCR cycles). If a gene is detected later than 30 cycles, it may be difficult or impossible to accurately quantify the gene of interest due to the low levels of template available for the qPCR reaction. On the other hand, if a gene is detected earlier than 20 cycles, we recommend that the template cDNA is diluted further in order to minimize waste of cDNA template (Table 10).

18. Always check for improved versions of QuantiFast real-time PCR primers before ordering qPCR primers.

Table 10
Potential of cDNA dilution in order to economize cDNA without compromising qPCR quality and accuracy

Detection Ct value (qPCR cycle number)	cDNA dilution factor to reach Ct 20	cDNA dilution factor to reach Ct 25
15	32	1,024
16	16	512
17	8	256
18	4	128
19	2	64
20	1	32
21		16
22		8
23		4
24		2
25		1

If the pilot qPCR returns Ct values lower than 20–25 cycles, it is possible to dilute the template cDNA without compromising qPCR quality and accuracy

Acknowledgments

The author would like to thank the Swedish Research Council and Prof. David Erlinge at Lund University, Sweden, for their support.

References

1. Amisten S, Braun OO, Bengtsson A, Erlinge D. Gene expression profiling for the identification of G-protein coupled receptors in human platelets. Thromb Res. 2008;122(1):47–57.

2. Bugert P, Dugrillon A, Gunaydin A, Eichler H, Kluter H. Messenger RNA profiling of human platelets by microarray hybridization. Thromb Haemost. 2003 Oct;90(4):738–48.

3. Gnatenko DV, Dunn JJ, McCorkle SR, Weissmann D, Perrotta PL, Bahou WF. Transcript profiling of human platelets using microarray and serial analysis of gene expression. Blood. 2003 Mar 15;101(6):2285–93.

4. McRedmond JP, Park SD, Reilly DF, Coppinger JA, Maguire PB, Shields DC, et al. Integration of proteomics and genomics in platelets: a profile of platelet proteins and platelet-specific genes. Mol Cell Proteomics. 2004 Feb;3(2): 133–44.

5. Rox JM, Bugert P, Muller J, Schorr A, Hanfland P, Madlener K, et al. Gene expression analysis in platelets from a single donor: evaluation of a PCR-based amplification technique. Clin Chem. 2004 Dec;50(12):2271–8.

6. Rolf N, Knoefler R, Suttorp M, Kluter H, Bugert P. Optimized procedure for platelet RNA profiling from blood samples with limited platelet numbers. Clin Chem. 2005 Jun;51(6): 1078–80.

7. Available from: http://www.pall.com/medical_48187.asp.

8. RNeasy MinElute Cleanup Handbook2007.

9. Kendziorski C, Irizarry RA, Chen KS, Haag JD, Gould MN. On the utility of pooling biological samples in microarray experiments. Proc Natl Acad Sci USA. 2005 Mar 22;102(12): 4252–7.

10. Kendziorski CM, Zhang Y, Lan H, Attie AD. The efficiency of pooling mRNA in microarray experiments. Biostatistics. 2003 Jul;4(3):465–77.

11. Mary-Huard T, Daudin JJ, Baccini M, Biggeri A, Bar-Hen A. Biases induced by pooling samples in microarray experiments. Bioinformatics. 2007 Jul 1;23(13):i313-8.

12. Zhang W, Carriquiry A, Nettleton D, Dekkers JC. Pooling mRNA in microarray experiments and its effect on power. Bioinformatics. 2007 May 15;23(10):1217–24.

13. Wilson CL, Pepper SD, Hey Y, Miller CJ. Amplification protocols introduce systematic but reproducible errors into gene expression studies. Biotechniques. 2004 Mar;36(3):498–506.

14. McCloskey C, Jones S, Amisten S, Snowden RT, Kaczmarek LK, Erlinge D, et al. Kv1.3 is the exclusive voltage-gated K+ channel of platelets and megakaryocytes: roles in membrane potential, Ca2+ signalling and platelet count. J Physiol. 2010 May 1;588(Pt 9):1399–406.

Part II

Additional Protocols for the Study of Megakaryocyte Function

Chapter 13

Characterization of Megakaryocyte Development in the Native Bone Marrow Environment

Anita Eckly, Catherine Strassel, Jean-Pierre Cazenave, François Lanza, Catherine Léon, and Christian Gachet

Abstract

Differentiation and maturation of megakaryocytes occur in close association with cellular and extracellular components in the bone marrow. Thus, direct examination of these processes in the native environment provides important information regarding the development of megakaryocytes. In this chapter, we present methods applied to mouse bone marrow to (1) examine the ultrastructure of megakaryocytes and their state of maturation in situ in fixed bone marrow sections and (2) study the dynamics of proplatelet formation by real-time observation of fresh bone marrow explants where megakaryocytes have matured in their natural physiological context. Combining these two approaches allows detailed investigation of in situ megakaryocyte differentiation, including proplatelet formation, which is the final maturation step before platelet release.

Key words: Bone marrow, Electron microscopy, Explants, In situ megakaryocyte ultrastructure, Proplatelet formation

1. Introduction

Since the cloning and availability of thrombopoietin (Tpo), the major growth factor for megakaryopoiesis, substantial effort has gone into developing megakaryocyte culture systems in order to better characterize the mechanisms that control thrombopoiesis (1, 2). Our understanding of the processes of platelet production has been significantly improved by the combined use of such cell culture systems and genetically manipulated mice (3, 4). Nevertheless, in vitro systems have not resolved all aspects of megakaryocyte differentiation and platelet production due to

Jonathan M. Gibbins and Martyn P. Mahaut-Smith (eds.), *Platelets and Megakaryocytes:*
Volume 3, Additional Protocols and Perspectives, Methods in Molecular Biology, vol. 788,
DOI 10.1007/978-1-61779-307-3_13, © Springer Science+Business Media, LLC 2012

their inability to reproduce the complexity of the bone marrow. This three-dimensional space is composed of a variety of cell types (e.g., blood cells and their precursors, stromal cells, endothelial cells, and osteoblasts) bound together by a meshwork of extracellular matrix (ECM) molecules (5), all of which could influence megakaryocyte development through cytokine secretion or cell–cell and cell–ECM interactions. For example, many of the cytokines that are critical for normal megakaryopoiesis (stem cell factor, IL-6, GM-CSF) are produced by osteoblasts and stromal fibroblasts (6–8). It has also been reported that the chemokine SDF-1 and the growth factor FGF-4 are necessary for megakaryocyte recruitment to sinusoid endothelial cells, thus contributing to platelet biogenesis and release (9). In addition, matrix proteins such as vitronectin, collagen, and fibrinogen have been proposed to regulate proplatelet formation (10–12). These data illustrate the influence of the bone marrow environment on megakaryopoiesis and emphasize the importance of directly studying megakaryocytes in a physiological context.

This chapter focuses on qualitative and quantitative methods to study megakaryocyte development using bone marrow from the mouse, although the approaches can be easily adapted to other species. We describe methods: (1) for bone marrow recovery allowing optimal examination; (2) for ultrastructural observation of megakaryocytes in fixed bone marrow sections and evaluation of their number and maturation stage; and (3) for analysis of the dynamics of proplatelet formation by living cells in fresh bone marrow explants where megakaryocytes have differentiated in a physiological context. These approaches are complementary, allowing exploration in situ throughout the major stages of megakaryopoiesis, a complex process of proliferation, maturation, and finally proplatelet formation, the final step before platelet release.

2. Materials

2.1. In Situ Observations of Megakaryocytes in Fixed Marrow Sections

2.1.1. Removal and Fixation of Bone Marrow

1. Anesthetic: A mixture of 100 μL 2% xylazin (Rompun® Bayer santé, France) and 100 μL 10% ketamin (Imalgene® 1000, Merial, France) in 800 μL 0.9% NaCl saline. Prepared daily.

2. Dissection instruments.

3. Syringe pump (Harvard Apparatus, Holliston MA, USA), 60 mL syringe and Surflo i.v. catheter 24-gauge (TERUMO, Polylabo, France) for whole-body fixation.

4. A combined gas/vapors particular respirator for protection against fixative vapor and projections.

5. 5-mL syringes with 21-gauge needles and 15-mL plastic tube for flushing of femurs.

6. 0.2-μm syringe filters (Pall Corporation, USA) for filtration of all solutions.

7. Cacodylate buffer: dissolve 21.4 g dimethylarsinic acid sodium salt trihydrate (Merck, Germany) and 20 g sucrose (Merck, Germany) in 1 L H_2O. Under constant agitation add 1 mL of 1 M $CaCl_2$ (Merck, Germany) and 1 mL of 1 M $MgCl_2$ (Merck, Germany). The pH is adjusted to 7.3 and the osmolarity to 306 mOsm/L with an approximate equal amount of 1 M $MgCl_2$ and 1 M $CaCl_2$. The final solution is clear and once filtrated (0.2 μm filter) is stable at +4°C for up to several weeks.

8. 25% glutaraldehyde (Electron Microscopy Sciences-EMS, USA), stored at +4°C.

9. Fixative: 2.5% glutaraldehyde in cacodylate buffer. Warning: fixative vapors and projections are very harmful to the eyes, nose, and throat.

2.1.2. Preparation of Marrow for Histology and Electron Microscopy

1. Agarose type LM-3 Low Melting Point Agar (Electron Microscopy Sciences, USA).

2. 2-mL plastic transfer pipettes for transfer of marrow.

3. Flat embedding silicone mold (Oxford Instrument, Agar Scientific, England) for use as capsules for histology and electron microscopy (EM).

4. Osmium tetroxide 2% (Merck, Germany) is dissolved at 1% in cacodylate buffer in a fume cupboard with good ventilation. Warning: Osmium tetroxide is highly volatile at room temperature and its vapors are very harmful to the eyes, nose, and throat.

5. Uranyl acetate 2% (Ladd Research Industries, USA) and ethanol 75, 90, and 100%, (VWR International, France) are dissolved in distilled water.

6. Propylene oxide (1.2-epoxypropane: Sigma, France) is purchased as a ready to use solution.

7. Epon: All chemicals to make Epon resin are from Ladd Research Industries (USA) and stored at room temperature. Components are mixed under constant agitation in the following order: LX-112 resin 217.4 g, dodecenyl succinic anhydride (DDSA) 104.4 g, nadic methyl anhydride (NMA) 102.8 g, and benzyldimethylamine (BDMA) 7.9 g. Resin components are toxic, some are carcinogenic and must be handled with care. The resin is drawn into 10-mL plastic syringes and stored at –20°C.

2.1.3. Histology of Bone Marrow

1. An Ultracut UCT ultramicrotome (Leica Microsystems GmbH, Austria) equipped with a Histo Diamond Knife 45° (Diatome, Switzerland) for obtaining histological sections.

2. Toluidine blue solution: 1 g toluidine blue (Ladd Research Industries, USA) and 1 g sodium borate (Sigma, France) are

dissolved in 100 mL distilled water. Toluidine blue solution is filtered (0.2 μm) and is kept at room temperature in a 10-mL syringe fitted with a 0.2-μm filter.

3. Mounting medium is Biomount (British BioCell, England) and is stored at room temperature.

2.1.4. Evaluation of the Megakaryocyte Density and Maturation Stages by EM

1. An Ultracut UCT ultramicrotome (Leica Microsystems GmbH, Austria).

2. 200-mesh thin-bar copper grids (Oxford Instrument, Agar Scientifics, England), on which sections are deposited.

3. Uranyl acetate (Ladd Research Industries, USA): 4% in distilled water.

4. Lead citrate (Ultrostain 2®: Leica Microsystems GmbH, Austria) purchased as a ready to use solution.

5. CM120 BioTWIN Transmission Electron Microscope (FEI, The Netherlands).

2.1.5. Ultrastructural Analysis of Demarcation Membrane Development

1. Tannic acid (Merck, Germany): dissolved at 1% (w/v) in distilled water.

2. Megaview camera (Olympus SIS, Germany) for capture of images and MetaMorph software (Version 5; Universal Imaging, Downingtown, PA) for image analysis.

2.2. Bone Marrow Explant

1. 13-mm diameter incubation chambers (EMS, USA).

2. Phase contrast inverted microscope (Leica Microsystems SA, Westlar, Germany) coupled to a video camera (DAGE MTI, USA). Metamorph™ software (MetaMorph software, Version 5; Universal Imaging, Downingtown, PA) for image analysis and processing of videos.

3. A heating incubator (PeCon GmbH, Germany) to maintain the chambers at 37°C.

4. 0.22-μm pore filters for sterilization of medium.

5. Stock solutions for filtered Tyrode's buffer:

   (a) Stock I: 160 g (2.73 M) NaCl, 4 g (53.6 mM) KCl, 20 g (238 mM) $NaHCO_3$, and 1.16 g (8.6 mM) NaH_2PO_4, H_2O made up to 1 L in distilled water and stored at 4°C.

   (b) Stock II: 20.33 g (0.1 M) $MgCl_2 \cdot 6H_2O$ made up to 1 L.

   (c) Stock III: 21.9 g (0.1 M) $CaCl_2 \cdot 6H_2O$ made up to 1 L.

   (d) HEPES stock: 0.5 M (N-[2-hydroxyethyl]piperazine-N'-[2-ethanesulfonic acid]) sodium salt (119 g) made up to 1 L.

   (e) Human serum albumin (HSA) stock: 200 g/L pasteurized human serum albumin for intravenous injection (Etablissement Français du Sang, Strasbourg, France) (purity >98%).

6. Tyrode's buffer containing albumin (0.35% albumin): 5 mL stock I, 1 mL stock II, 2 mL stock III, 1 mL HEPES stock, 1.75 mL HSA stock, and 0.1 g anhydrous D(+) sucrose in 100 mL of distilled water. Adjust the pH to 7.3 with 1N HCl and the osmolarity to 295 mOsm/L.

7. Antibiotics: PenicillinG at 10 U/mL final concentration and streptomycin sulfate at 0.29 mg/mL final concentration (GIBCO Invitrogen Corporation, USA) are added to prevent bacterial growth during explants culture.

8. Dissection instruments and ethanol (70% v/v) for sterilization.

9. Razor blades.

10. Preparation of mouse serum: Blood was collected from anesthetized mice in a glass tube without anticoagulant and allowed to clot for 2 h at 37°C and overnight at 4°C. The supernatant was taken and centrifuged at $2,000 \times g$ for 5 min. Serum aliquots were stored at −20°C until use.

11. Sterile transfer pipettes.

3. Methods

3.1. In Situ Observations of Megakaryocytes in Fixed Marrow Sections

3.1.1. Removal and Fixation of Bone Marrow (Fig. 1)

Attention must be paid to preserve as much of the native bone marrow structure as possible. For this purpose, a double fixation is performed, firstly by whole-body perfusion of the fixative followed by immersion of the bone marrow in the fixative (see Note 1) (Fig. 1a–c). Bone marrow flushing has to be performed carefully and immediately after femur dissection. The method is derived from that reported by Behnke (13). Local and national regulations for animal care, anesthesia, and euthanasia are followed at all times.

1. Anesthetize an adult mouse (8–12 weeks) by intraperitoneal injection with 10 mL/kg of a mixture of 0.2% xylazin and 1% ketamin.

2. Place the anesthetized mouse in a supine position on its back and expose the abdominal aorta by laparotomy (it is easier to expose the aorta when a 1-mL syringe is placed under the back of the animal). Open the diaphragm to expose the heart.

3. Connect a 60-mL syringe filled with 20 mL of fixative (2.5% glutaraldehyde in cacodylate buffer) at 37°C to a catheter and place in a perfusion syringe pump. Take care to avoid air bubbles.

4. Quickly insert the catheter into the aorta and make a small perforation using a 25-gauge needle in the right ventricle to allow blood and perfusate to escape.

5. Perfuse with fixative at the rate of 5 mL/min. A volume of 15 mL fixative is normally required for perfusing the entire

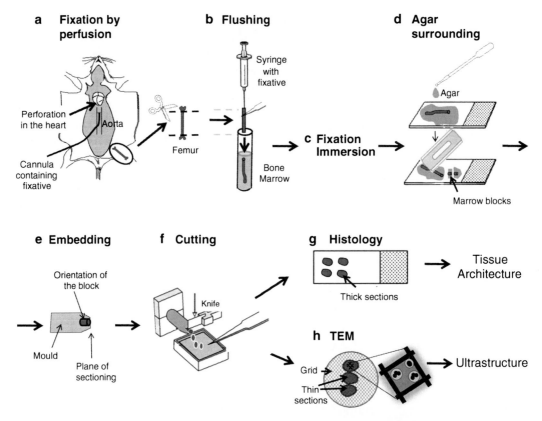

Fig. 1. Preparation of bone marrow for histological and EM observation. (**a**) The mouse is fixed by whole-body perfusion and the femur is removed. (**b**) The marrow is flushed with fixative using a 21-gauge needle. Care is taken to introduce only the bevel of the needle into the lumen of the bone. (**c**) The bone marrow cylinder is fixed further by immersion in fixative for 60 min. (**d**) The marrow is surrounded with agar to ensure tissue integrity and is cut into small cubes. (**e**) The blocks are postfixed, dehydrated, and infiltrated in Epon. The cubes are then positioned into flat silicone molds filled with epon before polymerization. (**f**) The blocks are cut in (**g**) thick sections for histological observation or in (**h**) thin sections for EM observation where the squares of the grid are defined as area for examination in order to facilitate quantification.

animal (see Note 2). The perfusion is stopped when the outflow from the ventricle is colorless. The time from skin incision to the end of the perfusion is normally about 4 min.

6. Marrow collection must then be performed as soon as possible. Rapidly dissect out the femur and remove adherent tissue. Gently cut away the epiphyses using a sharp razor blade.

7. While holding the femur with forceps, immediately flush the marrow into a 15-mL tube containing fixative using a 5-mL syringe filled with fixative and equipped with a 21-gauge needle. To achieve this, the extremity of the needle (just the bevel) is introduced into the opening of the bone and the plunger slowly pressed until the marrow has been extracted (Fig. 2a).

8. Typically, a bone marrow cylinder of 4–6 mm in length is obtained (Fig. 2b). Carefully immerse this in 1 mL of fresh fixative for a further 60 min.

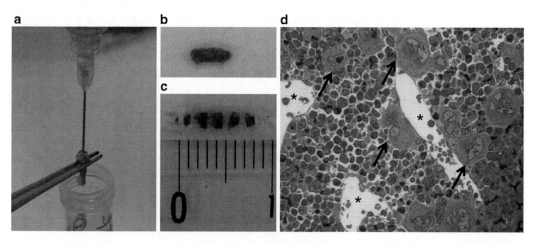

Fig. 2. Flushing of the bone marrow. (**a**) Photograph illustrating the flushing procedure. Only the extremity of the needle is introduced in the femur opening. (**b**) Typical appearance of the bone marrow cylinder after flushing. (**c**) Picture showing the size and the agar surrounding of marrow blocks (scale: 1 cm). (**d**) Normal appearance of bone marrow (image taken using a 40× objective lens). Cells are tightly packed with little extracellular space. The giant cells correspond to mature megakaryocytes (*arrows*), some being in contact with sinusoids (*asterisk*).

3.1.2. Preparation of Marrow for Histology and Electron Microscopy

Detailed description of standard dehydration and embedding methods can be found in many useful sources such as "Practical Electron Microscopy: A beginner's Illustrated Guide – Second Edition" from Elaine Hunter (14) (see Note 3). Here, we describe only some useful steps that improve sample preparation.

Marrow tissue is not cohesive enough to remain intact during the several washing steps and material can be easily lost. To maximize tissue integrity, the marrow is surrounded with a gel of agar before dehydration (Fig. 1d).

1. Prepare a 2% (w/v) solution of agar by dissolving agar powder in boiling cacodylate buffer.

2. Place the agar solution in a water bath at 45°C. At this temperature, the agar solution remains as a liquid.

3. Wash the fixed marrow from Subheading 3.1.1, step 8 in cacodylate buffer and carefully transfer to a glass slide using a plastic transfer pipette.

4. Quickly transfer a drop of 2% liquid agar using a warm pipette onto the top of the marrow sample (see Note 4).

5. Immediately place the slide on ice until the agar has solidified (1 or 2 min).

6. Under a microscope, cut the solidified agar containing the marrow into 1 mm³ cubes with a sharp razor blade (Fig. 2c). Discard the blocks corresponding to the extremities of the bone marrow cylinder because of possible tissue compression in these areas. Place the other blocks in 1.5-mL eppendorf tubes containing cacodylate buffer.

7. Postfix the blocks with 1% osmium tetroxide for 1 h at 4°C (see Note 5), stain with 2% uranyl acetate for 1 h, and dehydrate through a standard graded series of ethanol before infiltration with the embedding medium (see Note 6).

8. Epon resins are viscous, and thus to obtain uniform infiltration and polymerization inside the marrow, we recommend the use of the intermediate solvent propylene oxide. The samples are placed on a slow rotary shaker at room temperature and passed gradually from the intermediate solvent to the embedding epon according the following schedule:

100% propylene oxide	15 min
100% propylene oxide	15 min
1:1 mixture of 100% propylene oxide and Epon	60 min
100% Epon	Overnight (see Note 7)
100% Epon	2 h

9. Exact orientation of the marrow blocks is necessary to permit subsequent transversal sectioning of the entire bone marrow (Fig. 1e). This is done under a microscope where each block is placed and orientated into a flat silicone mold for embedding. The molds are filled with epon and placed at 60°C for 48 h. After that, the blocks are ready for cutting. Proceed to Subheading 3.1.3 for processing for histological examination and to Subheading 3.1.4 for processing for examination by EM.

3.1.3. Histology of Bone Marrow

Observation of the marrow sections under a light microscope before EM gives an excellent overview of the tissue architecture (e.g., cell compactness, cellularity, structure of vascular sinuses). The most mature megakaryocytes can be easily identified with a 40× objective due to their giant size and nuclear lobulation; they also appear as round cells, some of them being in contact with sinusoids (Fig. 2d). Observation of thick sections also allows selection of suitable blocks that contain megakaryocytes from which ultrathin sections will be obtained for TEM.

The marrow blocks from Subheading 3.1.2, step 9, are processed as follows (Fig. 1f, g):

1. Cut thick sections (0.5 μm) of epon-embedded marrow and deposit on a glass slide.

2. Immerse slides in 1% toluidine blue and then heat on a hot plate (60°C) for 1–2 min. Wash the slides thoroughly with distilled water.

3. Mount sections on coverslips with a drop of mounting medium and examine under a light microscope (Fig. 2d).

3.1.4. Evaluation of the Megakaryocyte Density and Maturation Stages by EM

Determination of the number of megakaryocytes present in the bone marrow gives indications concerning the regeneration capacity of megakaryocytes and the platelet production. Mouse megakaryocytes are usually counted under a light microscope after acetylcholinesterase staining and correspond to 0.02% of total marrow cells. Observation of megakaryocytes by TEM allows precise quantification of all megakaryocyte stages. They are ranked into three classes (stages I, II, and III) on the basis of cell size, nuclear morphology (bilobed or multilobed), presence of organelles, and development of platelet territories (15). Stage I megakaryocytes or megakaryoblasts represent the earliest recognizable cells by TEM. They correspond to cells of 10–15 μm in diameter with a large nucleus (see Note 8). Stage II or promegakaryocytes are cells of 15–30 μm in diameter containing platelet-specific granules. Stage III are mature megakaryocytes having a well-developed demarcation membrane system (DMS) with clearly defined platelet territories and a peripheral zone devoid of organelles (Fig. 3). Additional morphological characteristics can be studied and quantified, such as the interactions of fully mature megakaryocytes with sinusoids (which is informative regarding the ability of the megakaryocyte to migrate toward the sinusoids), apoptotic megakaryocytes or emperipolesis (presence of hematopoietic cells within the lumen of the demarcation membrane system of megakaryocytes). The samples are processed as follows (Fig. 1f, h):

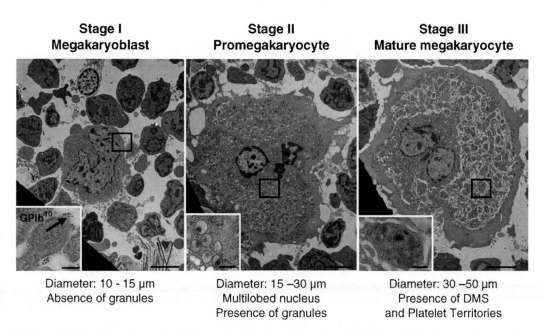

| Stage I Megakaryoblast | Stage II Promegakaryocyte | Stage III Mature megakaryocyte |

| Diameter: 10 - 15 μm Absence of granules | Diameter: 15 –30 μm Multilobed nucleus Presence of granules | Diameter: 30 –50 μm Presence of DMS and Platelet Territories |

Fig. 3. Classification of megakaryocytes in three maturation stages. Stage I, a cell 10–15 μm in diameter with a large nucleus expressing the major platelet membrane glycoproteins on their surface. To identify these cells, we recommend to perform immunostaining. This is shown in the *inset* where an immunogoldlabeling was performed using an anti-GPIb antibody conjugated with 10 nm gold particles (GPIb[10]). The *arrows* point to the gold particles located at the surface of stage I megakaryocyte; stage II, a cell 15–30 μm in diameter containing platelet-specific granules (see *inset*); stage III, a 30–50 μm cell containing a well-developed demarcation membrane system (DMS) defining platelet territories (one territory is shown in *inset*) and a thick membrane-free peripheral zone. Bars in main images: 5 μm; bars in inset images: 200 nm.

1. Cut thin (100 nm) transverse sections of epon-embedded marrow from Subheading 3.1.2, step 9 (see Note 9).

2. Deposit sections on 200-mesh thin-bar copper grids.

3. Double stain sections with uranyl acetate and lead citrate to obtain a good contrast. To achieve this, thin sections are floated on drops of the following solutions at room temperature: 4% uranyl acetate for 5 min, three washes of distilled water for 5 min each, lead citrate for 3 min, and three washes of distilled water for 5 min each.

4. Thin sections are examined at a constant magnification of 5,800× (necessary to distinguish the stage of development).

5. Megakaryocytes from each stage of development are counted manually on whole transversal sections. To facilitate quantification, each square of the grids is defined as an area for examination (which equals 16,000 μm^2) (Fig. 1h).

6. For each square, the number of stage I, II, or III megakaryocytes is scored. Using this procedure, the whole marrow transversal section is fully and systematically screened, square by square.

7. Using this procedure for wild-type C57BL/6 mice, the average number of mouse megakaryocytes per square is typically 1.7 ± 0.1 for a total of 50 megakaryocytes per section. The distribution expressed as a percentage of megakaryocyte in each stage is $8.8 \pm 1.3\%$ for stage I, $20.2 \pm 1.1\%$ for stage II, and $71.2 \pm 1.8\%$ for stage III. This is similar to what is found in a normal human marrow, where stage I megakaryocytes represent only 12% of total marrow megakaryocytes (16).

3.1.5. Ultrastructural Analysis of Demarcation Membrane Development

The most mature megakaryocytes (stage III) are characterized by the presence of a DMS within the cytoplasm that delineates platelet territories. The DMS is continuous with the megakaryocyte plasma membrane and appears as a network of membranes that separate the intracellular platelet territories. The difference in contrast between the cytoplasm (gray zone) and the extracellular medium (white zone) is used for image analysis (Fig. 4). To facilitate DMS visualization, we recommend the use of tannic acid. Through its action as a mordant for heavy metal and because it does not penetrate into the cells, tannic acid is particularly useful to enhance the contrast of extracellular structures such as plasma membrane and DMS. The following procedure is employed:

1. Fix and flush marrow as described in Subheading 3.1.1. The marrow cylinder is directly cut into 1 mm^3 cubes with a razor blade.

2. Incubate the blocks for 1 h in 1% tannic acid.

3. Surround the blocks with agar as described in Subheading 3.1.2, steps 2–6 and process for classical EM as described in Subheading 3.1.2, step 8.

+ Tannic Acid   + Image Analysis

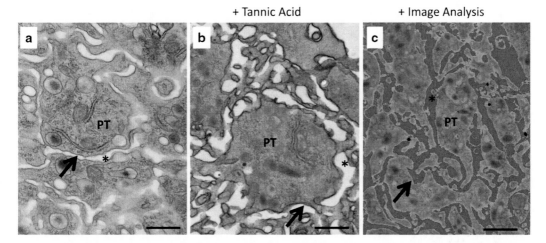

Fig. 4. Quantification of the demarcation membrane system (DMS) perimeter. (**a**) Normal ultrastructure of the DMS (*arrow*) appearing as a network of membranes delineating platelet territories (PT). (**b**) Treatment with tannic acid enhances the contrast of the DMS thereby facilitating its identification. (**c**) Analysis of the DMS perimeter in a region of interest. The lumen of the DMS is colored in *red* and the perimeter is calculated using Metamorph software. *Asterisk*: extracellular medium, *bars*: 500 nm.

4. Cut thin sections (100 nm) and examine at a constant magnification of 13,500 (necessary to clearly distinguish the membrane limits of the DMS).

5. Photograph the cytoplasm of fully mature megakaryocytes using a SIS Megaview Camera (1,280 × 1,024 pixels).

6. Transfer the images to the image analysis package MetaMorph.

7. Draw a region of interest (ROI) around the DMS-containing cytoplasm.

8. Apply a threshold to the image using Metamorph software to delimitate the DMS (colored in red, see Fig. 4 right panel) and unwanted background noise pixels are removed (e.g., electron lucent areas in dense granules).

9. Quantify the DMS in the ROI by measuring the perimeter (in pixels) of the pseudocolored zones which corresponds to the demarcation membranes.

3.2. Bone Marrow Explants and Proplatelet Visualization

3.2.1. Preparation of Marrow for Explants (Fig. 5a)

The final stage of megakaryocyte development includes proplatelet formation. Currently, most studies use cultured megakaryocytes to study the mechanism of proplatelet formation. However, in vitro culture does not represent the native milieu and it is possible that the absence of the bone marrow environment introduces artificial events or eliminates important factors required for normal megakaryocyte maturation. To evaluate proplatelet formation in a more native context, small pieces of fresh bone marrow (explants) are bathed in a physiological buffer and observed for 6 h by time-lapse microscopy. Over time, megakaryocytes are progressively visible at

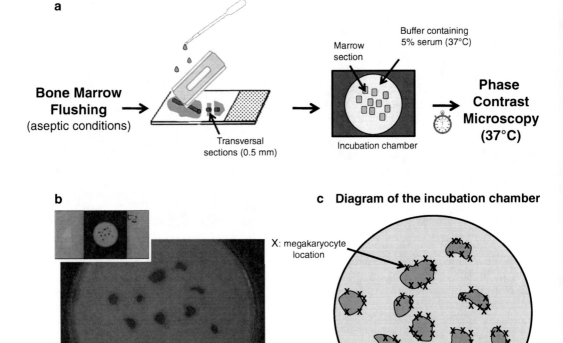

Fig. 5. Preparation of bone marrow explants. (**a**) Fresh bone marrow is obtained by flushing a mouse femur. The marrow is cut into 0.5-mm thick sections. Care is taken to avoid tissue dry out. Ten sections are transferred to an incubation chamber and are observed at 37°C under an inverted microscope. (**b**) Pictures illustrating the bone marrow sections in the incubation chamber. (**c**) To facilitate the location of megakaryocytes, a drawing recapitulates their position in each section.

the periphery of the explants and change their shape, allowing qualitative and quantitative investigations. Proplatelet formation in marrow explants was originally described by Thiery and Bessis in 1956 (17, 18).

1. Before starting, prepare and filter Tyrode's buffer containing albumin (0.35%) and antibiotics and warm to 37°C. The heating incubator containing the phase contrast microscope is also warmed to 37°C.

2. Use decerebration to euthenase mice and remove femurs under aseptic conditions. Disinfect all instruments in 70% (v/v) ethanol to avoid microbial contamination and immerse the mouse carcuss in 70% ethanol before laparotomy.

3. Obtain intact marrow by flushing femurs as described in Subheading 3.1.1, step 7, but using Tyrode's buffer containing albumin and antibiotics.

4. Using a sterile plastic transfer pipette, carefully transfer the marrow in a drop of Tyrode's buffer containing albumin to a glass slide and quickly flood with buffer. It is important not to let the tissue dry out.

5. Under a binocular microscope cut the marrow into 0.5-mm thick transverse sections with a sharp razor blade (see Note 10).

6. Gently pick up ten sections in a drop of buffer containing 5% mouse serum and transfer to an incubation chamber. Take care that the sections do not stick to each other or the walls of the chamber (Fig. 5b). Aspirate the buffer and replace with 45 μL fresh buffer containing 5% mouse serum at 37°C. It is possible to add various pharmacological agents to visualize their impact on the process of proplatelet formation. Start a timing device and consider this point as the start of the experiment.

7. Examine tissue sections by phase contrast under an inverted microscope coupled to a video camera and maintained at 37°C.

8. At the beginning of the experiment, no individual cells are visible due to the thickness of the sections, but after 30 min a monolayer of cells is formed at the periphery of the explants and megakaryocytes become recognizable from their large size (see Note 11). The cell layer extends with time and the number of visible megakaryocytes also increases with time (Fig. 6).

9. To facilitate megakaryocyte location, a diagram of the incubation chamber is drawn showing the position of megakaryocytes for each section (Fig. 5c). The number of megakaryocytes typically observed per section is 10.2 ± 1.3 ($n = 12$).

3.2.2. Morphology and Quantification of the Megakaryocytes Extending Proplatelets

1. Megakaryocytes are classified according to their morphology, some of them being spherical (type A), others developing thick protrusions (type B), and others extending numerous thin extensions (type C) (Fig. 7a).

2. Proplatelet morphology is similar to that observed when megakaryocytes are grown in culture from stem cells, with long, thin branched proplatelet shafts, proplatelet buds and intermediate swellings (19). A megakaryocyte is classified as a "proplatelet-forming cell" when at least one thin extension presenting a proplatelet bud is observed.

3. Based on the scheme, each megakaryocyte can be easily located and investigated over time according to its morphology (e.g., ability to extend proplatelets, quality and complexity of proplatelets, etc.).

4. Quantification is performed at 1 h 30 min, 3, and 6 h after sectioning of the explants. Results are expressed as the proportion

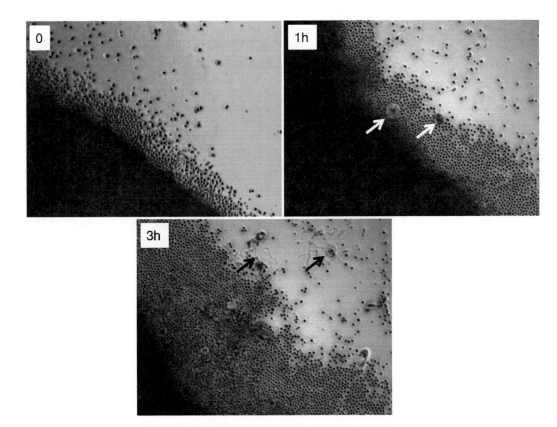

Fig. 6. Time-course observation of the cell layer at the periphery of a bone marrow explant. Cells progressively migrate out of the bone marrow explant, forming a layer in which megakaryocytes become visible. At the beginning, no megakaryocytes are observed. After 1 h, giant round cells (*white arrows*) can be observed. After 3 h, the number of megakaryocytes increases and several extend long extensions (*black arrows*).

of megakaryocytes in each class at the different times of observation (Fig. 7b). For wild-type C57BL/6 mice, typical results are $46 \pm 6.5\%$ in type A, $9 \pm 0.5\%$ in type B, and $45 \pm 6.3\%$ in type C at 6 h for a total of 850 megakaryocytes examined ($n = 5$). Using this technique, a decreased number of megakaryocytes forming proplatelets has been quantified in GPIbbeta and *MYH9* knock out mice, two models of human pathologies with macrothrombocytopenia ((20), Eckly et al., unpublished observation).

3.2.3. Time-Lapse Acquisition

Videomicroscopy recording is useful to follow the morphological changes of an individual megakaryocyte over time. Images are acquired sequentially at 10 s intervals and processed with the Metamorph™ software.

The spherical megakaryocyte first becomes distorted and then extends pseudopodia which elongate into proplatelets. Once started, the process is highly dynamic and the proplatelets are constantly remodeled, proplatelet extension usually being maximal within less than 2 h.

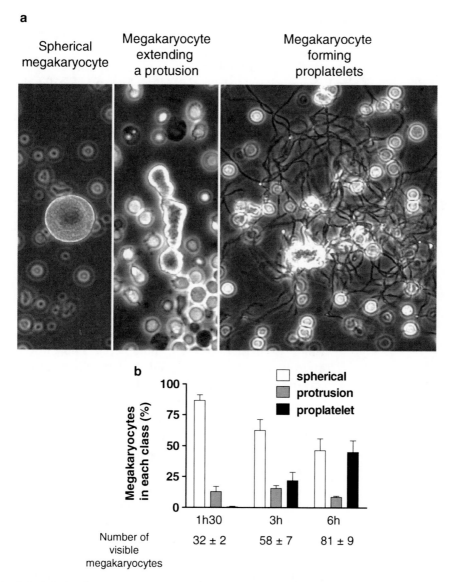

Fig. 7. Morphological classification of the megakaryocytes in explants. (**a**) Megakaryocytes are classified as "spherical," with one protrusion or as megakaryocytes extending proplatelets. (**b**) The proportion in each class was determined at 1 h 30 min, 3 h, and 6 h showing that the proportion of spherical megakaryocytes decreases while those extending proplatelets increase with time. At the different times of observation, the number of megakaryocytes counted per experiment is indicated ($n=5$).

3.3. Conclusion

Direct examination of megakaryocytes in the native environment is essential for a complete and valid understanding of megakaryopoiesis and platelet formation. In this chapter, we have described two complementary methods providing qualitative and quantitative information regarding in situ megakaryocyte development. Of interest, the revisited explant method, originally described in 1956 (17), has the advantage of being simple, rapid, and reproducible. Numerous proplatelets are formed within 6 h as compared to 4 days for megakaryocyte cultures. Moreover, this technique visualizes megakaryocytes that have fully matured in their physiological context avoiding artifactual in vitro differentiation. This needs to be considered in view of differences observed, for example in proplatelet formation between $MYH9^{-/-}$ cultured megakaryocytes and megakaryocytes from $MYH9^{-/-}$ explants. Finally, the explant method allows examination of the impact of pharmacological agents exclusively on the process of proplatelet extension, without interfering with the differentiation process as is likely for in vitro cultures.

4. Notes

1. We tested three different methods to collect the bone marrow (1) dissection of the mouse femur followed by fixation of the bone by immersion and EDTA decalcification, (2) dissection of the mouse femur and flushing of the marrow, which is then fixed by immersion, and (3) whole-body fixation by perfusion, dissection of the femur, and flushing of the bone marrow, which is further fixed by immersion. The optimum bone marrow morphology (i.e., tightly packed hematopoietic cells) was obtained with the third method.

2. The fixative must be made fresh prior to use and be filtered. Its temperature should be near the body temperature of mice (37–38°C). Fixatives are harmful and must be handled with care. Thus, animals must be perfused under an extraction hood or using a combined gas/vapor particular respirator. The excess of outflow fixative has to be collected with a cotton pad placed near the heart. The cotton should be discarded in a specific biohazard container in accordance with safety guidelines.

3. Almost all reagents used for EM are toxic and should be handled carefully using an extraction hood, latex gloves and air-tight goggles. Their storage in specific safety cabinets and their disposal should follow safety guidelines.

4. It is important to ensure that the agar remains warm until the moment it is poured onto the slide or it will start to solidify before the bone marrow is totally surrounded.

5. Before being discarded, 2% osmium tetroxide is neutralized by adding it to twice its volume of vegetable oil.

6. The blocks remain in the same eppendorf tubes throughout incubation in the different baths of osmium, acetate uranyl, and ethanol. Each solution is removed carefully with a Pasteur pipette before the next one is poured on. Care should be taken to ensure that the volume of solution used is at least tenfold greater than the volume of the specimen.

7. The eppendorf tubes containing the marrow blocks are placed opened on a slow rotary shaker overnight to allow any remaining propylene oxide to evaporate.

8. Stage I megakaryocytes are the rarest and the most difficult to recognize megakaryocytes. The cytological features of these cells are not specific, although they are slightly larger than other cells from the bone marrow. Immunostaining could help in the identification of these young species of megakaryocytes by using antibodies to platelet membrane glycoproteins (GPs) (GP IIb–IIIa, GPIb) (see inset in the left panel of Fig. 3).

9. Ultrathin sections can be cut for EM immediately following a thick section. This will give close correspondence between thick and thin sections, a desirable feature if corresponding areas are studied or photographed.

10. Sections must be rather thin to allow a detailed observation, but thick enough, however, not to damage megakaryocytes by compression (usual thickness between 0.4 and 0.7 mm).

11. Formation of the cell layer in bone marrow explants results from migration of bone marrow cells out of the tissue, a process which is prevented by incubation with cytochalasin D (10^{-5} M) which depolymerizes actin filaments.

Acknowledgments

The authors wish to thank Patricia Laeuffer, Fabienne Proamer, Monique Freund, Catherine Ziessel, and Jean-Yves Rinckel for technical assistance. This work was supported by INSERM, EFS-Alsace, ARMESA, and ANR (Agence Nationale pour la Recherche, Grant N ANR-07-MRAR-016-01).

References

1. Kaushansky, K., 2005, Thrombopoietin and the hematopoietic stem cell. Ann N Y Acad Sci, 1044: p. 139–41.

2. Shivdasani, R.A., and H. Schuze., 2005, Culture, expansion, and differentiation of murine megakaryocytes. Curr Protoc Immunol, Chapter 22: p. Unit 22F 6.

3. Battinelli, E.M., et al., 2007, Delivering new insight into the biology of megakaryopoiesis and thrombopoiesis. Curr Opin Hematol, 14(5): p. 419–26.

4. Liu, J., et al., 2008, Genetic manipulation of megakaryocytes to study platelet function. Curr Top Dev Biol, 80: p. 311–35.

5. Travlos, G.S., 2006, Normal structure, function, and histology of the bone marrow. Toxicol Pathol, 34(5): p. 548–65.

6. Kacena, M.A., et al., 2006, A reciprocal regulatory interaction between megakaryocytes, bone cells, and hematopoietic stem cells. Bone, 39(5): p. 978–84.

7. Debili, N., et al., 1993, Effects of the recombinant hematopoietic growth factors interleukin-3, interleukin-6, stem cell factor, and leukemia inhibitory factor on the megakaryocytic differentiation of CD34+ cells. Blood, 82(1): p. 84–95.

8. Avraham, H., et al., 1992, Effects of the stem cell factor, c-kit ligand, on human megakaryocytic cells. Blood, 79(2): p. 365–71.

9. Avecilla, S.T., et al., 2004, Chemokine-mediated interaction of hematopoietic progenitors with the bone marrow vascular niche is required for thrombopoiesis. Nat Med, 10(1): p. 64–71.

10. Leven, R.M., and F. Tablin., 1992, Extracellular matrix stimulation of guinea pig megakaryocyte proplatelet formation in vitro is mediated through the vitronectin receptor. Exp Hematol, **20**(11): p. 1316–22.

11. Sabri, S., et al., 2004, Differential regulation of actin stress fiber assembly and proplatelet formation by alpha2beta1 integrin and GPVI in human megakaryocytes. Blood, **104**(10): p. 3117–25.

12. Larson, M.K., and S.P. Watson., 2006, Regulation of proplatelet formation and platelet release by integrin alpha IIb beta3. Blood, **108**(5): p. 1509–14.

13. Behnke, O., 1968, An electron microscope study of the megacaryocyte of the rat bone marrow. I. The development of the demarcation membrane system and the platelet surface coat. J Ultrastruct Res, **24**(5): p. 412–33.

14. Hunter, E., 1993, Practical Electron Microscopy. A Beginner's Illustrated Guide, ed. C.U. Press. Vol. 2: Hunter E, Maloney P and Bendayan M. 1–170.

15. Zucker-Franklin, D., 1988, Megakaryocytes and platelets, ed. A.o.B.C.F.a. Pathology. Vol. 2: Zucker-Franklin D, Greaves MF, Grossi CE, Marmont AM, eds. 621–93.

16. Harker, L.A., and C.A. Finch., 1969, Thrombokinetics in man. J Clin Invest, **48**(6): p. 963–74.

17. Thiery, J.P., and M. Bessis., 1956, (Mechanism of platelet genesis; in vitro study by cinemicrophotography.). Rev Hematol, **11**(2): p. 162–74.

18. Thiery, J.P., and M. Bessis., 1956, (Genesis of blood platelets from the megakaryocytes in living cells.). C R Hebd Seances Acad Sci, **242**(2): p. 290–2.

19. Italiano, J.E, Jr., et al., 1999, Blood platelets are assembled principally at the ends of proplatelet processes produced by differentiated megakaryocytes. J Cell Biol, **147**(6): p. 1299–312.

20. Strassel, C., et al., 2009, Intrinsic impaired proplatelet formation and microtubule coil assembly of megakaryocytes in a mouse model of Bernard-Soulier syndrome. Haematologica, **94**(6): p. 800–10.

Chapter 14

Culture of Murine Megakaryocytes and Platelets from Fetal Liver and Bone Marrow

Harald Schulze

Abstract

Megakaryocytes (MKs) are the largest hematopoietic cells in bone marrow. Since the mid-1990s, recombinant thrombopoietin has been commercially available. Together with the emerging knowledge of mouse models, this has provided an unprecedented contribution to the understanding of external and internal factors that orchestrate the transition as one large MK gives rise to thousands of discoid platelets that are indistinguishable from one another. Taking advantage of the murine fetal liver as a source of MKs allows for analysis at the single cell level and also serves to provide enough cells for classical biochemical analyses. Other sources of MK progenitors are adult bone marrow and spleen, although the yields are normally lower. This chapter describes methods to establish a standard culture of murine MKs from fetal liver and bone marrow along with approaches to purify MKs using bovine serum albumin gradients and magnetic-activated cell sorting. These culture systems allow investigations of MK development (ploidy, surface markers, etc.) and platelet biogenesis, including effects of external factors such as structural proteins or feeder layers.

Key words: Megakaryocytes, Fetal liver, Thrombopoietin

1. Introduction

This chapter describes the isolation of megakaryocytes from murine fetal livers and alternate sources. In the mouse, the fetal liver is the prominent hematopoietic organ from day 12.5 of gestation up to birth, and the abundant presence of hematopoietic stem cells with broad plasticity allows their differentiation into mature MKs in the presence of thrombopoietin (TPO). The first Subheading (3.1) deals with the preparation of a culture of dissociated fetal liver cells (referred to as a single-cell culture), the differentiation of these cells into megakaryocytes, and some typical differentiation readouts. The second Subheading (3.2) describes the major procedures for

Jonathan M. Gibbins and Martyn P. Mahaut-Smith (eds.), *Platelets and Megakaryocytes: Volume 3, Additional Protocols and Perspectives*, Methods in Molecular Biology, vol. 788, DOI 10.1007/978-1-61779-307-3_14, © Springer Science+Business Media, LLC 2012

MK enrichment. The bone marrow and spleen of adult animals can be used as alternate sources, and culture of MKs from bone marrow is described in the third Subheading (3.3). The fourth and final Subheading (3.4) explains how magnetic-activated cell sorting (MACS) can be used to further purify megakaryocytes. Some additional aspects will be addressed in the notes, including methods for analyzing the influence of non cell-autonomous factors (e.g., stroma proteins or stromal feeder layers) on megakaryocyte maturation, proplatelet formation, and platelet biogenesis.

2. Materials

2.1. Single-Cell Culture Derived from Fetal Liver

1. Pregnant mice at a period of gestation of 12.5–15.5 days. Either inbred strains (C57/BL6, 129 Sv, Balb/c) or out-bred strains (CD1) are commercially available (Charles River, Taconic, Jackson Laboratories).

2. Sterile surgical instruments including scissors, fine-tip forceps (e.g., Dumont type #5).

3. 10-mL syringes (Becton Dickinson) and needles (18, 21, 23, and 25 gauges).

4. DMEM (Gibco, Invitrogen) with 10% fetal bovine serum (FBS).

5. Antibiotics: 50 U/mL penicillin and 50 µg/mL streptomycin, final concentrations.

6. Recombinant murine thrombopoietin (rmTPO) (R&D Systems).

7. 15 mL conical centrifuge tubes (Becton Dickinson).

8. 10 cm plastic cell culture dishes (Techno Plastic Products, TPP, Switzerland).

9. Humidified incubator at 37°C with 5% CO_2.

10. Inverted light microscope with phase contrast and 20× and 40× objective lenses and a 10× eyepiece.

11. Cell strainer with 70 µm size pores (Becton Dickinson).

12. Syringe filters with 0.45 µm size pores (Millipore).

2.2. Enrichment and Purification of Megakaryocytes

1. Bovine serum albumin (BSA), fraction V, cell culture quality (Sigma).

2. Phosphate-buffered saline (PBS) without Ca^{2+} and Mg^{2+} (PBS) (Dulbecco, Biochrom).

3. 1.5 and 3% (w/v) BSA in PBS. The solutions are filtered through a 0.45-µm syringe filter.

4. 15 mL conical centrifuge tubes (Becton Dickinson).

5. DMEM (Gibco, Invitrogen) with 10% FBS.

2.3. Single-Cell Culture Derived from Bone Marrow

1. Sterile surgical instruments including scissors, 1 or 3-mL syringes (Becton Dickinson) and a 25-gauge needle.

2. Sterile PBS without Ca^{2+} and Mg^{2+}.

3. 15 mL conical centrifuge tubes (Becton Dickinson).

4. 6-cm cell culture dishes (Techno Plastic Products, TPP, Switzerland).

2.4. Magnetic-Activated Cell Sorting

1. 50-mL conical centrifuge tubes (Becton Dickinson).

2. Antibodies raised against epitopes of interest, here antimouse CD61-FITC (Becton Dickinson, or Emfret, Würzburg, Germany).

3. 500 mM Na_2EDTA (pH 8.0 with NaOH).

4. MACS buffer: 0.5% (w/v) BSA in Ca^{2+}- and Mg^{2+}-free PBS, supplemented with 2 mM EDTA (from 500 mM stock solution in item 3), no need to adjust the pH afterward.

5. Anti-FITC-microbeads (Miltenyi) or alternate products.

6. MACS depletion columns, type LD (Miltenyi) and magnetic holding device.

7. Hemocytometer (Neubauer Chamber) for cell counting under an inverted light microscope equipped with 20× or 40× objective lenses and 10× eyepieces.

3. Methods

3.1. Single-Cell Culture Derived from Fetal Liver and Differentiation to Megakaryocytes

This section describes the cultivation of fetal liver-derived cells in the presence of TPO as the only exogenous growth factor, followed by the isolation of megakaryocytes. Different organ systems are involved in hematopoiesis during ontogeny of mammals. During gestation days 12.5–15.5, the liver is the most prominent organ in the developing fetus and it has an enrichment of hematopoietic stem cells. During this time window, MKs are readily obtained in ample amounts by a simple cultivation procedure.

1. Use timed-pregnant mice between gestation days 12.5 and 15.5, sacrificed in compliance with institutional guidelines (for instance by carbon dioxide-induced asphyxiation). Gently lift the skin of the abdominal wall and cut a 5-mm stretch open. Tear the skin further open to expose the abdomen and pelvic region. This can be performed under nonsterile conditions; prewetting the fur with 70% ethanol reduces the scattered distribution of loose hairs. Cut the peritoneum open and surgically remove the uterus, placing it in a new dish with fresh medium (see Note 1). Individual fetuses are prepared by opening the uterus and removing the placenta. Allow a brief period (2–5 s)

for exsanguination of the fetuses through the cut umbilical cord prior to placing them in a new dish with fresh medium.

2. Continue with two pairs of sterile forceps (fine-tip) and hold the fetus on its left side. Carefully introduce one end of the forceps and pull the fetal skin open. Continue to the front of the fetus without harming the liver. Albeit still small on day E12.5, the dark red-colored liver is readily recognizable as the prominent organ. Isolate the liver by severing all connecting tissue, including all segments of the gut. If necessary, make use of a binocular microscope (e.g., for fetuses of gestation days E12.5–E13.5). Place the livers in a new dish containing medium supplemented with 10% FBS. We recommend the use of antibiotics (50 U/mL penicillin and 50 µg/mL streptomycin).

3. Single livers can be pooled in 10 cm dishes or kept individually in 12- or 24-well dishes when derived from a heterogenic breeding. If genotyping is required, remove a fraction of a limb or tail of the remaining fetus for DNA preparation. For the standard protocol, pool 4–6 livers from E14.5 mice.

4. Fetal liver tissue is made into a single-cell suspension by passage through needles with decreasing diameter. Start with an 18-gauge needle attached to a sterile 10 mL syringe and aspirate and expel 8–10 times. Repeat the process with a 21-gauge needle attached to the same syringe. The quality of single-cell suspension should be controlled with the help of a standard cell culture light microscope. Depending on the gestation day, this step may be repeated, or performed using a 23-gauge needle. Use sterile conditions and a laminar flow hood from this step onward.

5. Most cells should be individualized or in small groups; larger clumps and connective tissue are removed by passage through a cell strainer (pore size 70 µm). Rinse the cell strainer subsequently with DMEM medium (containing FBS and antibiotics) to minimize cell loss. This procedure with the cell strainer is dispensable when individual livers are used. Cells are afterward transferred to a conical tube and centrifuged at ambient temperature for 3–5 min at $200 \times g$. Fetal erythropoiesis is enhanced on gestation day 15.5, and developing red blood cells should be lysed at this point (see Note 2).

6. Resuspend the cell pellet in 10 mL DMEM medium (+FBS and antibiotics), supplemented with 50 ng/mL thrombopoietin (see Note 3) and plate at a density of $3–4 \times 10^7$ cells (equals 3–4 fetal livers day 13.5) per sterile 10 cm cell culture dish or 75 cm² flask. Incubate cells in a standard humidified incubator at 37°C with 5% CO_2. We consider the day of plating as day 0. Cell differentiation can be followed daily, but we recommend minimizing the handling time outside the incubator for optimal differentiation results.

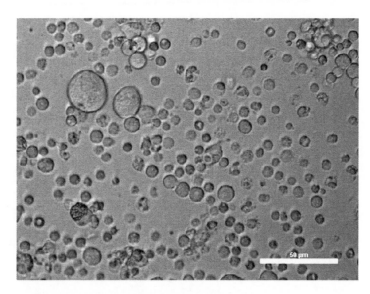

Fig. 1. A fetal liver-derived cell culture in the presence of thrombopoietin on day 1 of culture. Most cells are small, 3–5 μm, but larger cells can also be distinguished which are megakaryoblasts and MKs of 10–15 μm at an early stage of development that will develop into larger, mature MKs during culture. Scale bar 50 μm.

7. On day 0, most cells are uniformly small (3–5 μm). The first larger cells (10–15 μm) are readily detected on day 2 of culture (see Fig. 1), and their fraction increases over time. Around day 4, about 20–25% of cells are 20–30 μm in diameter or larger. Proplatelet formation also typically starts on day 4 and usually peaks on day 5. Quantification of the proplatelet-forming MKs should only be performed in areas where cells are not clustered. A typical example of a proplatelet-forming MK is depicted in Fig. 2. Cellular fragments derived from these cells harbor all features of platelets isolated from the peripheral blood and can be analyzed, e.g., by flow cytometry (1).

3.2. Enrichment and Purification of Megakaryocytes

Subheading 3.1 describes a simple and very robust procedure to generate MKs for analysis at the single cell level. Using all fetal livers from 1–2 timed-pregnant mice can also provide ample cells for biochemical analysis, which can be adapted to heterogenic breeding when each liver is treated separately. However, since the culture system also supports the growth of other hematopoietic and nonhematopoietic cells, enrichment of MKs is therefore often required. Application of two- or three-step BSA density gradients has been widely used to isolate a fraction with enriched MKs and a fraction depleted of MKs (often referred to as the "non-MK" fraction). In addition, fluorescence-activated cell sorting (FACS) or MACS (see Subheading 3.4) are feasible methods when more pure populations are required.

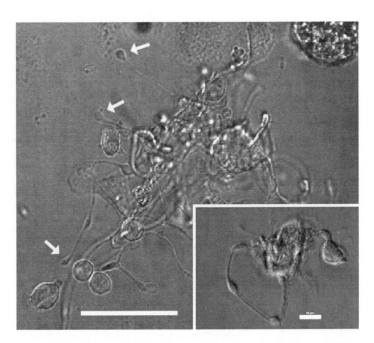

Fig. 2. Mature megakaryocyte undergoing proplatelet formation on day 5 of culture. An elaborated network of shafts is readily recognized with bead-like swellings along the shafts and tear-shaped tips (*white arrows*) that mark the nascent platelet. These structures are released into the medium and feature all characteristics of peripheral platelets. One severed proplatelet with a tear-shaped tip on each end and a central swelling will yield future platelets (*inset*). Scale bar 50 μm, inset scale bar 10 μm.

1. Pipet 1.5 mL of 3% BSA (w/v in PBS) solution into the bottom of a 15-mL conical tube without wetting the tube wall. Carefully tilt the tube and overlay the liquid with the same volume of a 1.5% BSA solution. The interface should be visible as a change in diffraction (see Note 4).

2. Pipet all cells from the MK cell culture dish into a 15-mL conical tube. Rinse the dish again to maximize retention of adherent cells and assess the success in an inverted light microscope. When late cultures are used, proplatelet integrity is best conserved by minimizing mechanical stress and through the use of pipets with wide openings. Centrifuge cells for 3–5 min at 200×*g*, aspirate the clear supernatant and resuspend the cell pellet in 1 mL PBS.

3. Take the tube with the one-step BSA gradient, tilt it slightly and carefully pipet the cell suspension onto the 1.5% BSA layer (typically there will be some mixture between the 1.5% layer and the PBS). Place the tube vertically in a rack. After 10–15 min, the first white cells should become detectable as they migrate into the 3% BSA layer. Depending on the experimental question to be asked, the incubation time can be extended, which will result in a higher megakaryocyte count at

the expense of purity. In general, 20–30 min yields a good ratio of mass to purity (see Note 5). Do not exceed 45 min.

4. Carefully pipet the top 2 mL of the combined PBS/1.5% BSA phase into a new tube. This fraction contains cells with a low density; they are mostly hepatocytes, hematopoietic precursors, and early megakaryocytes (see Note 6). The next ≈1 mL (the "interphase") can be removed and used separately or all remaining solution (2 mL) used as a single fraction. Although the overall cell count in the interphase is rather low, it is enriched for mature, proplatelet-forming MKs. This fraction is typically used to analyze single MKs by immunofluorescence staining. However, if a greater cell number is required at the expense of purity, the experimenter can combine the remaining 2 mL. To the interphase or 2 mL lower fraction add 10 mL of PBS and centrifuge for 5 min at $200 \times g$. The washed pellets can be used for further cultivation, e.g., to assess proplatelet or platelet formation, for ploidy analysis, for immunophenotyping, or for any other biochemical manipulation. Inspection using an inverted light microscope should reveal 80% large cells in the lower phase (see Note 7). The purity can be increased by applying the cells to a second BSA two-step gradient. Alternatively, magnetic-cell sorting (see Subheading 3.4) can be used.

3.3. Single-Cell Culture Derived from Bone Marrow

The fetal liver is the best choice if high yields of MKs are needed. However, due to the laborious and time-consuming timed-pregnancies, alternate sources can be used. The murine spleen is a hematopoietic organ that is enriched in MKs, however, the overall yield of mature MKs is limited, and cells often refrain from generating proplatelets. In contrast, bone marrow can be readily isolated from both femurs, and if required also from both tibia, and does yield reasonable numbers of mature proplatelet-generating MKs (see Note 8).

1. Mice are sacrificed according to institutional guidelines. Use sharp scissors to carefully prepare the femur by removing adherent muscles and connective tissue. First, sever the bone at the hip joint, preferably using the inner part of the blade, so that it is easy to move. Then cut the bone at the patella and remove the femur. The red marrow should be readily detectable in the center of the bone. Hold the femur with tweezers and carefully flush 0.5 mL sterile PBS through the lumen of the bone and collect the flow-through in a conical tube. We use a 25-gauge needle attached to a 1- or 3-mL syringe. Check for residual marrow and if necessary repeat the procedure (either from the same or the opposite end) for increased yield, and collect cells in the same conical tube.

2. Marrow of several bones can be collected in one tube before finally filling it up to 15 mL with sterile PBS. Centrifuge at

$200 \times g$ for 5 min. Expect 1×10^7 cells from one femur of an adult mouse and less from the tibia. Plate cells at a density of $1–2 \times 10^7$ per 6 cm cell culture dish with DMEM + 10% FBS (and antibiotics) containing rmTPO, or purify cells with magnetic beads as described in Subheading 3.4.

3.4. Magnetic-Activated Cell Sorting of Megakaryocytic Cells

Enrichment of MKs by a BSA-step density gradient from TPO-stimulated fetal liver cell cultures (Subheading 3.2) results in a population clearly positive for late megakaryocytic markers, including expression of the p45 subunit of NF-E2 transcription factor, beta1-tubulin, or Rab27b (2–4). However, as the purification step is solely based on cell density, early MK precursors are retained in the upper fraction due to low density, whereas a significant number of neutrophils are found in the high density (i.e., the lower) fraction due to their high granule content. It can therefore be desirable to use MK surface markers for enrichment instead. While flow cytometric sorting (FACS) is often considered the gold standard, in many laboratories these cytometers are not easily accessible. During the last decade, magnetic cell sorting (MACS), described below, has become a common technology exhibiting comparable purity although less megakaryocytes are recovered compared to FACS.

1. In a 50-mL conical tube place up to 2×10^8 cultivated fetal liver cells (days 2–4) and centrifuge for 5 min at $200 \times g$. Resuspend the pellet in 200 µL MACS-buffer. Add 20 µL FITC-conjugated anti mouse-CD61 antibody (see Note 9) and incubate for 20 min in the dark at 4°C.

2. Add 1 mL MACS-buffer and carefully resuspend cells, preferentially without plastic pipets to avoid cell clumping. Small clumps can be dissociated by pipetting up and down using a pipet with a wide opening. Fill the tube up to 50 mL with MACS buffer, invert 2–3 times and centrifuge for 5 min at $200 \times g$. Aspirate the supernatant.

3. Resuspend the cell pellet in 200 µL cold MACS-buffer and add 20 µL anti-FITC-microbeads, mix gently and incubate for another 20 min in the dark at 4°C.

4. Wash cells by repeating step 2 and resuspend in 1 mL MACS-buffer. In between, attach a magnetic column type LD (see Note 10) to the holder, place a tube underneath and equilibrate the column with 2 mL MACS-buffer. Apply the cells to the column and allow flow-through by gravity, which should take no longer than 5 min. Discard the liquid, place a new tube below for collecting the "negative fraction" and wash the column twice with 2 mL MACS buffer.

5. Place a new 15 mL conical tube under the column, detach it from the magnetic device and add 5 mL of MACS-buffer.

Insert the plunger into the column and carefully increase pressure so that liquid comes out dropwise. Up to 5 mL of liquid is recovered.

6. After removing an aliquot for analysis, centrifuge cells for 3 min at $200 \times g$. In the meantime, count cells in a Neubauer chamber using an inverted light microscope. Expect recovery rates of 2–3% CD61-positive cells for day 2 cultures and up to 15% when day 4 cultures are used.

7. Aspirate the supernatant and resuspend for further purposes appropriately. For further culture, plate CD61+ MKs on cell culture dishes. Overall, purifed MKs are more synchronized, and initial proplatelet formation can be increased from 30 to 35% using a BSA gradient to over 50% in MACS-sorted cells (see Note 11).

4. Notes

1. The number of fetuses per mouse varies substantially between different strains. The outbred strain CD1 typically has 12–15 fetuses, while inbred strains have lower numbers. C57/BL6 mice have 8–10 fetuses and 129/Sv often less than 6. Individual pregnancies vary, but in general, more timed-pregnant mice from inbred strains are required for the same outcome.

2. For red blood cell lysis, resuspend the cell pellet in 3 mL ACK lysis buffer (150 mM ammonium chloride, 1 mM potassium bicarbonate, 0.1 mM EDTA) and incubate for 1.5 min on ice. Fill the tube up to 15 mL with PBS, gently mix, and centrifuge for 5 min at $200 \times g$. The supernatant should be of reddish color while the cell pellet has now an overall white appearance. Remaining red blood cell progenitors do not interfere with the culture conditions. Aspirate and discard the supernatant and continue with the protocol.

3. Both human and murine recombinant thrombopoietin are effective on murine fetal liver cells for megakaryocytic differentiation. The initial concentration of thrombopoietin is effective for the incubation time of 5–6 days. Note that if the medium is changed after large MKs have already been formed, high TPO concentrations can inhibit proplatelet formation (5), but typically, we do not see reduced proplatelets when low TPO concentrations are maintained. Conditioned medium derived from TPO-producing cell lines can also be used.

4. Some bulk BSA preparations contain substantial amounts of endotoxins that might interfere with the maturation or proplatelet formation. For best results use cell culture quality, and

prewarm solutions, as the density is temperature-dependent. Some groups use three-step BSA gradients (6) to enrich MKs. However, we did not see a substantial improvement and rather suggest to use two consecutive rounds of the two-step gradient instead.

5. Sorting of MKs derived from a single fetal liver by flow or beads (see Subheading 3.4) might not be feasible due to the small amount of starting material. If enrichment of MKs is desirable, perform the BSA gradient in an Eppendorf tube using 0.5 mL of each solution, and overlay with 0.1–0.2 mL of cell suspension.

6. This cell population, albeit depleted for mature MKs, still has a considerable potential to differentiate into MKs. Cells can be resuspended in DMEM (supplemented with FBS, antibiotics, and rmTPO) and plated again for generation of "second wave" MKs. We often seed the cells back into the original dishes or flasks with double the original density for good results.

7. For unclear reasons, there can be significant variability between different cultures. We strongly recommend to control and compare the quality of each preparation. In addition to cell number and morphology, we suggest using a small aliquot for flow cytometric analysis or preparation of cytospins followed either by Wright–Giemsa staining or by immunofluorescence staining using MK-specific antibodies. In contrast to human MKs, mature murine MKs express acetylcholinesterase, and standard staining protocols are available (7).

8. If more cells are required, the humerus can be used as well, though cell yields are usually markedly lower.

9. Depending on the desired degree of maturation, any MK-specific antibody can be used (most preferably CD41, CD61, or CD42). Fluorophore-conjugated antibodies allow specific staining that can be followed using immunofluorescence on an inverted microscope or by flow cytometry. In addition, these chromophores can be removed from the cell surface (together with the beads) using specific proteases.

10. MKs are very large cells. Although special "large cell columns" are commercially available, in our hands recovery rates were extremely low. We use "LD depletion columns" for purification of MKs with best results. Increased pore size and an extended magnetic passage might contribute to that observation. However, other investigators might provide better results with other systems.

11. MKs can be plated on untreated cell culture dishes indicating that MK maturation and proplatelet formation is a cell-autonomous process (8). In addition, cells can be plated on matrix proteins including fibrinogen, fibronectin, or collagen

to study their influence on the process (9). In addition stromal feeder layers such as OP9-cells or OP9-DL1 cells have been described (10). This procedure has the potential to answer significant scientific questions when well-controlled internally. However, outcomes between different investigators differ substantially due to reagents and read-outs. We thus suggest to refrain from using these matrices unless a specific question is to be addressed.

Acknowledgments

The author would like to thank Silke Schwiebert for overall excellent technical assistance, Imke Meyer for images, and Stefan Kunert for help with the chapter on bone marrow isolation. Some sections in this chapter are based on an earlier article co-authored by Ramesh A. Shivdasani (7). This work was supported by a grant from the Deutsche Forschungsgemeinschaft (SCHU 1421/5-1).

References

1. Choi ES, Nichol JL, Hokom MM et al. (1995) Platelets generated in vitro from proplatelet-displaying human megakaryocytes are functional, Blood **85**: 402–413

2. Lecine P, Italiano JE, Jr., Kim SW et al. (2000) Hematopoietic-specific beta 1 tubulin participates in a pathway of platelet biogenesis dependent on the transcription factor NF-E2, Blood **96**: 1366–1373

3. Schwer HD, Lecine P, Tiwari S et al. (2001) A lineage-restricted and divergent beta-tubulin isoform is essential for the biogenesis, structure and function of blood platelets, Curr Biol **11**: 579–586

4. Tiwari S, Italiano JE, Jr., Barral DC et al. (2003) A role for Rab27b in NF-E2-dependent pathways of platelet formation, Blood **102**: 3970–3979

5. Choi ES, Hokom MM, Chen JL et al. (1996) The role of megakaryocyte growth and development factor in terminal stages of thrombopoiesis, Br J Haematol **95**: 227–233

6. Tajika K, Nakamura H, Nakayama K et al. (2000) Thrombopoietin can influence mature megakaryocytes to undergo further nuclear and cytoplasmic maturation, Exp Hematol **28**: 203–209

7. Shivdasani RA, Schulze H. (2005) Culture, Expansion and Differentiation of Murine Megakaryocytes., in *Current Protocols in Immunology*, 22 F.26.21-22 F.26.12, John Wiley & Sons, Inc.

8. Lecine P, Villeval JL, Vyas P et al. (1998) Mice lacking transcription factor NF-E2 provide in vivo validation of the proplatelet model of thrombocytopoiesis and show a platelet production defect that is intrinsic to megakaryocytes, Blood **92**: 1608–1616

9. Larson MK, Watson SP. (2006) A product of their environment: do megakaryocytes rely on extracellular cues for proplatelet formation?, Platelets **17**: 435–440

10. Mercher T, Cornejo MG, Sears C et al. (2008) Notch signaling specifies megakaryocyte development from hematopoietic stem cells, Cell stem cell **3**: 314–326

Chapter 15

In Vitro Generation of Megakaryocytes and Platelets from Human Embryonic Stem Cells and Induced Pluripotent Stem Cells

Naoya Takayama and Koji Eto

Abstract

Human embryonic stem cells (hESCs) represent a potential source of blood cells for transfusion therapies and a promising tool for studying the ontogeny of hematopoiesis. Moreover, human-induced pluripotent stem cells (hiPSCs), recently established by defined reprogramming factors expressed in somatic cells, represent a further source for the generation of hematopoietic cells. When undifferentiated hESCs or hiPSCs are cultured on either mesenchymal C3H10T1/2 cells or OP-9 stromal cells, they can be differentiated into a hematopoietic niche that concentrates hematopoietic progenitors, which we named "embryonic stem cell-derived sacs" (ES-sacs). We have optimized the in vitro culture condition for obtaining mature megakaryocytes derived from the hematopoietic progenitors within ES-sacs, which are then able to release platelets. These in vitro-generated platelets display integrin activation capability, indicating normal hemostatic function. This novel protocol thus provides a means of generating platelets from hESCs as well as hiPSCs, for the study of normal human thrombopoiesis and also thrombopoiesis in disease conditions using patient-specific hiPSCs.

Key words: Human ESCs, Human iPSCs, ES-sac, Hematopoietic progenitor, Megakaryocyte, Platelet, Transfusion

1. Introduction

Platelets, whose adhesive and signaling functions are essential for normal hemostasis, are generated from megakaryocytes within the bone marrow (BM). A number of different methods for obtaining megakaryocytes from various sources have been described; for example, hematopoietic stem cells of BM (1), cord blood (2, 3), or adipose tissue (4). However, it has proven difficult to generate megakaryocytes and platelets in vitro on a large scale, probably due

Jonathan M. Gibbins and Martyn P. Mahaut-Smith (eds.), *Platelets and Megakaryocytes: Volume 3, Additional Protocols and Perspectives*, Methods in Molecular Biology, vol. 788, DOI 10.1007/978-1-61779-307-3_15, © Springer Science+Business Media, LLC 2012

to the limited ex vivo expansion capability of BM cells and cord blood cells. Human embryonic stem cells (hESCs) are pluripotent cells that can proliferate almost infinitely in vitro (5), and thus represent a potential source of platelets for transfusion and also for studying the ontogeny of hematopoiesis (6–10). It is well known, however, that repeated transfusion induces the production of antibodies against allogenic human leukocyte antigen (HLA) on the transfused platelets (11). To establish a supply of identical platelet concentrates without loss of responsiveness due to immunorejection, in particular for patients with a rare HLA, we focused on platelet generation from human-induced pluripotent stem cells (hiPSCs) as well as from hESCs. In this chapter, we describe methods for the in vitro culture of hESCs (or hiPSCs) to yield megakaryocytes that extend proplatelets and release platelets.

2. Materials

2.1. Culturing of C3H10T1/2 Feeder Cells

1. The mouse C3H10T1/2 cell line (RIKEN Bio-Resource Center, Tsukuba, Ibaraki, Japan).

2. Eagle's basal medium (BME) (Invitrogen).

3. 0.05% Trypsin–EDTA (Sigma).

4. Fetal bovine serum (FBS) for mouse feeder cells (Biological Industries).

5. C3H10T1/2 cell medium: BME supplemented with 10% FBS for mouse feeder cells and 2 mM L-glutamine.

6. 15 ml conical tubes and 10 cm culture dishes (or 6 cm plates) (TPP®, Techno Plastics Products AG, Switzerland).

7. Phosphate buffer saline (PBS): PBS used in the protocols within this chapter is Ca^{2+}- and Mg^{2+}-free.

8. Gelatin from porcine skin, Type A (Sigma): 0.1% gelatin PBS is made by dissolving 0.1 g gelatin in 100 ml of PBS, which is then autoclaved.

9. 0.1% Trypan blue: dissolve the powder in PBS.

2.2. Differentiation of Multipotent Hematopoietic Progenitors from hESCs via "ES-Sacs"

1. Human ES cell line (KhES-3) (8, 12) can be obtained from Kyoto University (Kyoto, Japan). Human iPSCs are available from Tokyo University (Tokyo, Japan) (see Note 1). These are maintained in culture as described previously (12). Briefly, they are cultured on irradiated mouse embryonic fibroblasts in a 1:1 mixture of Dulbecco's modified Eagle's medium and Ham's F-12 medium (Sigma) supplemented with 0.1 mM nonessential amino acids (Invitrogen), 2 mM L-glutamine (Invitrogen), 20% knockout serum replacement (KSR) (Invitrogen), 0.1 mM

2-mercaptoethanol, and 5 ng/ml basic fibroblast growth factor (bFGF, Upstate).

2. Iscove's modified Dulbecco's medium (IMDM) (Sigma).

3. FBS (JRH Biosciences) (see Note 2).

4. Insulin/transferrin/selenite (ITS) 100× stock solution (purchased as a premixed solution): a cocktail of 10 μg/ml human insulin, 5.5 μg/ml human transferrin, 5 ng/ml sodium selenite (Sigma).

5. 200 mM L-glutamine (Invitrogen).

6. Ascorbic acid (Sigma): to prepare the stock solution (50 mg/ml), dissolve 0.1 g ascorbic acid in 2 ml water, pass through a 0.22-μm filter, and store at 4°C for no longer than 1 month.

7. α-Monothioglycerol (MTG) (Sigma): to prepare the stock solution (450 mM), dissolve 780 μl MTG in 20 ml PBS, pass through a 0.22-μm filter, aliquot at 500 μl/tube, and store at −20°C. Avoid repeated freeze–thaw cycles.

8. PBS: PBS used in the protocols within this chapter is Ca^{2+}- and Mg^{2+}-free.

9. Human recombinant vascular endothelial growth factor (VEGF) (1 mg) (R&D systems): to prepare the stock solution (100 ng/μl), add PBS containing 1% bovine serum albumin to the bottle, pipet to mix, aliquot at 50 μl/tube, and store at −20°C for up to 12 months. Working solutions (20 ng/μl) are prepared by the addition of 200 μl of IMDM to the tube and kept at 4°C for no longer than 1 month. Avoid repeated freeze–thaw cycles.

10. 2.5% Trypsin (Invitrogen).

11. 0.25% Trypsin–EDTA (Invitrogen).

12. KSR (Invitrogen).

13. 100 mM $CaCl_2$: $CaCl_2$ is dissolved in water and stored at 4°C for up to 12 months.

14. 6- and 10-cm culture dishes and 50 ml tubes (TPP®, Techno Plastics Products AG, Switzerland).

15. 40-μm nylon cell strainer (BD Bioscience).

16. Human ESC dissociation buffer: Ca^{2+}- and Mg^{2+}-free PBS supplemented with 0.25% trypsin, 1 mM $CaCl_2$, and 20% KSR. Usually we add 20 ml of 2.5% trypsin (Subheading 2.2, item 10), 40 ml of KSR (Subheading 2.2, item 12), and 2 ml of 100 mM $CaCl_2$ (Subheading 2.2, item 13) in 138 ml of PBS. Aliquot at 10 ml/15 ml tube, and store at −20°C for up to 12 months.

17. Hematopoietic differentiation medium (for hematopoietic progenitors from hESCs): IMDM supplemented with 10 μl/ml

ITS 100× stock solution, 2 mM L-glutamine, 0.45 µM MTG, 50 µg/ml ascorbic acid, and 15% FBS (Subheading 2.2, item 3).

18. 0.1% Trypan blue: dissolve the powder in PBS.

2.3. Differentiation of Megakaryocytes and Platelets from Hematopoietic Progenitors Within "ES-Sacs"

1. Human recombinant thrombopoietin (TPO) (1 mg) (R&D systems): to prepare a stock solution (100 ng/µl), add PBS containing 1% bovine serum albumin to the bottle, pipet to mix, aliquot 100 µl/tube, and store at –20°C for up to 12 months. Working solutions (10 ng/µl) are prepared by the addition of 900 µl of IMDM to the tube and kept at 4°C for no longer than 1 month. Avoid repeated freeze–thaw cycles.

2. Human recombinant stem cell factor (SCF) (1 mg) (R&D systems): to prepare a stock solution (500 ng/µl), add PBS containing 1% bovine serum albumin to the bottle, pipet to mix, aliquot at 100 µl/tube, and store at –20°C up to 12 months. Working solutions (50 ng/µl) are prepared by the addition of 900 µl of IMDM to the tube and kept at 4°C for no longer than 1 month. Avoid repeated freeze–thaw cycles.

3. Heparin sodium (AJINOMOTO, Japan).

4. Six-well plates, 50 ml tubes (TPP®).

5. GM6001 (EMD Bioscience); to prepare a stock solution (100 mM), dissolve 5 mg GM6001 in 128 µl DMSO, aliquot at 10 µl/tube, and store at –20°C. Avoid repeated freeze–thaw cycles.

2.4. Characterization of Hematopoietic Progenitors, Megakaryocytes, and Platelets Generated from hESCs In Vitro

1. Staining medium for flow cytometer (SM): PBS supplemented with 3% FBS.

2. Allophycocyanin (APC)-conjugated antihuman CD41a (integrin αIIb) monoclonal antibody (MAb) (BD Bioscience).

3. PE-conjugated antihuman CD42b (GPIbα) MAb (BD Bioscience).

4. PE-conjugated antihuman CD9 MAb (BD Bioscience).

5. Fluorescein isothiocyanate (FITC)-conjugated antihuman CD42a (GPIX) MAb (BD Bioscience).

6. Alexa 405-conjugated antihuman CD45 MAb (CALTAG LABORATORIES).

7. PE-conjugated antihuman CD31 (PECAM1) MAb (Bio Legend).

8. APC-conjugated antihuman VEGF-R (KDR/Flk-1) MAb (BD Bioscience).

9. FITC-conjugated PAC1 antibody; antihuman-activated integrin $\alpha_{IIb}\beta_3$ MAb (BD Bioscience) (13).

10. True count tube (BD Bioscience).

11. Hemacolor (staining set for microscopy) (MERCK).

12. Propidium iodide: dissolve 1 mg Propidium iodide with 1 ml of PBS and store at 4°C in dark (caution: this is extremely toxic and the user should follow local guidelines for safe use and disposal).

13. Shandon Cytospin 4 Cytocentrifuge and accessories for cytospin: slide, a filter card, and a sample chamber (Thermo Fisher Scientific, USA). Please follow the manufacturer's instructions in detail.

14. 100% methanol.

15. Acid citrate dextrose solution: 85 mM sodium citrate, 104 mM glucose, and 65 mM citric acid.

16. Modified Tyrode's–HEPES buffer: 10 mM HEPES, 12 mM NaHCO$_3$, 138 mM NaCl, 5.5 mM glucose, 2.9 mM KCl, 1 mM MgCl$_2$, pH 7.4. The buffer is titrated with NaOH.

17. Tirofiban, a specific antagonist to human integrin $\alpha_{IIb}\beta_3$ (Merck, Whitehouse Station, NJ): 10 mM stock solution is made with water. To prepare a stock solution (10 mM), dilute with water and store at −20°C.

3. Methods

3.1. Culturing of C3H10T1/2 Stromal Cells and Their Preparation for Hematopoietic Differentiation

The C3H10T1/2 cell line was established from mesenchymal cells derived from C3H mouse embryo (8, 14). In the presence of an adequate combination of cytokines this cell line can support the expansion of multipotent hematopoietic progenitors and the development of mature hematopoietic cells (including megakaryocytes and platelets) from monkey ESCs (14) and human ESCs (8). The OP9 mouse stromal cell line may also be used to provide feeder cells for hematopoietic differentiation (see Note 3). C3H10T1/2 cells should be passaged at the point when the cells are 80% confluent (see Note 4).

3.1.1. Thawing of C3H10T1/2 Cells

1. Thaw a vial of frozen cells in a 37°C water bath with gentle agitation.

2. Transfer the cells into a 15-ml conical tube, add prewarmed C3H10T1/2 cell medium gradually to avoid cell damage by osmotic changes, and centrifuge at 400 × g for 5 min at room temperature. Aspirate the medium.

3. Resuspend the cells with 10 ml of C3H10T1/2 cell medium.

4. Transfer the cells into a 10-cm dish.

3.1.2. Maintenance of C3H10T1/2 Stromal Cells

1. Aspirate the medium and wash twice with 5 ml of PBS.

2. Add 1 ml of 0.05% trypsin–EDTA to the 10-cm dish and incubate at 37°C for 5 min.

3. Add 5 ml of C3H10T1/2 cell medium and pipet well to disperse into single cells.

4. Split the cells at 1:8 in a 10-cm dish with 10 ml of C3H10T1/2 cell medium.

5. Maintain the cells at 37°C in a 5% CO_2 incubator. Change the medium every 2 days.

3.1.3. Preparation of Feeder Cells for Hematopoietic Differentiation

1. Irradiate the C3H10T1/2 cells at 50 Gy (total) to stop the cell growth.

2. Aspirate the medium and wash twice with 5 ml of PBS.

3. Add 1 ml of 0.05% trypsin–EDTA to the 10-cm dish and incubate at 37°C for 5 min.

4. Add 5 ml of C3H10T1/2 cell medium and pipet well to make them into single cells.

5. Measure the density of viable cells with 0.1% trypan blue.

6. Resuspend single cells in the medium ($7–8 \times 10^4$ cells/ml).

7. Add 0.1% gelatin–PBS to a new 10-cm dish or six-well plate and leave at 37°C for 30 min to allow the surface to become coated with gelatin. Aspirate the remaining gelatin–PBS solution before use.

8. Transfer the irradiated feeder cells to the gelatin-coated dish (10 ml/10-cm dish or 12 ml/six-well plate).

9. Confirm that the cells are homogeneously spread (Fig. 1).

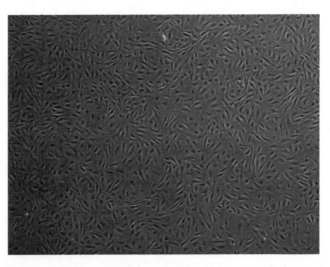

Fig. 1. Representative image of C3H10T1/2 cells before coculture with hESCs. C3H10T1/2 cells irradiated at 50 Gy are spread on a gelatin-coated dish and allowed to form a monolayer.

3.2. Differentiation of Hematopoietic Progenitors via an ES-Sac

It is important to culture hESCs at the correct density for differentiation. In addition, it should be noted that hESCs display very limited growth when they are dispersed into single cells in culture. In order to determine the number of hESCs available for differentiation, we usually prepare at least two 6-cm dishes containing the same number of hESCs. We use one hESC dish to measure the density of viable cells and the other hESC dish is used for hematopoietic differentiation. The peak of megakaryocyte and platelet generation is around days 22–26 (8). An identical protocol is followed for generation of MKs and platelets from hiPSCs.

3.2.1. Determination of hESC Density Prior to Differentiation Experiments

1. Split hESCs from a 6-cm dish into two or three 6-cm dishes (i.e., 1:2–3) 5–6 days before differentiation experiments. Change the medium every day.

2. To calculate the number of hESCs, aspirate the medium from one dish and add 1 ml of 0.25% trypsin–EDTA.

3. Incubate at 37°C for 5 min.

4. Pipet well with a P-1000 pipette to disperse into a single cell suspension.

5. Accurately determine the number of viable cells after adding 0.1% trypan blue to stain dead cells.

3.2.2. Preparation of hESCs for Differentiation

The protocol to prepare hESCs for differentiation is shown schematically in Fig. 2.

1. Aspirate the medium and wash once with 5 ml of PBS.

2. Add 1 ml of hESC dissociation buffer (Subheading 2.2, item 16) and incubate at 37°C for 5–6 min.

3. When the edges of the colonies are rolled up, aspirate the dissociation buffer.

4. Add 2 ml of hematopoietic differentiation medium (Subheading 2.2, item 17).

5. Scrape off the hES colonies with a P-1000 pipet. Take care not to disperse into extremely small-sized colonies (see Note 5).

6. Resuspend the small clumps of hESC colonies in the hematopoietic differentiation medium ($5–10 \times 10^3$ cells/ml).

3.2.3. Differentiation of Human ESCs into ES-Sacs Containing Hematopoietic Progenitors

A schematic of the protocol is shown in Fig. 3.

1. Transfer small clumps of hESCs onto irradiated C3H10T1/2 cells and culture with hematopoietic cell differentiation medium containing 20 ng/ml recombinant human VEGF.

2. Change the medium every 2–3 days. The culture medium should contain 20 ng/ml recombinant human VEGF up to 14–15 days of culture.

3. ES-sacs emerge on days 14–15 of culture.

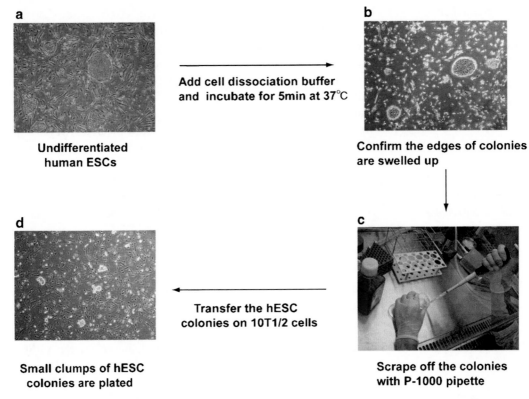

Fig. 2. Preparation of human ESCs for differentiation into megakaryocytes.

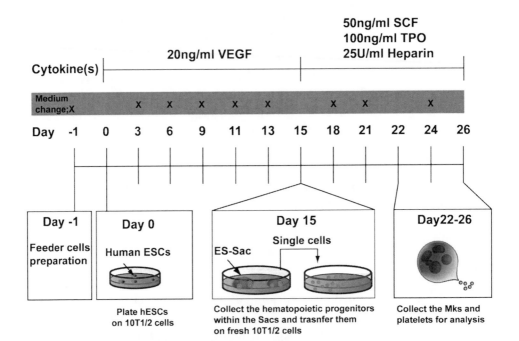

Fig. 3. Human ESC-derived megakaryocytes yielding platelets are generated via "sac"-like structures (ES-sacs). Schematic diagram of the in vitro differentiation protocol for hESC-derived megakaryocytes and platelets. Mature megakaryocytes are generated from cells within the ES-Sacs on day 14–15. *VEGF* vascular endothelial growth factor; *TPO* thrombopoietin; *SCF* stem cell factor.

1. Collect ES-sacs using a cell scraper and transfer into a 50-ml tube.

2. Centrifuge the cells at $400 \times g$ for 10 min and aspirate the medium to leave the cells at the bottom of the tube with approximately 1 ml of media remaining.

3. Gently break up the ES-sacs with P-1000 pipette and transfer the cells into a new 50 ml tube through a 40-μm cell strainer to obtain hematopoietic progenitors.

4. Centrifuge the cells at $400 \times g$ for 10 min.

5. Aspirate the medium and add 1 ml of fresh medium.

6. Count the hematopoietic progenitors in a sample stained with 0.1% trypan blue.

***3.3. Differentiation
of Megakaryocytes
and Platelets from
Hematopoietic
Progenitors***

To differentiate the hematopoietic progenitors harvested from Subheading 3.2.4 into megakaryocytes and platelets, the cells are grown in the presence of TPO, SCF, and heparin (see Fig. 3). We have recently demonstrated that GPIbα and GPV are shed from mouse ESC-derived platelets by a disintegrin and metalloproteinase (ADAM) family protein, ADAM17, under culture conditions at 37°C (15). Loss of GPIbα extracellular domain attenuates the ability of platelets to adhere to the extracellular matrix and to form thrombi. Moreover, in vivo, platelets lacking the GPIbα extracellular domain are quickly cleared from the circulation (16), leading to insufficient levels of circulating platelets following transfusion (15). To avoid this, without affecting platelet yield, GM6001, a broad inhibitor of metalloproteases in applied 2 days before analysis of platelets.

1. Resuspend the hematopoietic progenitors from Subheading 3.2.4 in hematopoietic differentiation medium (Subheading 2.2, item 17) supplemented with 100 ng/ml human TPO, 50 ng/ml human SCF, and 25 U/ml heparin sodium at 1×10^4 cells/ml.

2. Transfer the cells onto fresh and irradiated feeder cells in a six-well plate with differentiation medium (4×10^4 cells/4 ml/well).

3. Half medium change is performed every 2–3 days (Fig. 3). 2 ml of medium are collected in a 15-ml tube and centrifuged at $400 \times g$ for 5 min. Aspirate the medium and resuspend with 2 ml of fresh medium containing 200 ng/ml human TPO, 100 ng/ml human SCF, and 50 U/ml heparin sodium (final concentration of the growth factors should be 100 ng/ml human TPO; 50 ng/ml human SCF, and 25 U/ml heparin sodium).

4. If the purpose of the experiment is to analyze the function of released platelets, 50 μM GM6001, a broad metalloprotease

inhibitor that prevents shedding of GPIbα and GPV by ADAM 17, is administered to the culture medium 2 days before analysis (see Note 6).

5. Harvest the nonadherent cells and analyze on days 22–26 of culture.

3.4. Characterization of hESC-Derived Hematopoietic Progenitors, Megakaryocytes and Platelets

3.4.1. Flow Cytometric Analysis of Surface Molecules of Hematopoietic Progenitors and Megakaryocytes

1. Resuspend the hematopoietic cells from ES-sacs (day 15 of culture), or from floating cells (days 22–26 of culture), in 100 μl of SM (Subheading 2.4, item 1) containing 1×10^5 cells.

2. Add 3 μl of MAbs conjugated with APC, FITC, or PE fluorochromes to 100 μl of cell suspension and mix briefly by pipetting.

3. Incubate for 30 min at 4°C in the dark.

4. Wash the cells once with 5 ml of SM ($400 \times g$, 5 min).

5. Resuspend the cells in 300 μl of SM containing 1 μg/ml propidium iodide as a marker of dead cells.

6. Analyze cells by flow cytometry. The percentages of positive cells are determined as compared to isotype controls that establish the background levels of nonspecific staining.

3.4.2. Morphological Characterization of hESC-Derived Megakaryocytes

1. Re-suspend the floating cells in 100 μl PBS at $1–2 \times 10^4/100$ μl.

2. Position a cytospin slide, filter card, and sample chamber in the slide holder.

3. Load into a cytocentrifuge rotor.

4. Add 100 μl of cell suspension to the bottom of the chamber.

5. Centrifuge at $400 \times g$ for 2 min.

6. Gently separate the slide with the filter card and the chamber. Leave slide at room temperature for 10 min.

7. Fix the slide with 100% methanol for 5 s.

8. Stain with red Hemacolor solution for 5 s.

9. Stain with blue Hemacolor solution for 5 s.

10. Rinse once with water.

11. Dry in air.

3.4.3. Flow Cytometric Analysis of Surface Molecules and Numbers of the Platelets Generated

1. Collect 200 μl of the culture supernatant containing hESC-derived platelets.

2. Add 2 μl of APC-conjugated anti-CD41a, FITC-conjugated anti-CD42a, or PE-conjugated CD42b MAbs to 200 μl of culture supernatant and mix briefly by pipetting.

3. Incubate for 30 min at room temperature in the dark.

4. To measure the number of platelets in the culture supernatant, add the sample into a true count tube that contains a precise number of microbeads (Subheading 2.4, item 10).

5. Analyze cells by flow cytometry. The percentages of positive cells are determined by comparison with isotype controls that establish the background levels due to nonspecific staining.

3.4.4. Functional Analysis of hESC-Derived Platelets by Studies of Agonist-Mediated Integrin $\alpha_{IIb}\beta_3$ Activation

Stimulating platelets with an agonist changes the conformation of the integrin $\alpha_{IIb}\beta_3$ and promotes its clustering. This inside-out activation of $\alpha_{IIb}\beta_3$ promotes the binding of its ligands (17), principally fibrinogen, but also von Willebrand factor. These ligands in turn promote outside-in signaling to induce the cytoskeletal changes and spreading required for production of stable platelet thrombi in vivo (17).

1. Gently collect platelets from the culture medium and add a one-ninth volume of acid citrate dextrose solution (Subheading 2.4, item 12).

2. Centrifuge at $150 \times g$ for 10 min to precipitate any large cells.

3. Transfer the supernatant containing platelets to a new tube and centrifuge at $400 \times g$ for 10 min to precipitate a platelet pellet.

4. Wash the pellet once with 5 ml modified Tyrode's–HEPES buffer at pH 7.4 (Subheading 2.4, item 13).

5. Resuspend the pellet in an appropriate volume of modified Tyrode's–HEPES buffer at pH 7.4 and add 1 mM $CaCl_2$.

6. Incubate some samples of hESC-derived platelets with 10 µM tirofiban, a specific antagonist to human integrin $\alpha_{IIb}\beta_3$ (18) for 10 min, to act as a control (see Note 7).

7. Incubate 50 µl aliquots of platelets in the absence or presence of 1 U/ml of human thrombin or 50 µg/ml of ADP for 10 min.

8. Incubate platelets with 2 µl of PE-conjugated anti-CD42b and 5 µl of FITC-conjugated PAC-1 at room temperature for 10 min in the dark.

9. Quantify the binding of PAC-1 to platelets using flow cytometry. Nonspecific binding is determined in the presence of 10 µM tirofiban.

3.5. Perspective

Megakaryocytes are a rare population of cells that develop in the bone marrow (BM), which limits use of BM specimens for studies of megakaryopoiesis and platelet generation. Thus, we have attempted to establish a protocol to obtain megakaryocytes capable of generating platelets from ESCs of mouse (15, 19) or human (8) in vitro. This method has allowed us to relatively large numbers of megakaryocytes and platelets from hESCs, and also hiPSCs (20). Megakaryocytes derived from patient-specific hiPSCs represent a powerful tool to understand the unresolved aspects of the mechanisms underlying thrombocytopenia and also to screen novel therapeutic agents to treat diseases of platelet production and function.

4. Notes

1. Human iPSCs are generated from dermal fibroblasts transduced with four factors (*OCT3/4, SOX2, KLF4,* and *c-MYC*) or three factors (without *c-MYC*). The differentiation capacity differs between the different hESC and hiPSC clones. We recommend that several human ESC clones or human iPSC clones should be examined for megakaryocyte differentiation.

2. The differentiation efficiency can vary with the batch of FBS and therefore different lots of FBS should be assessed in the experiments.

3. We previously compared the ability of C3H10T1/2 cells and OP9 cells to support megakaryocyte differentiation from hESCs, and confirmed no significant difference between these two types of feeder cells.

4. Avoid allowing the cells to become over-confluent as they may lose the ability to support hematopoietic differentiation from hESCs.

5. In our experience, small size hESC colonies show poor differentiation into hematopoietic cells. Also restrict the period of trypsinization of hESCs to a minimum.

6. As GM6001 is cytotoxic for megakaryopoiesis (15), we usually administer GM6001 2 days before analysis, which does not affect the platelet yield.

7. Tirofiban should be incubated with the platelet suspension for at least 10 min to block the activation of integrin $\alpha_{IIb}\beta_3$.

References

1. Debili, N. et al. The Mpl-ligand or thrombopoietin or megakaryocyte growth and differentiative factor has both direct proliferative and differentiative activities on human megakaryocyte progenitors. *Blood* **86**, 2516–25 (1995).

2. Schipper, L.F. et al. Differential maturation of megakaryocyte progenitor cells from cord blood and mobilized peripheral blood. *Exp Hematol* **31**, 324–30 (2003).

3. Matsunaga, T. et al. Ex vivo large-scale generation of human platelets from cord blood CD34+ cells. *Stem Cells* **24**, 2877–87 (2006).

4. Matsubara, Y. et al. Generation of megakaryocytes and platelets from human subcutaneous adipose tissues. *Biochem Biophys Res Commun* **378**, 716–20 (2009).

5. Thomson, J.A. et al. Embryonic stem cell lines derived from human blastocysts. *Science* **282**, 1145–7 (1998).

6. Vodyanik, M.A., Bork, J.A., Thomson, J.A. & Slukvin, II. Human embryonic stem cell-derived CD34+ cells: efficient production in the coculture with OP9 stromal cells and analysis of lymphohematopoietic potential. *Blood* **105**, 617–26 (2005).

7. Wang, L. et al. Endothelial and hematopoietic cell fate of human embryonic stem cells originates from primitive endothelium with hemangioblastic properties. *Immunity* **21**, 31–41 (2004).

8. Takayama, N. et al. Generation of functional platelets from human embryonic stem cells in vitro via ES-sacs, VEGF-promoted structures that concentrate hematopoietic progenitors. *Blood* **111**, 5298–306 (2008).

9. Ma, F. et al. Generation of functional erythrocytes from human embryonic stem cell-derived definitive hematopoiesis. *Proc Natl Acad Sci USA* **105**, 13087–92 (2008).

10. Yokoyama, Y. et al. Derivation of functional mature neutrophils from human embryonic stem cells. *Blood* **113**, 6584–92 (2009).

11. Schiffer, C.A. Diagnosis and management of refractoriness to platelet transfusion. *Blood Rev* **15**, 175–80 (2001).

12. Suemori, H. et al. Efficient establishment of human embryonic stem cell lines and long-term maintenance with stable karyotype by enzymatic bulk passage. *Biochem Biophys Res Commun* **345**, 926–32 (2006).

13. Shattil, S.J., Hoxie, J.A., Cunningham, M. & Brass, L.F. Changes in the platelet membrane glycoprotein IIb.IIIa complex during platelet activation. *J Biol Chem* **260**, 11107–14 (1985).

14. Hiroyama, T. et al. Long-lasting in vitro hematopoiesis derived from primate embryonic stem cells. *Exp Hematol* **34**, 760–9 (2006).

15. Nishikii, H. et al. Metalloproteinase regulation improves in vitro generation of efficacious platelets from mouse embryonic stem cells. *J Exp Med* **205**, 1917–27 (2008).

16. Bergmeier, W. et al. Metalloproteinase inhibitors improve the recovery and hemostatic function of in vitro-aged or -injured mouse platelets. *Blood* **102**, 4229–35 (2003).

17. Shattil, S.J. & Newman, P.J. Integrins: dynamic scaffolds for adhesion and signaling in platelets. *Blood* **104**, 1606–15 (2004).

18. Peerlinck, K. et al. MK-383 (L-700,462), a selective nonpeptide platelet glycoprotein IIb/IIIa antagonist, is active in man. *Circulation* **88**, 1512–7 (1993).

19. Eto, K. et al. Megakaryocytes derived from embryonic stem cells implicate CalDAG-GEFI in integrin signaling. *Proc Natl Acad Sci USA* **99**, 12819–24 (2002).

20. Takayama, N. et al. Transient activation of c-MYC expression is critical for efficient platelet generation from human induced pluripotent stem cells. *J Exp Med* **207**, 2817–30 (2010).

Chapter 16

Megakaryocyte and Platelet Production from Human Cord Blood Stem Cells

Amélie Robert, Valérie Cortin, Alain Garnier, and Nicolas Pineault

Abstract

The cloning of thrombopoietin together with advances in the culture of hematopoietic stem cells have paved the way for the study of megakaryopoiesis, ongoing clinical trials and, in the future, for the potential therapeutic use of ex vivo produced blood substitutes, such as platelets. This chapter describes a 14-day culture protocol for the production of human megakaryocytes (MKs) and platelets, and assays that can be used to characterize the functional properties of the platelets produced ex vivo. CD34⁺ cells isolated from cord blood cells are grown in a serum-free medium supplemented with newly developed cytokine cocktails optimized for MK differentiation, expansion, and maturation. Detailed methodologies for flow cytometry analysis of MKs and platelets, for the purification of platelets and functional assays, are presented together with supporting figures. The chapter also provides a brief review on megakaryocytic differentiation and ex vivo MK cultures.

Key words: Megakaryocytes, Platelets, Hematopoietic stem cells, Cord blood, Flow cytometry, Ex vivo cell culture, Culture medium optimization, Stem cells expansion, Cytokines

1. Introduction

Human hematopoietic stem cells (HSCs) and megakaryocytes (MKs) are both extremely rare cell types, representing approximately 0.05 and 0.4% of whole marrow, respectively. As a consequence of their low frequency, these cells are often studied after enrichment using various cell surface antigens. For instance, MKs are commonly enriched by selection of CD41a (GPIIb) or CD61 (GPIIIa) positive cells. Conversely, MKs at varying levels of purity can be generated in larger numbers using simple culture protocols such as the one described herein. Culture systems are commonly used to study the development of human MKs, their maturation, and platelet biogenesis. Most key events occurring during MK

Jonathan M. Gibbins and Martyn P. Mahaut-Smith (eds.), *Platelets and Megakaryocytes:*
Volume 3, Additional Protocols and Perspectives, Methods in Molecular Biology, vol. 788,
DOI 10.1007/978-1-61779-307-3_16, © Springer Science+Business Media, LLC 2012

maturation appear to be well-replicated ex vivo, except for the regulatory activities normally provided by the microosteoblastic and microvascular niches on immature and mature MKs, respectively (1–3).

MKs and platelets can now be produced ex vivo (recently reviewed in ref. 4) from various sources, such as cell lines (5), embryonic stem cells (6–8), and HSCs (9–13) maintained in suspension cultures (9, 10, 12, 14) or with feeder layers (6, 7, 11) or in bioreactors (13). Hematopoietic progenitor cells, which are enriched in the CD34+ cell fraction, can be isolated from bone marrow, mobilized peripheral blood, or umbilical cord blood (CB). CB cells present logistical and safety advantages compared to other CD34+ sources. However, it is important to point out that the maturation of CB-derived MK ex vivo differs from that reported for bone marrow-derived MKs. The major differences include a greater proliferative potential ex vivo, reduced terminal cell size, and reduced polyploidization potential for CB-derived MKs (15, 16).

1.1. Overview of Megakaryocyte Differentiation and Maturation

MKs are derived from HSCs following a number of differentiation events (Fig. 1a, reviewed in ref. 17) that lead to the emergence of a bipotent progenitor with MK and erythroid potential (18). Differentiation of this progenitor along the MK lineage is first apparent by the expression of CD41a (GPIIb) and CD61 (GPIIIa/β3), the earliest markers of MK differentiation. CD41a and CD61 are the most abundant glycoproteins on the surface of platelets and form a functional dimeric complex that recognizes and binds to fibrinogen and von Willebrand factor (vWF). During the course of MK maturation, the MK expresses various cell surface molecules also present on platelets. These include a second vWF receptor composed of three chains: GPIbα, GPV, and GPIX. GP1bα, also referred to as CD42b, is commonly used to discriminate between immature and mature MKs. The expression of these markers during the course of megakaryopoiesis is shown schematically in Fig. 1a. As MK maturation progresses, the expression of MK receptors is enhanced along with an increase in cell size and DNA content. Flow cytometry is perfectly suited to monitor the kinetics of MK differentiation and maturation, as shown in Fig. 2. The last stage of MK maturation is the formation of proplatelet filaments that occurs asynchronously in MK cultures (Fig. 1c). The release of platelets from proplatelet-bearing MKs is normally accompanied by a complex process of compartmentalized apoptosis (19, 20). Thus, the viability of MK cell culture is necessarily low once the MKs are fully matured and platelets are being released.

The maturation of MKs is a complex process essentially characterized by two important events: polyploidization and cytoplasmic maturation (21). Polyploidization occurs through successive cycles of a specialized process called endomitosis. Endomitotic MKs undergo normal cell cycle progression up to the telophase, but fail

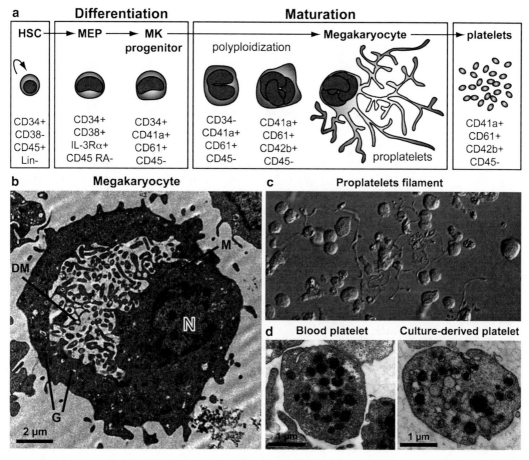

Fig. 1. An overview of megakaryocyte development and platelet production. (**a**) The changes in expression of the major cell surface antigens used to characterize and isolate cells at various stages of MK differentiation are shown. *HSC* hematopoietic stem cell, *MEP* common megakaryocyte and erythroid progenitor, *MK* megakaryocyte. (**b**) Mature culture-derived MK viewed by transmission electron microscopy (TEM). *N* nucleus, *G* granule, *DM* demarcation membrane, *M* mitochondria (5,000×). (**c**) Proplatelet extensions in a representative MK culture at day 11 (phase-contrast picture). (**d**) TEM of blood and culture-derived platelets (30,000×).

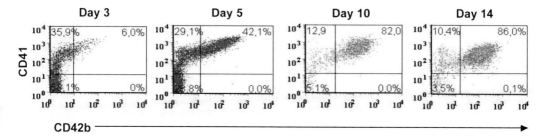

Fig. 2. Flow cytometry analysis of CB CD34+ cell culture undergoing MK differentiation. The expression of CD41a (*y*-axis) and CD42b (*x*-axis) is presented in dot-plot analysis as a function of time. Note that the majority of MKs at day 3 are immature since they do not express CD42b. As megakaryocytic maturation progresses, the majority of the MKs gain expression of CD42b and the intensity of both markers increases. These cultures were maintained in the BS1-supplemented medium as described in the methods. Reproduced from Cortin et al. (42) with permission from Humana Press.

to complete cell division due to a late interruption of cytokinesis at the cleavage furrow ingression stage (22). This results in the formation of a polyploidic cell with DNA content greater than 2n (4–128n) contained in a single polylobulated nucleus or in separated nuclei (22). Cytoplasmic maturation is characterized by an increase in size of the cytoplasm, increased organelle synthesis, and invagination of the plasma membrane leading to the formation of the demarcation membranes (DMs) (Fig. 1b). The DM represents a reservoir of cell membrane required for the formation and extension of proplatelets (23). The latter occurs following a profound reorganization of the cytoskeleton with microtubule involvement that leads to the elongation of organelle-containing pseudopods (23, 24). Proplatelets appear as long, thin filament extensions (2–4 μM in diameter) composed of uniform beads linked by thin cytoplasmic bridges that extend in total length up to 250–500 μM (19, 25). Organelles are actively loaded into the proplatelet ends by an intricate mechanism of microtubule-based bidirectional trafficking (24). This process is terminated by the release of newly assembled platelets from the protruding proplatelet filaments in the marrow-intravascular sinusoidal space. Multiphoton intravital microscopy has recently shown that proplatelets extend from mature MKs through endothelial cells into the blood vessel, and such studies suggest that platelet release and maturation are completed in part in the blood flow under conditions of shear stress (2), which may significantly improve the yield of platelet production (3).

1.2. Induction of Megakaryocyte Differentiation in Ex Vivo Cultures

As a consequence of their low frequency and the need for large cell numbers for functional and biochemical studies, numerous articles have been published on the expansion of MKs ex vivo (10, 26–33). Many conditions described in the literature are based on serum-free medium supplemented with various cytokine cocktails, all of which include thrombopoietin (TPO). The use of pleiotropic cytokines, such as IL-3 and IL-6, and early acting cytokines, such as stem cell factor (SCF) and Flt-3 ligand (FL), usually increase MK expansion but at the expense of purity (12, 26, 33). In contrast, high MK purity (>90% CD41$^+$) can be obtained simply by culturing CD34$^+$ cells with TPO alone (10–100 ng/mL) in serum-free media, though MK yields are then substantially reduced. Recent studies have also shown that adhesion molecules, allowing MK interactions with bone marrow components, may also be involved in MK expansion (34) and maturation (1).

Due to the potential therapeutic value of MK progenitors and MKs generated in culture, for example to reduce the period of thrombocytopenia in patients undergoing CB transplantation, much effort has been directed toward the development of cytokine cocktails optimized for the expansion of these cells ex vivo (14, 27, 29, 31, 35). Cytokines, such as TPO, SCF, FL, IL-3, IL-6, GM-CSF, and IL-11, have all been shown to support the expansion of MK progenitors to different extents.

1.3. Overview of the Chapter

A simple and efficient two-step culture protocol developed in our laboratory is described for the differentiation and maturation of CD34+ cells along the MK lineage. The proposed culture conditions and cytokine cocktails (BS1 and OMP) were developed with CB CD34+ cells. However, efficient MK differentiation of CD34+ cells from peripheral blood (mobilized cells) and BM can also be achieved with this protocol, although MK and platelet yields are significantly greater with CB CD34+ cells. The cytokine cocktail BS1 was developed by statistical design of experiment, which included the screening of 13 cytokines and optimization of the concentration of the selected cytokines by a response surface methodology (26). OMP was recently optimized using a similar strategy for the large-scale production of MKs and platelets ex vivo (36). This chapter also includes a simple method to purify platelet-like particles (PLPs) and adapted functional assays to investigate the quality and function of platelets produced ex vivo. Although not described herein, genetic engineering (e.g., via lentiviruses) of CD34+ or CD41+ cells could also be applied to this culture system in order to investigate the role of genes of interest in formation and function of MKs and platelets.

2. Materials

2.1. CB Mononuclear Cell and CD34+ Cell Preparation

CB CD34+-enriched cells are used in this method as a starting source of HSCs and multipotent progenitors to derive MK progenitors, MKs, and platelets ex vivo. CD34+ cells can be enriched by any means available; the details provided below describe the technique of negative immunomagnetic purification currently used in our laboratory.

1. Ficoll-Hypaque (1.077 g/mL) isotonic solution (Ficoll-Paque™ PLUS, GE Healthcare) for isolation of MNC from CB (see Note 1).

2. A complete EasySep separation unit (Easy Sep, Stem Cell Technologies, Vancouver, BC, Canada) for enrichment of CD34+ cells.

3. The Human Progenitor Enrichment Cocktail kit with CD41 depletion (Stem Cell Technologies); this isolates hematopoietic cells using antibody mixtures consisting of anti-CD2, -CD3, -CD11b, -CD11c, -CD14, -CD16, -CD19, -CD24, -CD41, -CD56, -CD66b, and glycophorin A.

4. Cryoprotective medium: 40% Fetal Bovine Serum (FBS, qualified, Invitrogen) and 10% Dimethylsulfoxide (DMSO; Sigma–Aldrich) in Iscove's modified Dulbecco medium (IMDM; Invitrogen).

5. 15- and 50-mL polypropylene centrifuge tubes (Falcon, Beckton Dickinson Labware, Franklin Lakes, NJ, USA).

6. 2-mL propylene cryotubes (VWR International, Mississauga, Ont., Canada).

7. DNase A (Sigma, D4513): Reconstitute at 1 mg/mL with H_2O, keep frozen at $-35°C$ in a 1-mL aliquot.

8. 0.4% Trypan blue solution (Invitrogen).

9. Hemacytometer (Hausser Scientific, Horsham, PA, USA).

10. Phosphate-Buffered Saline (PBS; this is Mg^{2+}- and Ca^{2+}-free for all protocols in this chapter), pH 7.4 (Gibco BRL).

11. EDTA 200 mM stock solution; 7.44 g EDTA (Fisher) in 100 mL of H_2O, adjust to pH 8.0 with NaOH and filter sterilize (0.22 µM).

12. PBS-glucose with 2% FBS and 1 mM EDTA: 2 g/L Glucose (Sigma–Aldrich) and 0.5% Phenol red (Sigma–Aldrich) in PBS (item 10, this section).

13. IMDM (Invitrogen, Burlington, Ontario, Canada) with 20% FBS.

2.2. CD34⁺ Cell Proliferation and Differentiation into MK

1. Enriched CD34⁺ cell suspension, obtained following the procedure 3.1.

2. 2-Mercaptoethanol (2-ME), 5×10^{-2} M containing 50 µM EDTA: 35 µL 14.3 M 2-ME (Sigma–Aldrich) diluted in 9 mL, sterile-filtered (0.22 µM) PBS. Add 2.5µL 200 mM EDTA and adjust to pH 6.0 with 0.1 N HCl, and then complete to 10 mL with PBS before filtration (0.22 µM). Keep at 4°C for 2 weeks (see Note 2).

3. PBS–1% Bovine serum albumin (BSA) (w/v): 1.33 mL 7.5% (w/v) BSA, fraction V, (Invitrogen) added to 8.67 mL PBS.

4. Human recombinant cytokines (SCF, TPO, FLT-3, IL-9, and IL-6 (Peprotech, Rocky Hill, NJ)). Aseptically reconstitute each cytokine at 10 µg/mL in PBS–1% BSA. Mix and dissolve well before aliquoting. Store at $-35°C$ (see Note 3).

5. Basal medium: IMDM aseptically supplemented with 20% BSA/insulin/transferin (BIT, serum substitute solution; Stem Cell Technologies), 20 µg/mL Low density lipoprotein (LDL; Sigma–Aldrich) (see Note 4), 100 µM 5×10^{-2} M 2-ME. For instance, to prepare 25 mL of medium, add 5 mL BIT, 100 µL LDL, and 50 µL 5×10^{-2} M 2-ME to 19.9 mL IMDM.

6. BS1- and OMP-supplemented culture medium: Basal medium (item 5, this section) supplemented with the required volumes of cytokines to obtain the desired concentrations. We recommend the use of the BS1 cytokine cocktail for the one-phase culture protocol (final concentration of 1 ng/mL SCF, 30 ng/mL TPO, 13.5 ng/mL IL-9, and 7.5 ng/mL IL-6; the so-called BS1-supplemented medium). For large-scale production (50–200 mL), cytokine consumption can be reduced by using the cytokine cocktail OMP (2 ng/mL TPO, 11 ng/mL FLT-3, 7.5 ng/mL SCF; the so-called OMP-supplemented medium) to seed the cells, and the BS1-supplemented medium for successive culture dilutions (see Notes 5–6).

7. GM6001 (Calbiochem, 364205): Prepare a 40 mM stock solution in DMSO. Keep in small aliquots at –20°C.

8. PBS–glucose solution: 2 g/L Glucose (Sigma–Aldrich) and 0.5% Phenol red (Sigma–Aldrich) in PBS, sterile filtered (0.22 μM).

9. 0.4% Trypan blue solution (Invitrogen).

10. Hemacytometer (Hausser Scientific, Horsham, PA, USA).

11. 6-well and 24-well cell culture microplates (Costar, Corning Incorporated, Corning, NY, USA or Beckton Dickinson Labware).

12. 25 and 75 cm^2 cell culture flasks (Corning, NY).

13. 100×20 mm tissue culture dishes (Beckton Dickinson Labware).

14. 15- and 50-mL polypropylene centrifuge tubes (Falcon, Beckton Dickinson Labware).

15. MegaCult™ kit (StemCell Technologies, Vancouver, Canada).

2.3. Flow Cytometry

1. 10× MK buffer: 20 mM theophylline and 272 mM Na-citrate in PBS, dissolved in a water bath at 37°C and sterile filtered (0.22 μM). For instance, add 360 mg theophylline (Sigma–Aldrich) and 7.643 g Na-citrate (Fisher Scientific) in 200 mL PBS. Stable for 3 months at 4°C (see Note 7).

2. 1× MK buffer: For 200 mL, mix 20 mL 10× MK buffer with 26.8 mL 7.5% BSA (w/v) and bring volume to 200 mL with PBS (see Note 7).

3. Fluorochrome-coupled human antibodies and their respective isotypic murine controls: Phycoerythrin (PE)-coupled anti-CD34 (Immunotech, Beckman Coulter Co., Marseille, France), PE-coupled anti-CD62 (Immunotech), allophycocyanin (APC)-coupled anti-CD41a (Becton Dickinson), and fluorescein isothiocyanate (FITC)-coupled anti-CD42b (Becton Dickinson). Isotypic controls: Mouse IgG1 isotype control conjugated to APC, PE, and FITC.

4. 1 mg/mL propidium iodide (PI): Dissolve 25 mg PI (Sigma–Aldrich) in 25 mL deionized water. Prepare a 1-mL aliquot and store at –20°C.

5. Flow cytometer (FACS-Calibur, Becton Dickinson Immunocytometry Systems, San Jose, CA, USA).

6. Centrifuge microtubes made of homopolymer, 1.5 mL (Axygen Scientific, Union City, CA, USA).

7. 5-mL Polystyrene FACS tubes (Bio-Rad Laboratories).

8. 3% (v/v) acetic acid (Glacial Acetic Acid (Fisher)).

9. Blood collection tubes; Citrate Tubes (16×100 mm $\times 8.5$ mL BD Vacutainer® glass whole blood tube, ACD solution A, BD Vacutainer Cat Number 364606).

2.4. BSA Gradient

1. BSA fraction V (Fisher, BP1605).

2. PBS Mg and Ca free (Gibco BRL).

3. 10× CGS buffer: 100 mM sodium citrate, 300 mM D-Glucose, 1.2 M NaCl, pH 6.5. For instance, add 2.9 g Na-citrate (Fisher Scientific), 5.4 g d-Glucose (Sigma–Aldrich), and 7 g NaCl (Fisher) to 90 mL H_2O. Adjust pH to 6.5 with citric acid (Fisher) and bring volume to 100 mL with H_2O. Sterile filter (0.22 µM) and keep at 4°C.

4. 1× CGS buffer: Mix 10 mL of 10× CGS buffer with 90 mL of sterile water. Keep at 4°C.

5. PGE_1 (Sigma, P7527): Prepare 500 µM stock solution in ETOH 99%. Store in small aliquots at –20°C.

6. Filtered (0.22 µM), heat-inactivated (1 h at 55°C) human AB serum (Sigma–Aldrich, H4522).

7. Platelet isolation medium: Basal medium (item 5, see Subheading 2.2) supplemented with the BS1 cytokine cocktail (1 ng/mL SCF, 30 ng/mL TPO, 13.5 ng/mL IL-9, and 7.5 ng/mL IL-6). For storage of platelets (>8 h), 1% inactivated human serum and 25 µM GM6001 (see Subheading 2.2) should be added before use (see Note 8).

8. 15- and 50-mL polypropylene centrifuge tubes (Falcon, Beckton Dickinson Labware).

2.5. Additional Material for Functional Assays

1. Platelet agonists stock solutions: To prepare 1 mg/mL fibrinogen, add 15 mg fibrinogen from human plasma (Calbiochem) to 15 mL sterile PBS. Dissolve at 37°C for 12 h, sterilize by filtration (0.22 µM), and keep at 4°C. For 13.4 U/mL Thrombin, add 1 mL of sterile deionized water to one vial of Thrombin (13.4 U, Sigma). ADP (Cat. No. 364) and Collagen type I (Cat. No. 385) are used as recommended by the manufacturer (ChronoLog Corporation). Store thrombin and ADP

stock solutions in small aliquots at –20°C to avoid repeated freeze/thaw cycles.

2. Flow cytometer (FACS-Calibur, Becton Dickinson Immunocytometry Systems, San Jose, CA, USA).

3. Calcein AM (Molecular Probes, Invitrogen): Prepare 2 mg/mL stock solution in DMSO. Store in a desiccated environment at –20°C, protected from light (see Note 9).

4. 6-well and 96-well cell culture microplates (Costar, Corning Incorporated, Corning, NY, USA or Beckton Dickinson Labware).

5. Titertek microplate shaker (Flow Laboratories).

6. 22 mm glass coverslips, Square No. 1- 0.13–0.17 thick (Fisherbrand Scientific).

7. Formaldehyde 37% (Sigma–Aldrich).

8. 10% (v/v) Triton X-100: Dissolve 1 volume of pure Triton X-100 (Bio-Rad Laboratories, Life Science Research, Hercules, CA, USA) in 9 volumes of PBS.

9. Tween 20 (Fisher).

10. 4′,6-Diamidino-2-phenylindole (DAPI; Molecular Probes, Invitrogen): For using at 1:1,000 final dilution, prepare 5 mg/mL stock solution in deionized water. Stock solution can be aliquoted and stored at –20°C. Thawed aliquots can be kept at 4°C for 6 months protected from light.

11. ALEXA fluor conjugated phalloidin (Molecular Probes, Invitrogen): Add 1.5 mL of methanol to one vial of ALEXA fluor-conjugated phalloidin. Keep solution at –20°C in desiccate, protected from light. Use at 1:50 final dilution.

12. ProLong® Gold antifade reagent (Molecular Probes, Invitrogen).

13. TE-2000-s inverted microscope (Nikon, Melville, NY, USA) equipped with a 100× oil, 1.25 NA objective and an LWD 40×, 0.55 NA objective.

3. Methods

CB mononuclear cells (MNCs) are isolated on a Ficoll-Hypaque density gradient and cryopreserved (see Note 1). MNCs from three to six CB samples are mixed prior to CD34+ enrichment in order to reduce the interdonor variability and to allow production of larger cell numbers. Following enrichment using negative immunomagnetic selection, CD34+-enriched cells (purity≥75%) are aliquoted and cryopreserved.

3.1. Purification of CD34+ Cells

3.1.1. MNC Preparation

MNCs are isolated on a Ficoll-Hypaque density gradient following the manufacturer's instructions. Then, cells are cryopreserved (follow Subheading 3.1.2) or CD34+ cells can be purified immediately (follow Subheading 3.1.4).

3.1.2. MNC Cryopreservation

1. Centrifuge the cells obtained in Subheading 3.1.1 for 10 min at $514 \times g$.

2. Resuspend the cell pellet in cryoprotective medium, maintained at 4°C, at a density not higher than 300×10^6 cells/mL. Prepare a 1-mL aliquot of this suspension into cryotubes.

3. Place the cryotubes at −80°C for 24 h, and then transfer to liquid nitrogen.

3.1.3. MNC Thawing

1. Thaw cryopreserved MNC in a 37°C water bath without mixing.

2. Transfer cells into a 15-mL tube. Add IMDM containing 20% FBS at 4°C to bring to approximately 15 mL.

3. Centrifuge for 10 min at $228 \times g$ and room temperature (RT).

4. Remove the supernatant and resuspend the pellet in 10 mL IMDM containing 20% FBS at 4°C.

5. Add DNase A (final concentration of 100 μg/mL) and incubate for 15 min at room temperature.

6. Take a sample to count the cells and assess viability (see Note 10).

7. Centrifuge for 10 min at $228 \times g$ and room temperature.

8. Decant the supernatant and resuspend cells to obtain a density between 2×10^7 and 8×10^7 cells/mL in PBS–glucose containing 2% FBS with 1 mM EDTA for magnetic labeling (Subheading 3.1.4).

3.1.4. Magnetic Labeling and CD34+ Cell Separation

CD34+-enriched cells can be purified by several means, and several commercial kits are available for this purpose. The culture protocols described here were developed with CD34+-enriched cells negatively immunoselected with the kit offered by Stem Cell Technologies (Easy-Sep, Human Progenitor Enrichment Cocktail kit, Stem Cell Technologies) and used according to the manufacturer's instructions. Once CD34+-enriched cells are purified, they are cryopreserved in aliquots (~5×10^5/vials), as described in Subheading 3.1.2.

3.2. Ex Vivo Expansion and Maturation of MK Starting from CB CD34+ Cells

Cultures are initiated in a serum-free BS1- or OMP-supplemented culture medium. TPO alone at 10–100 ng/mL is often reported in the literature since it does not support the growth of non-MK cells, resulting in pure MK cell cultures. However, we strongly recommend to use the BS1 cytokine cocktail developed in our laboratory for the entire 14 days of culture (one-phase culture protocol), as it supports both high MK purity and cell expansion. For large-scale production of MKs and platelets, we also provide a two-phase, resource-efficient protocol that combines the newly developed cocktail OMP with BS1 (see Note 6).

Cultures are usually performed in 24-well tissue culture plates with volumes ranging from 0.8 to 1 mL. We recommend using only the 8 central wells and to fill the 16 external wells with PBS to reduce the risk of overestimation of cell density due to evaporation, especially when culturing cells at 39°C (see Note 11). The cultures can also be performed in 6-well to 96-well plates or in larger flasks by keeping the same volume-to-culture surface ratio (see Note 12). The whole cell culture lasts 2 weeks, where the cells mainly proliferate and differentiate into MK progenitors during the first week while they mature and produce platelets during the second week with a peak of platelet production around day 14. Flow cytometry (Subheading 3.3.1) and cell counts are usually performed on days 0, 7, 10, and 14 to assess cell differentiation and expansion and to evaluate platelet production (Subheading 3.3.3). Platelets are usually purified at day 14 (Subheading 3.4).

1. Thaw the cryopreserved CD34$^+$ cells in a 37°C water bath without mixing.

2. Transfer the cells to a 15-mL tube. Make up to 15 mL with PBS–glucose solution at 37°C by adding the first 5 mL very slowly.

3. Centrifuge for 10 min at $228 \times g$ and discard the supernatant.

4. Resuspend the cells in culture medium (BS1- or OMP-supplemented medium; see Note 6) to a density of around 3–6×10^5 cells/mL.

5. Take a sample to measure the cell density and cell viability (see Note 10).

6. Dilute the CD34$^+$ cell suspension to 40,000 cells/mL in BS1 or OMP culture medium. Incubate at 39°C, 10% CO_2, in a humidified incubator (see Note 11). Greater cell density can be used if analyses of the cultures are performed within the first few days of culture.

7. After 4 days, gently mix the contents of each well, remove half of the cell suspension volume, and add an equivalent volume of fresh BS1 or OMP culture medium (see Note 12). For large-scale platelet production, just add an equivalent volume of fresh OMP- or BS1-supplemented culture medium (see Note 6).

8. Dilute the cells with fresh BS1-supplemented culture medium on day 7 to reach 2×10^5–3×10^5 cells/mL and incubate the cell culture at 37°C (see Notes 11 and 12).

9. As metabolic waste can affect platelet integrity, replace the culture medium at day 10 or 11. Take a sample to measure the cell density and transfer the entire culture to 50-mL tube(s).

Centrifuge for 10 min at $228 \times g$ and discard the supernatant. Resuspend the culture in fresh BS1-supplemented culture medium to reach 3×10^5 cells/mL. For large-scale culture, greater platelet yield can be reached if the culture is transferred to 100-mm dishes instead of flasks. In this case, do not exceed 10–12 mL per dish for a better oxygenation of newly formed platelets.

10. Add 25 μM of the metalloproteinase inhibitor GM6001 to all culture dishes on day 12 (see Note 8).

11. At day 14, harvest cell cultures using the following procedure (Note 13):

   (a) For small cultures (<5 mL), mix the cultures thoroughly by pipetting up and down (10–15 times) with a P1000 micropipette.

   (b) For large cultures (>5 mL):

   • Fix a P1000 tip at the end of a serological 5- or 10-mL pipette.

   • Mix the cell culture well by pipetting the cell suspension up and down 10–20 times (depending on the size of the culture).

3.3. Culture Analysis

MK differentiation can be observed directly by using a phase-contrast, inverted microscope; mature MKs are discernible as larger cells and can be observed within 7 days while proplatelet filaments are clearly visible by day 10. Note that the proplatelets are extremely fragile, and great care must be taken when handling the cultures. For a quantitative monitoring of megakaryopoiesis, cells should be characterized by flow cytometry (see Subheading 3.3.1, Fig. 3) while MK progenitors can be measured using colony assays (see Subheading 3.3.3).

Phenotypic analysis by flow cytometry is carried out using fluorochrome-conjugated antibodies against specific MK markers, such as CD41a, CD61, and CD42b. Other markers can also be used, such as CD62 which is used to determine the activation status of platelet (see Note 14). Note that platelets do not express CD45. Recent results from our laboratory demonstrated that functional platelets produced in culture coexpress CD41a and CD42b and that those that fail to maintain CD42b expression are for the most part metabolically inactive (36). MK progenitors can also be estimated by flow cytometry as the CD34+CD41a+ population, although we recommend using a more robust assay, such as MegaCult™.

3.3.1. Flow Cytometry Acquisition Setup

The first step prior to the analysis of the samples is to establish a proper cell/platelet analysis acquisition template for the acquisition software used with the cytometer (such as CellQuest™). We recommend using a similar acquisition strategy to that described in Fig. 3d–h. A hematopoietic cell line can also be used to facilitate

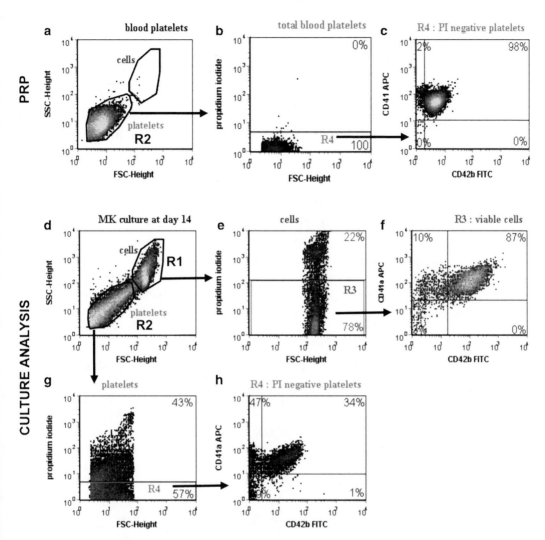

Fig. 3. Flow cytometry methodology used to analyze MKs and platelets. (**a–c**) Analysis of fresh platelets from platelet-rich plasma (PRP). (**d–i**) Analysis of ex vivo produced MKs and platelets. Further details: (**a**) Normal platelets derived from PRP are first used to set the analytical platelet region "R2" in a 2D dot-plot acquisition window showing the forward-side scatter (FSC) and side scatter (SSC) properties on the x- and the y-axis, respectively. (**b**) On an independent dot-plot window, platelet events from the R2 region are analyzed for their retention of propidium iodide (PI), which is null for normal platelets. (**c**) PI-negative platelet events (R2*R4 region) are then analyzed for the expression of CD41a and CD42b on the y- and x-axis, respectively. The use of fresh PRP platelets also ensures proper adjustments of the various settings of the flow cytometer. (**d**) FSC and SSC analysis window of cell cultures. Platelet events appear in the previously drawn R2 region, whereas MNCs appear as distinctive events of greater size (FSC) and greater granularity (SSC). The cell region "R1" is drawn around that second population. (**e** and **g**) The cell and platelet events from the regions R1 and R2 are then analyzed for their retention of PI. The PI-negative selection regions for the cells (R3) and platelet (R4) events are shown in **e** and **g**, respectively. (**f** and **h**) The PI-negative cell and platelet events (from R1*R3 and R2*R4 regions, respectively) are analyzed for CD41a and CD42b (or other markers of interest) expression in **f** and **h**, respectively. Reproduced from Cortin et al. (42) with permission from Humana Press.

the setup. Platelets are enumerated as PI-negative CD41a⁺CD42b⁺ events with scatter properties similar to those of control platelets prepared from peripheral blood. We strongly recommend to validate the platelet-gating strategy and cytometer settings with a normal platelet sample that can easily be isolated from a small blood sample (described in the next section).

1. Cell region R1: In a dot-plot acquisition window, with forward scatter (FSC) and side scatter (SSC) on the x- and y-axis, respectively, draw the R1 region around the nucleated cells, which have FSC and SSC properties significantly greater than platelets and cell debris (Fig. 3d). This region is used as a gate to specifically select cell events.

2. Propidium iodide-negative cell region R3: In a dot-plot acquisition window acquiring events from the R1 region only, select FSC and PI (read on FL3 channel) on the x- and y-axis, respectively, and draw the R3 region around the PI-negative cell events (Fig. 3e). This region is used as a gate to select viable cells.

3. Platelet region R2: Draw the platelet region R2 in the same dot-plot acquisition window in which the cell region was drawn (step 1, this section). Platelet events have reduced FSC and SSC properties. The platelet region is used as a gate to select platelet events and it must be validated using fresh platelets (see Subheading 3.3.2).

4. PI-negative platelet region R4: In a dot-plot acquisition window acquiring events from the R2 region only, select FSC and PI (on FL3 channel) on the x- and y-axis, respectively, and draw the R4 region around the PI-negative platelet events (Fig. 3g). This region is used as a gate to select events emanating from platelets specifically. Normal platelets do not stain positively with PI (Fig. 3b). PI$^+$ platelet events seen with culture samples are most likely cellular MK debris and/or MK-derived microvesicles.

5. In dot-plot acquisition window, select CD41a (if using an APC antibody, FL4 channel) and CD42b (if using an FITC antibody, FL1 channel) on the y- and x-axis, respectively, to analyze MKs (gating R1*R3 events only) or platelets (gating R2*R4 events only) as shown in Fig. 3f and h, respectively.

6. Run unstained and isotopic control-stained cells first to help adjust the cytometer settings and to measure background levels. Run samples afterward.

7. Once the flow cytometry acquisition template and flow cytometer settings are validated with culture samples and normal platelets, they should be saved for future use. In the latter case, new samples can be analyzed rapidly without the requirement of preparing fresh normal platelets.

3.3.2. Preparation of the Culture Samples for Flow Cytometry Analysis of the Cells and Platelets

This procedure is designed for the simultaneous analysis of cells and platelets produced in culture by flow cytometry. The first step can be skipped if platelets are not to be analyzed or if the flow cytometry acquisition template is already set up.

First step: Preparation of control platelet for validation of the platelet region.

1. Centrifuge citrated whole adult blood (1–10 mL) at $329 \times g$ for 10 min without a brake (see item 9, Subheading 2.3 for suitable blood collection tubes. Note that appropriate ethical permission and donor consent must be obtained).

2. Collect the platelet-rich plasma (PRP) with a pipette in the upper phase of the centrifuged sample. Centrifuge this PRP at $1{,}428 \times g$ for 15 min.

3. Discard the supernatant and resuspend platelets in MK buffer.

4. Sample an aliquot for platelet count; this can be done on a hematologic analyzer. Alternatively, counts can be estimated using a hemacytometer. If the latter method is used, dilute the sample in 3% acetic acid to lyse red cells; platelets appear as small dots since they are 10–25 times smaller than nucleated cells (for more details, see ref. 37).

5. Stain a small fraction of PRP samples by diluting $\sim 1 \times 10^6$ platelets in MK buffer prewarmed to room temperature, add antibodies against platelet antigens, such as CD41a, CD42b, and CD62, and incubate for 15–30 min at room temperature (see Note 15). CD62 is optional (see Note 14). Each specific stain should be accompanied by an isotopic control stain done in parallel at the same antibody dilution.

6. To the 0.2-mL stained aliquots, directly add 0.5 mL MK buffer prewarmed to room temperature containing 7 µg/mL PI in order to reach a final PI concentration of 5 µg/mL (see Note 16).

7. Analyze by flow cytometry as indicated in Subheading 3.3.1 as soon as possible since platelets are sensitive to spontaneous activation. Acquire a minimum of 10,000 events to identify the platelet region R2 (Fig. 3a) with specific FSC and SSC characteristics.

Second step: Analyses of cells and platelets produced in vitro.

1. Measure cell density and cell viability using a hemacytometer and trypan blue (see Note 10).

2. Gently resuspend the cell culture and transfer a 0.2-mL aliquot of cell suspension (100,000–500,000 cells) into two 5-mL FACS tube.

3. Add antibodies (CD41a, CD42b …) to each tube (see Note 15) and incubate for 15–30 min in the dark (on ice for cells only or room temperature otherwise as platelets are activated by low temperatures). Each specific stain should be accompanied by an isotopic control stain done in parallel at the same antibody dilution.

4. Option 1: For cell-only analysis, wash the cell/antibody mixtures by adding 1–2 mL of MK buffer. Centrifuge at $1,000 \times g$ for 5 min, discard the supernatants, and resuspend the pellets in 0.3–0.5 mL MK buffer containing 5 µg/mL PI.

5. Option 2: For cell and platelet simultaneous analysis: To the 0.2-mL stained aliquot, directly add 0.5 mL MK buffer pre-warmed to room temperature containing 7 µg/mL PI in order to reach a final PI concentration of 5 µg/mL (see Note 16).

6. Proceed to flow cytometry analysis as indicated in Subheading 3.3.1 as soon as possible since samples are not fixed and platelets are sensitive to spontaneous activation. Acquire a minimum of 10,000 PI-negative cell events (R1*R3 events).

3.3.3. Analysis of MK and Platelet Production

1. Production of MKs:

$$\text{Total MKs} = \text{TNC}(t) \times \%\text{CD41a}^+$$

$$\text{Total mature MKs} = \text{TNC}(t) \times \%\text{CD41a}^+\text{CD42b}^+$$

TNC (t): Total nucleated cells at time t = density of viable cells $(t) \times$ culture volume(t)

%CD41$^+$ and %CD41$^+$CD42b$^+$ cells: Derived from the cytometry analysis at time t

The theoretical cumulative (Σ) number of MKs produced per seeded cells at time t, when all cells have been maintained in culture from day 0 up to time "t":

$$\sum \text{Total MKs} = \text{DVC}(t) \times \text{CDF}(t) / \text{DVC}(day0) \times \%\text{CD41a}^+$$

DVC: Density of viable cell

CDF: Cumulative dilution factor up to time t.

2. Production of PLPs:

Platelet yields are estimated using the cell density and the platelet-to-cell ratio (R(P/C)).
The number of PLPs (Nb PLP) produced at time "t":

$$\text{Nb PLT(t)} = \text{DVC}(t) \times \text{CV}(t) \times \%\text{CD41}^+\text{CD42}^+ \text{PLPs} \times \text{R(P/C)}t$$

DVC: Density of viable cell at time t

CV: Culture volume at time t

%CD41$^+$CD42$^+$ PLPs: Percentage of platelet events that are CD41$^+$CD42b$^+$ (Fig. 3h, upper right quadrant)

R(P/C)t: Ratio at time "t" of the number of platelet events in R4 (Fig. 3g) and the number of cell events in R3 (Fig. 3e). These numbers are available in the statistical analysis of the dot plots.

The theoretical cumulative number of PLPs produced in culture at time "t" per s eeded cells (Σ) Nb PLPt):

$$\sum Nb\ PLPt = DVC(t) \times CDF(t) / DVC(day0)$$
$$\times \%CD41a^{+}CD42b^{+}PLPs \times R(P/C)\ t$$

CDF: Cumulative dilution factor up to time t

3.3.4. CFU-MK Titration

Culture conditions favoring MK expansion and differentiation also support the expansion of MK progenitors. However, the magnitude of the expansion is for the most part dictated by (1) the cytokine cocktail used (nature and concentration) and (2) the length of expansion. Thus, these two parameters must be carefully optimized to obtain the maximal expansion. The BS1 cocktail suggested in this chapter provides good MK progenitor expansion ranging from 5- to 75-fold for culture maintained for 3–10 days at 39°C (N. Pineault personal results, 2005). A number of assays are available to measure MK progenitors: the plasma-clot technique, the serum-free fibrin clot assay, and a commercially available collagen-based kit known as MegaCult™. More information on the fibrin-clot assay can be found in ref. (38).

We routinely use the MegaCult™ assay (2,250–8,000 cells per plate). In general, the frequency of colony-forming unit-megakaryocyte (CFU-MK) progenitors in culture decreases as a function of culture time, hence we suggest plating two different cell doses (1×–2×). We recommend plating 2,250 cells between day 0 and day 5 of culture, and 3,500 between day 6 and day 10. The MegaCult™ assay uses immunostaining to ensure proper enumeration of MK progenitors (referred to as CFU-MK). In contrast to most myeloid colonies, colonies originating from MK progenitors are small, both in total size and the number of cells they contain. The colonies are scored based on their size, with mature MK progenitor forming small colony (3–20 cells) and immature progenitor forming larger colony (≥50 cells). Background information, material, and methods are all supplied in the MegaCult™ kit (StemCell Technologies, Vancouver, Canada). To assess the expansion of MK progenitors in culture, MK progenitor frequency (f) should be determined at day 0 and at other times of interest (day X). MK progenitor expansion is then calculated as follows: the total cell expansion is multiplied by f on day X and divided by f on day 0.

3.4. Purification of Culture-Derived Platelet-Like Particles

The above flow cytometry technique can be used to assess platelet phenotype with or without activation in whole culture. Similar analysis and additional functional assays can be done on purified platelets to eliminate the MK and non-MK cells. Technically, the simplest method to isolate platelets from a human blood sample is the preparation of washed platelets by successive centrifugation (39).

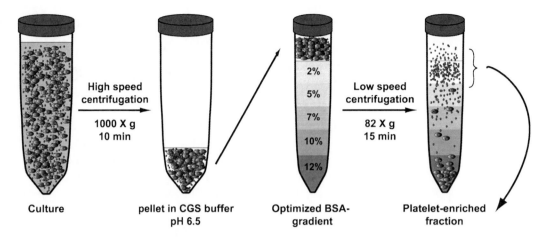

Fig. 4. Isolation of platelets from MK culture by centrifugation over a BSA density gradient; *see* Subheading 3.4.1 for details.

This technique can be easily used to isolate PLPs from the culture if purity is not an important criterion for your purpose. Usually, we obtain a purity of around 80% PLPs and 20% nucleated cells by using standard centrifugation. We tried to improve the purity of this preparation by gel filtration method on Sepharose 2B. The latter was found to provide excellent purity (~98%) but unreliable yields (~30%) due to cell debris that often obstructed the column. Finally, we had much better success in terms of yield, purity, and ease by adapting the albumin density gradient centrifugation technique described by Walsh et al. (40). Following optimization of the albumin concentration (Fig. 4) and centrifugation speed parameters, we have achieved a platelet purity of 98% with more than 60% of PLP recovery. This adapted method, described in detail below, can be used to isolate blood platelets and culture-derived PLPs.

3.4.1. BSA Gradient Preparation

1. Dissolve BSA in PBS at a concentration of 12% w/v and filter the solution through a 0.22-µM filter.

2. Use this stock to make the following BSA solutions (2 mL of each solution per gradient);10, 7, 5, and 2% w/v in PBS.

3. Add 1.5 mL of the 12% BSA solution in a 15-mL centrifuge tube and gently layer 1.5 mL 10% BSA on the top. Repeat the second part of this procedure for the other concentrations of BSA finishing with the 2% BSA solution (as depicted in Fig. 4).

3.4.2. Platelet Isolation from MK Culture over BSA Gradient

1. Gently resuspend the cell culture and transfer the cell suspension to 50-mL tube(s).

2. Centrifuge at $1,000 \times g$ for 10 min.

3. Discard the supernatant and resuspend cells and platelets in 2 mL of CGS buffer prewarmed to room temperature.

4. Gently layer the cell suspension in CGS on the top of the BSA gradient (created in Subheading 3.4.1).

5. Centrifuge the BSA gradient at $80 \times g$ for 15 min without a brake.

6. Discard the 0.5 mL upper phase of the gradient and collect the next 4.5 mL upper phase that contains the PLPs.

7. Dilute the collected PLPs with an equal volume of 1% BSA w/v in PBS and add 1 μM PGE_1. Centrifuge at $1,000 \times g$ for 10 min.

8. Discard the supernatant and resuspend PLPs in 0.5 mL of BS1-supplemented medium (see Note 17).

9. For long-term storage, adjust the PLP concentration to around 2×10^7 PLPs/mL in BS1 medium supplemented with 1% heat-inactivated human serum and 25 μM GM6001, and then maintain the PLPs in a humidified atmosphere at 37°C (see Note 8).

3.4.3. Platelet Isolation from PRP over BSA Gradient

1. Prepare PRP as described in Subheading 3.3.2 (steps 1–3).

2. Add 1 μM PGE_1 to PRP and then centrifuge at $1,000 \times g$ for 10 min.

3. Discard the supernatant and resuspend the platelets in 2 mL of CGS buffer.

4. Complete the isolation procedure by proceeding to step 4 of Subheading 3.4.2.

3.5. In Vitro Functional Assays for Culture-Derived PLPs

The assays described below have been adapted from classic in vitro platelet functional assays to investigate the function of PLPs. Each of these assays should be performed in parallel using isolated blood platelets as a control.

3.5.1. Platelet Vitality by Calcein Loading

Calcein acetoxymethyl ester (calcein AM) is membrane permeant and nonfluorescent until hydrolyzed by intracellular esterases present in viable cells and platelets. Generally, the vast majority of freshly washed blood platelets are positive for fluorescent calcein (Fig. 5a). In contrast, a significant proportion of the PLPs is composed of cell debris, microparticles, and dead platelets which remain nonfluorescent (Fig. 5b). Consequently, this simple labeling method allows one to distinguish viable and perhaps functional platelets in PLPs and to follow their stability over the time in storage. By using our culture conditions, all of the CD41a+CD42b+ PLPs are metabolically active according to the appearance of calcein fluorescence, suggesting that functional assays should be performed in this subpopulation of PLPs.

1. Add calcein AM from a stock of 2 mg/mL to the isolated platelets or PLPs to reach a final concentration of 2 μg/mL and incubate for 1 h at 30°C (see Note 19).

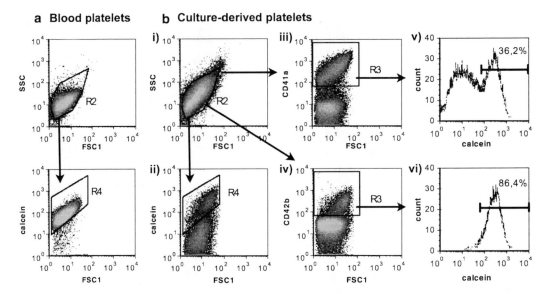

Fig. 5. Flow cytometry methodology used to analyze platelets' viability after calcein AM loading. (**a**) Analysis of fresh plate-lets from PRP. Normal platelets isolated from PRP are first used to set the analytical platelet region "R2" in a 2D dot-plot acquisition window showing the FSC and SSC properties on the *x*- and the *y*-axis, respectively. On a second dot-plot window, platelet events from the R2 region are analyzed for their capacity to convert calcein AM into a fluorescent com-pound, which reaches almost 100% in freshly isolated blood platelets. Viable platelet region "R4" is set in a 2D dot plot with the FSC on the *x*-axis and the calcein (usually, FL1 channel) on the *y*-axis. (**b**) Analysis of isolated culture-derived platelets. (i) FSC and SSC analysis of purified platelets. (ii) Platelets from the R2 region analyzed in a 2D plot showing the FSC and the calcein fluorescence. Note that the proportion of platelet-like particles (PLPs) that are metabolically active (R4 region) is inferior to that of control blood platelets. PLP events (R2 region) analyzed in a 2D plot for FSC on the *x*-axis and for the expression of CD41a (iii) or CD42b (iv) on the *y*-axis, respectively, to create the CD41a or CD42b platelet region "R3". (v–vi) On the histogram windows, platelet events from the R3 regions are analyzed for calcein fluorescence to exclude cell debris and microparticles. This analysis shows that more than 80% of the CD42b-expressing PLPs (vi) are viable while more than half of the CD41a-expressing PLPs (v) are metabolically inactive.

2. Add 1 µM PGE$_1$ (see Subheading 2.4, item 5) and centrifuge labeled PLPs at 1,000 × *g* for 10 min.

3. Discard the supernatant, wash PLPs with 1% BSA (w/v) in PBS supplemented with 1 µM PGE$_1$, and then centrifuge at 1,000 × *g* for 10 min.

4. Discard the supernatant and suspend PLPs in the desired buffer.

Calcein fluorescence can be observed by fluorescent microscopy with standard filters used for FITC or GFP or by flow cytometry using the FL1 channel. Calcein fluorescence is stable for several hours, so counterstaining with other compounds, for example phalloidin ALEXA-conjugated or APC-CD42b MoAb, is possible prior to the analysis. For flow cytometry analysis, it is important to use freshly washed blood platelets loaded with calcein to delineate the metabolically active platelet region.

3.5.2. Activation State upon Agonist Stimulation

The cytometry analysis described above depicts two distinct populations of CD41a⁺ PLPs present in culture. The first one is composed of nonviable CD41a⁺CD42b⁻ particles according to the low levels of calcein fluorescence (Subheading 3.5.1). The second one is composed of viable CD41a⁺CD42b⁺ PLPs (positive for calcein fluorescence, see Fig. 5b panel vi), which represents putative functional platelets. Investigation of the functionality of these PLPs can be performed by testing their capacity to become activated following agonist stimulation. A common in vitro assay to address the activation state of platelet is the monitoring of P-selectin (CD62P), which is exposed only on the surface of platelets that have released alpha-granules. This can be analyzed by flow cytometry using fluorochrome-conjugated antibody against CD62P following platelet stimulation with or without agonist. Other antibodies can also be used, such as PAC-1, which recognize the active conformation state of the extracellular domain of the integrin $\alpha_{IIb}\beta_3$ (CD41a/CD61 complex). Normally, the vast majority of CD41a⁺CD42b⁺ PLPs respond strongly to agonist stimulation.

1. Place 500,000–1 million platelets into a 1.6-mL Eppendorf tube and bring the volume to 0.2 mL with MK buffer (Subheading 2.3, item 2) prewarmed to room temperature.

2. Add agonist(s) (10–50 µM ADP, 1 U/mL thrombin, …) into each test tube and incubate at 37°C for 10 min. Keep one test tube without agonist as a control to indicate the basal level of activation.

3. Add antibodies (CD41a, CD42, and CD62P) to each tube and incubate for 15–30 min at room temperature. Each specific stain should be accompanied by an isotopic control stain done in parallel at the same antibody dilution.

4. Transfer the 0.2-mL stained aliquots into 5-mL FACS tubes and directly add 0.2 mL formaldehyde 0.5% (v/v in PBS). Fix the stained platelets for 1 h at room temperature.

5. Add 0.4 mL of MK buffer prewarmed to room temperature and proceed to flow cytometry analysis as indicated in Subheading 3.3.1 (see Note 18).

3.5.3. Evaluation of Platelet Adhesion and Shape Change Following Stimulation by Fluorescence Microscopy

Among the first steps leading to platelet thrombi formation is the tethering of circulating platelets and firm adhesion to the injury site. This is triggered in part by the integrin $\alpha_{IIb}\beta_3$ (CD41a/CD61 complex) that mediates cell adhesion and signaling. In resting platelets, $\alpha_{IIb}\beta_3$ has a low affinity for fibrinogen, its main ligand. Following platelet exposure to extracellular matrix or soluble agonists, platelet membrane receptors specific for each stimulus induce intracellular signals that are transmitted to the cytoplasmic domain of $\alpha_{IIb}\beta_3$. This inside-out signaling converts the integrin from an inactive to an active state through conformational change, which

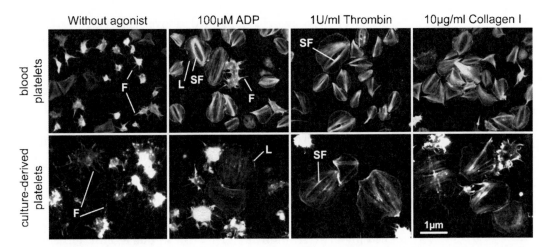

Fig. 6. Platelet adhesion and shape change following stimulation. Purified blood and culture-derived platelets were placed on immobilized fibrinogen for 30 min with or without the indicated agonists. Adherent platelets were then fixed with 3.7% formaldehyde, and F-actin filaments were stained with ALEXA 488-conjugated phalloidin as described in Subheading 3.5.3 to reveal filopodias (F), lamellipodias (L), and stress fibers (SFs) induced by the $\alpha_{IIb}\beta_3$ bidirectional signal response. The fixed platelets were visualized under fluorescence microscopy by using a 100× oil, 1.25 NA objective.

increases its fibrinogen affinity. Ligand binding to the $\alpha_{IIb}\beta_3$ causes integrin clustering and conformational changes of their cytoplasmic tails that promote interactions with intracellular proteins. This outside-in signaling leads to important actin cytoskeleton reorganization that supports platelet adhesion and thrombi formation. To monitor the bidirectional signaling capacity of culture-derived PLPs, evaluation of the capacity of platelets to undergo normal shape change and spreading following adhesion to immobilized fibrinogen can be performed by fluorescence microscopy. Subsequent to adhesion, normal platelets first induce filopodia extension (see F, Fig. 6) followed by lamellipodia formation (see L, Fig. 6) supported by stress fibers (see SF, Fig. 6). These changes are accelerated by the addition of soluble agonists, namely, ADP, thrombin, or collagen, and can be revealed by actin filament staining using fluorochrome-conjugated phalloidin.

1. Place sterile coverslips in the bottom of the wells of a six-well plate (see Note 20).

2. Coat coverslips with 100 μg/mL fibrinogen in PBS for 1 h at 37°C or for O/N at 4°C.

3. Aspirate fibrinogen solution and block coverslips with 1% BSA (w/v) in PBS (see Subheading 2.4, items 1 and 2) for 1 h at 37°C.

4. Rinse coverslips once with PBS.

5. Aspirate PBS and add around 5×10^6 platelets (PLPs or from PRP) diluted in platelet isolation medium (with or without

serum) (see item 7, Subheading 2.4) in the absence or presence of agonists (0.5–1 U/mL thrombin; 50–100 μM ADP; 5–10 μg/mL collagen I). Put no more than 100 μL of platelet suspension on the center of each coverslip.

6. Incubate for 30 min at 37°C. Gently wash off the culture medium with 1 mL PBS preheated to 37°C.

7. Aspirate the PBS, add 1 mL of preheated 3.7% formaldehyde (v/v) in PBS, and incubate for 15 min at 37°C.

8. Rinse twice with PBS and proceed to actin filament staining using the immunofluorescence protocol described below.

Immunofluorescence for platelets and MKs.

1. Permeabilize platelets for 5 min at room temperature with 0.1% Triton X-100 (v/v) in PBS.

2. Rinse twice with PBS.

3. Incubate for 30 min at 37°C in blocking solution (1% BSA (w/v) in PBS).

4. Incubate for 1 h at 37°C with primary antibody diluted in the blocking solution. Example: For GPIbα receptor, use CD42b moAb diluted 1:50.

5. Rinse five times with washing solution (0.05% Tween (v/v) in PBS).

6. Incubate for 30 min at 37°C with secondary antibody conjugated with fluorochromes diluted in the blocking solution. ALEXA Fluor-conjugated phalloidin and DAPI can be added at this stage. Example: Anti-mouse ALEXA 555 (final dilution 1:1,000) could be mixed with phalloidin ALEXA 488 (final dilution 1:50) and DAPI (final dilution 1:1,000) to reveal simultaneously CD42b moAb, actin filaments, and DNA (see Note 21).

7. Wash five times with washing solution and one time with deionized water.

8. Mount glass coverslips (cell side down) in an appropriate mounting medium. We recommend to use ProLong Gold from Molecular Probes that contains antifading agents to decrease photobleaching during observations.

3.5.4. Coated Microwell Aggregation Assay

The capacity of PLPs to aggregate can be evaluated using classic light transmittance aggregometry. In that case, purified PLPs must be resuspended in filtered platelet-poor plasma (PPP) at a concentration greater than 200×10^6 PLP/mL. This assay is complicated by the fact that it requires a very large quantity of PLPs. In addition, since an important proportion of PLPs are metabolically compromised, the aggregation response of CD42b$^+$ PLPs is often masked by the greater amount of debris. Hence, we designed an assay to evaluate the aggregation properties of CD41a$^+$CD42b$^+$

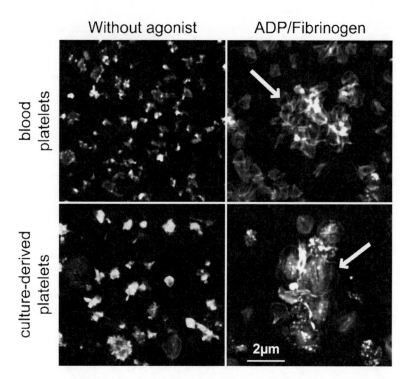

Fig. 7. Aggregation response of platelets in fibrinogen-coated microwells. Purified blood and culture-derived platelets bound to immobilized fibrinogen were incubated in the absence or presence of soluble fibrinogen (300 μg/mL) and ADP (20 μM) under vigorous shaking for 15 min. Fixed aggregates (*arrows*) were observed under fluorescence microscopy following phalloidin A488 staining. For more details, see Subheadings 3.5.3 and 3.5.4.

PLPs. The latter can be highly enriched by incubating the PLP mixture on fibrinogen-coated microwells. PLPs and cell debris that fail to adhere to fibrinogen are washed away before the induction of aggregation. Because fibrinogen coating causes platelets to spread, the aggregates look like an agglutination of flat platelets (see Fig. 7). To obtain more compact aggregates, poly-L-lysine-coated wells (10 μg/mL) may be used to enrich viable PLPs.

1. Coat the number of wells needed of a 96-well plate with 50 μL of 100 μg/mL fibrinogen solution in PBS for 1 h at 37°C or O/N at 4°C. Calculate one well per sample (PLPs or blood platelets) per condition (without agonist, with agonist 1, with agonist 2, …).

2. Discard fibrinogen and incubate the coated wells with 50 μL of 1% BSA (w/v) in PBS for 1 h at 37°C.

3. Wash once with PBS.

4. Add around 375,000 GPIbα⁺ isolated PLPs or 750,000 isolated blood platelets resuspended in culture medium in each treated well.

5. Centrifuge the plate for 5 min at $80 \times g$ to favor rapid contact between platelets and fibrinogen coating.

6. Wash once with PBS preheated at 37°C to remove nonadherent debris and dead platelets.

7. Remove PBS and add 50 μL of culture medium with or without the agonist. The addition of 20 μM ADP together with 300 μg/mL fibrinogen and 0.5 U/mL thrombin provides good reproducible results.

8. Incubate the plate at 37°C in a Titertek microplate shaker, speed 5 for 15 min.

9. Incubate the plate for another 10 min without shaking.

10. Gently wash the platelets with PBS and fix them with 50 μL of 3.7% formaldehyde (v/v) in PBS for 20 min at 37°C.

11. Rinse twice with PBS and proceed to actin filament staining using the immunofluorescence protocol (see Subheading 3.5.3), but skip step 8.

12. Observe platelet aggregates directly in the plate with an inverted fluorescence microscope using an LWD 40×, 0.55 NA objective.

4. Notes

1. CB was collected after informed consent from the mother and approved protocols from Héma-Québec and Québec City Saint-François d'Assise hospital ethical committees. MNCs were isolated within 12 h following their collection in aseptic conditions.

2. 2-ME is added in culture to prevent oxidation of LDL. Predilution of 2-ME is prepared following Sigma–Aldrich recommendations which specify that antioxidant properties of the compound are better preserved at pH 6 in the presence of EDTA.

3. Cytokine aliquots should not be submitted to more than one freeze/thaw cycle.

4. Due to the susceptibility of LDL to become oxidized, basal or cytokine-supplemented medium should be used within 7 days or fresh on the day that the cell cultures are diluted. The type and source of LDL have an important impact on the platelet production. Our MK culture protocol was developed with presolubilized LDL solution added to our medium.

5. For the one-phase culture protocol: Use the cytokine cocktail BS1 (26) for the entire length of culture, which gives the same

MK purity (>90% CD41a⁺ cells on day 14), but increased MK and platelet yields (approximately fivefold) when compared to TPO (10–100 ng/mL) alone.

6. For the two-phase culture protocol: Use the OMP cocktail from day 0 to day 4 (or up to day 6), and the BS1 cocktail from day 4 to day 14 (or from day 6 to day 14) of culture. This combination is suggested for the large-scale production of MKs and platelets. MK purity is reduced to 70%, but MK and platelet yields are usually improved. MK purity can be increased if cells are washed during the medium change for the second phase of culture at day 4 (or day 6).

7. It is important to keep the MK buffer (10× and 1×) sterile even though cytometry analysis does not require aseptic manipulations. Indeed, bacterial contaminated MK buffer results in numerous false platelet events (R2 region, Fig. 3) that perturb flow cytometry analysis. Moreover, MK buffer concentrated more than 10× crystallizes.

8. GM6001 is a broad-range metalloproteinase inhibitor added to avoid ectodomain shedding of receptors (such as GPV, GPVI, and GPIbα) on newly released platelets (7, 36). GM6001 is added at day 12 during the culture protocol and daily to purified platelets during storage.

9. As calcein AM solution is very sensitive to moisture, we recommend using the special packaging offered by Molecular Probe and the use of freshly diluted calcein AM for each experiment.

10. For cell counts and viability assessment, a 20 μL cell suspension sample is mixed with 20 μL of 0.4% trypan blue solution. Further dilution is made in order to count no more than 200 events in the hemacytometer.

11. The Héma-Québec Laboratory has previously demonstrated that CB cells cultured at 39°C experience an accelerated MK differentiation kinetics and improved MK and platelet yields (14), but the culture protocol described herein can also be performed at 37°C. Recent optimization of the culture period length at 39°C led to the discovery that MK and platelet yields are even greater if the cultures are switched from 39°C back to 37°C on day 4 (41).

12. The culture size depends principally on the quantity of cells desired or the number of different conditions needed to be tested. Thus, early large cell requirements necessitate larger culture volume or greater seeding cell density (cell expansion ratio is around fourfold at day 4), whereas analysis done after 10 days of culture requires fewer starting cells due to increasing cell expansion (expected final cell expansion ratio of 150–250- and 300–600-fold at day 10 and 14, respectively). For large-scale platelet production, start with 300,000 CD34⁺ cells

to obtain 150–250 mL of culture at day 14 with a PLP yield between 12 and 54 million.

13. This procedure is used to mimic shear stress to ensure proper dissociation of PLPs bound to MKs and to favor the release of PLPs from proplatelet filaments (2). This procedure improves PLP yields.

14. Platelet activation can be conveniently monitored by flow cytometry using an antibody recognizing the P-selectin CD62. This antigen is present in the alpha granules of platelets, and is exposed to their surface following activation of the platelets.

15. Routinely, all antibodies listed in Subheading 2.3 are diluted 1:40, with the exception of the anti-CD41a which is used at 1:200. We recommend titrating all antibody solutions.

16. For platelet analyses, centrifugation is omitted in order to avoid platelet activation. In our experience, no significant difference has been noticed between washed and unwashed stained cells in regards to nonspecific staining.

17. We recommend to resuspend isolated PLPs in BS1-supplemented medium since all of the functional assays described above were set up in this medium. However, PLPs could be resuspended in other physiologic buffers usually used for platelet storage as tyrode–Hepes buffer.

18. Although the cells are fixed with formaldehyde at this point, the flow cytometry analysis should be performed within 4 h after the sample treatment. After this time, a loss of the marker intensity is observed, which leads to an underestimation of the MK population.

19. If a humidified incubator at 30°C is not available, calcein AM incubation can be achieved at 37°C without any problems.

20. For sterilization, simply immerge coverslips in a 70% (v/v) ETOH solution for 30 min and then rinse twice with deionized sterile water.

21. For labeling of actin and/or nucleus, avoid steps 3–5. Then, ALEXA-conjugated phalloidin (dilution 1:50) and DAPI (dilution 1:1,000) can be used directly in PBS without the need for a blocking solution.

Acknowledgments

The authors wish to thank Lucie Boyer for helpful discussions and revision of this chapter. This work was supported in part by a Strategic Project grant from the Canadian Natural Science and Engineering Research Council (NSERC). A. Robert owns an Industrial R&D Fellowship from NSERC.

References

1. Avecilla, S. T., Hattori, K., Heissig, B., Tejada, R., Liao, F., Shido, K., Jin, D. K., Dias, S., Zhang, F., Hartman, T. E., Hackett, N. R., Crystal, R. G., Witte, L., Hicklin, D. J., Bohlen, P., Eaton, D., Lyden, D., de Sauvage, F., and Rafii, S. (2004) Chemokine-mediated interaction of hematopoietic progenitors with the bone marrow vascular niche is required for thrombopoiesis *Nat Med* **10**, 64–71.

2. Junt, T., Schulze, H., Chen, Z., Massberg, S., Goerge, T., Krueger, A., Wagner, D. D., Graf, T., Italiano, J. E., Jr., Shivdasani, R. A., and von Andrian, U. H. (2007) Dynamic visualization of thrombopoiesis within bone marrow *Science* **317**, 1767–70.

3. Dunois-Larde, C., Capron, C., Fichelson, S., Bauer, T., Cramer-Borde, E., and Baruch, D. (2009) Exposure of human megakaryocytes to high shear rates accelerates platelet production *Blood* **114**, 1875–83.

4. Reems, J. A., Pineault, N., and Sun, S. (2010) In vitro megakaryocyte production and platelet biogenesis: state of the art. *Transfus Med Rev* **24**, 33–43.

5. Gandhi, M. J., Drachman, J. G., Reems, J. A., Thorning, D., and Lannutti, B. J. (2005) A novel strategy for generating platelet-like fragments from megakaryocytic cell lines and human progenitor cells *Blood Cells Mol Dis* **35**, 70–3.

6. Fujimoto, T. T., Kohata, S., Suzuki, H., Miyazaki, H., and Fujimura, K. (2003) Production of functional platelets by differentiated embryonic stem (ES) cells in vitro *Blood* **102**, 4044–51.

7. Nishikii, H., Eto, K., Tamura, N., Hattori, K., Heissig, B., Kanaji, T., Sawaguchi, A., Goto, S., Ware, J., and Nakauchi, H. (2008) Metalloproteinase regulation improves in vitro generation of efficacious platelets from mouse embryonic stem cells *J Exp Med* **205**, 1917–27.

8. Takayama, N., Nishikii, H., Usui, J., Tsukui, H., Sawaguchi, A., Hiroyama, T., Eto, K., and Nakauchi, H. (2008) Generation of functional platelets from human embryonic stem cells in vitro via ES-sacs, VEGF-promoted structures that concentrate hematopoietic progenitors *Blood* **111**, 5298–306.

9. Choi, E. S., Nichol, J. L., Hokom, M. M., Hornkohl, A. C., and Hunt, P. (1995) Platelets generated in vitro from proplatelet-displaying human megakaryocytes are functional *Blood* **85**, 402–13.

10. Ungerer, M., Peluso, M., Gillitzer, A., Massberg, S., Heinzmann, U., Schulz, C., Munch, G., and Gawaz, M. (2004) Generation of functional culture-derived platelets from CD34⁺ progenitor cells to study transgenes in the platelet environment *Circ Res* **95**, e36–44.

11. Matsunaga, T., Tanaka, I., Kobune, M., Kawano, Y., Tanaka, M., Kuribayashi, K., Iyama, S., Sato, T., Sato, Y., Takimoto, R., Takayama, T., Kato, J., Ninomiya, T., Hamada, H., and Niitsu, Y. (2006) Ex vivo large-scale generation of human platelets from cord blood CD34⁺ cells *Stem Cells* **24**, 2877–87.

12. Boyer, L., Robert, A., Proulx, C., and Pineault, N. (2008) Increased production of megakaryocytes near purity from cord blood CD34⁺ cells using a short two-phase culture system *J Immunol Methods* **332**, 82–91.

13. Sullenbarger, B., Bahng, J. H., Gruner, R., Kotov, N., and Lasky, L. C. (2009) Prolonged continuous in vitro human platelet production using three-dimensional scaffolds *Exp Hematol* **37**, 101–10.

14. Proulx, C., Dupuis, N., St-Amour, I., Boyer, L., and Lemieux, R. (2004) Increased megakaryopoiesis in cultures of CD34-enriched cord blood cells maintained at 39°C *Biotechnol Bioeng* **88**, 675–80.

15. Schipper, L. F., Brand, A., Reniers, N. C., Melief, C. J., Willemze, R., and Fibbe, W. E. (1998) Effects of thrombopoietin on the proliferation and differentiation of primitive and mature haemopoietic progenitor cells in cord blood *Br J Haematol* **101**, 425–35.

16. Mattia, G., Vulcano, F., Milazzo, L., Barca, A., Macioce, G., Giampaolo, A., and Hassan, H. J. (2002) Different ploidy levels of megakaryocytes generated from peripheral or cord blood CD34⁺ cells are correlated with different levels of platelet release *Blood* **99**, 888–97.

17. Pang, L., Weiss, M. J., and Poncz, M. (2005) Megakaryocyte biology and related disorders *J Clin Invest* **115**, 3332–8.

18. Debili, N., Coulombel, L., Croisille, L., Katz, A., Guichard, J., Breton-Gorius, J., and Vainchenker, W. (1996) Characterization of a bipotent erythro-megakaryocytic progenitor in human bone marrow *Blood* **88**, 1284–96.

19. Italiano, J. E., Jr., Lecine, P., Shivdasani, R. A., and Hartwig, J. H. (1999) Blood platelets are assembled principally at the ends of proplatelet processes produced by differentiated megakaryocytes *J Cell Biol* **147**, 1299–312.

20. Clarke, M. C., Savill, J., Jones, D. B., Noble, B. S., and Brown, S. B. (2003) Compartmentalized megakaryocyte death generates functional platelets committed to

caspase-independent death *J Cell Biol* **160**, 577–87.

21. Kikuchi, J., Furukawa, Y., Iwase, S., Terui, Y., Nakamura, M., Kitagawa, S., Kitagawa, M., Komatsu, N., and Miura, Y. (1997) Polyploidization and functional maturation are two distinct processes during megakaryocytic differentiation: involvement of cyclin-dependent kinase inhibitor p21 in polyploidization *Blood* **89**, 3980–90.

22. Lordier, L., Jalil, A., Aurade, F., Larbret, F., Larghero, J., Debili, N., Vainchenker, W., and Chang, Y. (2008) Megakaryocyte endomitosis is a failure of late cytokinesis related to defects in the contractile ring and Rho/Rock signaling *Blood* **112**, 3164–74.

23. Schulze, H., Korpal, M., Hurov, J., Kim, S. W., Zhang, J., Cantley, L. C., Graf, T., and Shivdasani, R. A. (2006) Characterization of the megakaryocyte demarcation membrane system and its role in thrombopoiesis *Blood* **107**, 3868–75.

24. Richardson, J. L., Shivdasani, R. A., Boers, C., Hartwig, J. H., and Italiano, J. E., Jr. (2005) Mechanisms of organelle transport and capture along proplatelets during platelet production *Blood* **106**, 4066–75.

25. Italiano, J. E., Jr., Patel-Hett, S., and Hartwig, J. H. (2007) Mechanics of proplatelet elaboration *J Thromb Haemost* **5** Suppl 1, 18–23.

26. Cortin, V., Garnier, A., Pineault, N., Lemieux, R., Boyer, L., and Proulx, C. (2005) Efficient in vitro megakaryocyte maturation using cytokine cocktails optimized by statistical experimental design *Exp Hematol* **33**, 1182–91.

27. De Bruyn, C., Delforge, A., Martiat, P., and Bron, D. (2005) Ex vivo expansion of megakaryocyte progenitor cells: cord blood versus mobilized peripheral blood *Stem Cells Dev* **14**, 415–24.

28. Bruno, S., Gunetti, M., Gammaitoni, L., Dane, A., Cavalloni, G., Sanavio, F., Fagioli, F., Aglietta, M., and Piacibello, W. (2003) In vitro and in vivo megakaryocyte differentiation of fresh and ex-vivo expanded cord blood cells: rapid and transient megakaryocyte reconstitution *Haematologica* **88**, 379–87.

29. Williams, J. L., Pipia, G. G., Datta, N. S., and Long, M. W. (1998) Thrombopoietin requires additional megakaryocyte-active cytokines for optimal ex vivo expansion of megakaryocyte precursor cells *Blood* **91**, 4118–26.

30. Shaw, P. H., Gilligan, D., Wang, X. M., Thall, P. F., and Corey, S. J. (2003) Ex vivo expansion of megakaryocyte precursors from umbilical cord blood CD34 cells in a closed liquid culture system *Biol Blood Marrow Transplant* **9**, 151–6.

31. Dolzhanskiy, A., Basch, R. S., and Karpatkin, S. (1997) The development of human megakaryocytes: III. Development of mature megakaryocytes from highly purified committed progenitors in synthetic culture media and inhibition of thrombopoietin-induced polyploidization by interleukin-3 *Blood* **89**, 426–34.

32. Tajika, K., Ikebuchi, K., Inokuchi, K., Hasegawa, S., Dan, K., Sekiguchi, S., Nakahata, T., and Asano, S. (1998) IL-6 and SCF exert different effects on megakaryocyte maturation *Br J Haematol* **100**, 105–11.

33. Proulx, C., Boyer, L., Hurnanen, D. R., and Lemieux, R. (2003) Preferential ex vivo expansion of megakaryocytes from human cord blood CD34$^+$-enriched cells in the presence of thrombopoietin and limiting amounts of stem cell factor and Flt-3 ligand *J Hematother Stem Cell Res* **12**, 179–88.

34. Fox, N. E., and Kaushansky, K. (2005) Engagement of integrin $\alpha 4\beta 1$ enhances thrombopoietin-induced megakaryopoiesis *Exp Hematol* **33**, 94–9.

35. Bertolini, F., Battaglia, M., Pedrazzoli, P., Da Prada, G. A., Lanza, A., Soligo, D., Caneva, L., Sarina, B., Murphy, S., Thomas, T., and della Cuna, G. R. (1997) Megakaryocytic progenitors can be generated ex vivo and safely administered to autologous peripheral blood progenitor cell transplant recipients *Blood* **89**, 2679–88.

36. Robert A, B. L., Pineault N (2010) *Stem Cells Dev* (TITLE TO BE CONFIRMED). Submitted publication.

37. Harrison, P., Briggs, C., and Machin, S. J. (2004) Platelet counting. *Methods Mol Biol* 2004 **272**, 29–46.

38. Debili, N., Louache, F., and Vainchenker, W. (2004) Isolation and culture of megakaryocyte precursors *Methods Mol Biol* **272**, 293–308.

39. Cazenave, J.-P., Ohlmann, P., Cassel, D., Eckly, A., Hechler, B., and Gachet, C. (2004) Preparation of washed platelet suspensions from human and rodent blood *Methods Mol Biol* **272**, 13–28.

40. Walsh, P. N. (1972) Platelet washing by albumin density gradient separation (ADGS) *Adv Exp Med Biol* **34**, 245–56.

41. Pineault, N., Boucher, J. F., Cayer, M. P., Palmqvist, L., Boyer, L., Lemieux, R., and Proulx, C. (2008) Characterization of the effects and potential mechanisms leading to increased megakaryocytic differentiation under mild hyperthermia *Stem Cells Dev* **17**, 483–93.

42. Cortin, V., Pineault, N., and Garnier, A. (2009) Ex vivo megakaryocyte expansion and platelet production from human cord blood stem cells *Methods Mol Biol* **482**, 109–26.

Chapter 17

Culture of Megakaryocytes and Platelets from Subcutaneous Adipose Tissue and a Preadipocyte Cell Line

Yumiko Matsubara, Mitsuru Murata, and Yasuo Ikeda

Abstract

The molecular mechanisms whereby stem cells develop into platelet-producing megakaryocytes (MKs) are not yet fully understood. Within this chapter we describe a two-step in vitro culture system in which MKs and platelets are generated from primary subcutaneous adipose tissues and the preadipocyte cell line 3T3L1. The cells are first cultured in an adipocyte induction medium for 10–12 days, followed by 8–14 days culture in a MK differentiation medium. Adipose tissue-derived MKs and platelets display a number of morphological and functional characteristics (e.g., secretory granules, open canalicular membranes) comparable with the native cell type. The use of subcutaneous adipose tissue to produce a large number of platelets is advantageous because this tissue is easily obtained and available in large quantities. Thus, this in vitro culture system may prove useful in both regenerative medicine, but it may also be used in understanding fundamental research questions within MK and platelet research, including further elucidation of the pathways that cause cells to differentiate along the MK lineage ultimately leading to platelet production.

Key words: Megakaryopoiesis, Thrombopoiesis, Stem cell, Preadipocyte, Adipose tissue

1. Introduction

Adipose tissue, particularly subcutaneous adipose tissue, has recently been highlighted as a source of preadipocytes, some with very restricted potential and others with multipotency for use in tissue engineering and regenerative medicine (1, 2). Experimental studies show that adipocytes, chondrocytes, and osteoblasts can differentiate from subcutaneous adipose tissues (1, 3). Although these cell types are derived from mesenchymal stem cells distinct from hematopoietic stem cells (4), we obtained megakaryocytes (MKs) and platelets from normal human subcutaneous adipose tissues in an in vitro culture system (5). The adherent cells from

Jonathan M. Gibbins and Martyn P. Mahaut-Smith (eds.), *Platelets and Megakaryocytes: Volume 3, Additional Protocols and Perspectives*, Methods in Molecular Biology, vol. 788, DOI 10.1007/978-1-61779-307-3_17, © Springer Science+Business Media, LLC 2012

adipose tissues are first cultured in adipocyte induction medium for 12 days, and these cells are then cultured in MK lineage induction medium for 12 days. Although in vitro MK differentiation from fresh murine adipose tissues can be observed without preculture, pilot studies showed that a two-step culture produces larger numbers of MKs and platelets from human adipose tissues. The molecular mechanism responsible for this effect remains unclear, although it may relate to the fact that adipocytes secrete various cytokines and growth factors with roles in megakaryopoiesis and thrombopoiesis (6). The mouse preadipocyte cell line 3T3L1, originally derived from mouse fibroblasts, was also found to differentiate into MK lineages in an in vitro culture system (7). Characterization of MK lineage cells derived from adipocyte precursor cells was examined by lineage-specific surface markers, DNA ploidy, morphological analysis, and platelet functional assays (5). This culture system may prove useful in studies of megakaryopoiesis and thrombopoiesis, especially considering the greater availability of adipose tissue compared to other tissue samples.

2. Materials

2.1. Cell Culture for Differentiation into Adipocytes

1. Human Subcutaneous Preadipocytes, Lonza, Basel, Switzerland (see Note 1). The mouse preadipocyte cell line 3T3L1, originally derived from mouse fibroblasts (8, 9), is from the Health Science Research Resources Bank (Osaka, Japan).

2. Cell maintenance medium: Preadipocyte Growth Media-2 Basal Medium (PGM-2) with 10% fetal bovine serum (SAFC Bioscience, Tokyo, Japan), 2 mM L-glutamine, and 37 ng/mL GA-1000 (Lonza, Basel, Switzerland). Dulbecco's modified Eagle's medium (DMEM; Invitrogen, Carlsland, CA) can be used in place of the PGM-2.

3. Dexamethasone (Sigma, St. Louis, MO): stock 500 μM in 100% ethanol. This is mixed with DMEM at a ratio of 1:4 just before use.

4. Indomethacin (Sigma): stock 1 M in 100% ethanol, made just before use.

5. 3-Isobutyl-1-methylxanthin (IBMX; Sigma): stock 1 M. Dissolve in 100% dimethylsulfoxide (DMSO) with gentle warming, and store in single-use aliquots at −20°C.

6. Insulin (sodium salt, human) (Sigma).

7. Adipocyte induction medium: Cell maintenance medium (item 2), supplemented with 1.6 μM insulin, 1 μM dexamethasone, 0.5 mM IBMX, and 0.2 mM indomethacin.

8. Trypsin (0.25%) and ethylenediamine tetraacetic acid (EDTA; 1 mM) solutions (Invitrogen).

9. Six-well plates.

2.2. Cell Culture for Differentiation into MKs and Platelets

MK lineage induction medium (a modification of that described by Zauli et al (10) and Kerrigan et al (11)): Iscove's modified Dulbecco's medium (Invitrogen) supplemented with 2 mM L-glutamine (Invitrogen), 100 U/mL penicillin G sodium (Invitrogen), 0.1 mg/mL streptomycin sulfate (Invitrogen), 0.5% bovine serum albumin (Sigma), 4 µg/mL LDL cholesterol (Sigma), 200 µg/mL iron-saturated transferrin (Sigma), 10 µg/mL insulin (Sigma), 50 µM 2-β-mercaptoethanol (Invitrogen), 20 µM each nucleotide (ATP, UTP, GTP, and CTP) (Invitrogen), and 50 ng/mL thrombopoietin (TPO; a gift from KIRIN Brewery Ltd., Tokyo, Japan).

2.3. Oil Red-O-Staining

1. Oil Red-O (3 mg/mL; Sigma) is dissolved in isopropanol (Wako, Osaka, Japan). Just before use, a 60% solution is prepared by dilution in water, and filter (0.45 µm)-sterilized. This staining solution is stored in a light-resistant tube at room temperature until required for the experiment.

2. PBS without divalent cations (Sigma).

3. Ten percent formalin (Wako).

2.4. Flow Cytometry Analysis

1. Antihuman CD41 antibody and antihuman CD42b antibody (Beckman Coulter, Fullerton, CA).

2. Antimouse CD41 antibody (BD Bioscience, San Jose, CA). Antimouse CD42b antibody (EMFRET Analytics Gmbh and Co., KG, Germany).

3. Propidium iodide (PI, 50 µg/mL; Sigma) is dissolved in sterile water, and stored in a light-resistant container at 4°C.

4. RNAase: RNase A is from QIAGEN (Valencia, CA).

5. Buffer: EDTA (5 mM; Invitrogen) in PBS (Sigma). Store at room temperature.

2.5. Fibrinogen Binding and P-Selectin Surface Expression

1. Divalent cation-free HEPES–Tyrode's buffer, pH 7.4: 140 mM NaCl, 0.4 mM NaH_2PO_4, 2.7 mM KCl, 10 mM $NaHCO_3$, 10 mM HEPES, and 5 mM glucose, titrated to pH 7.4 with NaOH.

2. $CaCl_2$ (1.5 M) is dissolved in sterile water. $MgCl_2$ (2 M) is prepared in sterile water.

3. Human thrombin (25 U/mL: Sigma) is dissolved in sterile water and stored in single use aliquots at −80°C.

4. Agonists: Adenosine 5′-diphosphate (ADP, 1 mM; Biopool, Ventura, CA) is dissolved in sterile water. Epinephrine (100 µM; Daiichi-Sankyo, Tokyo, Japan) is dissolved in sterile water.

Protease-activated receptor 4 (PAR4)-activating peptide (250 mM; Sigma) is dissolved in sterile water. Protease-activated receptor 1 (PAR1)-activating peptide (10 mM; Sigma) is dissolved in sterile water. These are stored in single-use aliquots at −20°C.

5. A 1.5 mg/mL stock solution of Alexa Fluor 488-labeled fibrinogen (Invitrogen) is prepared by reconstituting the conjugate in 0.1 M sodium bicarbonate (pH 8.3, adjustment not needed), and stored in single-use aliquots at −20°C.

6. Fluorescein isothiocyanate (FITC)-conjugated antihuman P-selectin antibody (BD Bioscience).

2.6. Electron Microscopy

1. 2% glutaraldehyde (Electron Microscopy Sciences, Hatfield, PA) is prepared in 0.1 M phosphate buffer (pH 7.4).

2. 4% osmium tetroxide (EM Sciences) is dissolved in distilled water and stored at 4°C. 1% osmium tetroxide is prepared in 0.1 M phosphate buffer (pH 7.4) at the time of use.

3. Epon is prepared by mixing Epon 812, DDSA, MNA and DMP-30 (TAAB Laboratories, Berkshire, UK).

4. 2% uranyl acetate (TAAB) is dissolved in distilled water and stored at 4°C. 0.1% lead citrate (TAAB) is dissolved in 0.1 M NaOH at the time of use.

3. Methods

3.1. Differentiation from Human Subcutaneous Adipose Tissue into MK Lineages In Vitro

To obtain MK lineage cells from adipose tissues, adherent cells (see Note 1) from adipose tissues are initially cultured in adipocyte induction medium and then in MK lineage induction medium (Fig. 1).

1. Culture primary human subcutaneous adipose tissue (see Notes 1 and 2) in cell maintenance medium (item 2, Subheading 2.1) in a six-well plate. Passage every 4 days using trypsin/EDTA.

2. At passage 2–3, grow the adherent cells at 1.5×10^5 cells/well in a six-well plate with 2 mL/well of adipocyte induction medium.

3. By day 12, approximately 80% of cells will have differentiated into mature adipocytes based on their morphology with lipid droplets and positive Oil Red-O-staining (see Subheading 3.3).

4. The cell population on day 12 (see Notes 3 and 4) will have mostly adherent cells, but also several floating cells. Collect both floating and adherent cells and resuspend (1.5×10^5 cells/

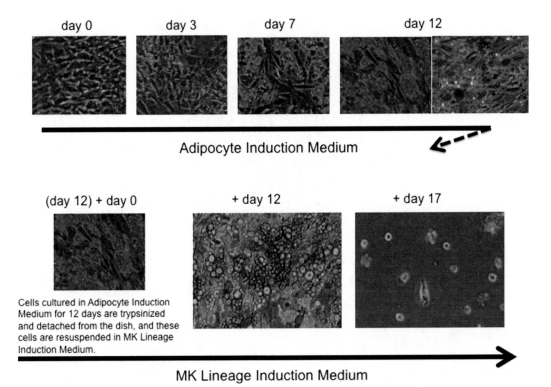

day 0   day 3   day 7   day 12

Adipocyte Induction Medium

(day 12) + day 0   + day 12   + day 17

Cells cultured in Adipocyte Induction
Medium for 12 days are trypsinized
and detached from the dish, and these
cells are resuspended in MK Lineage
Induction Medium.

MK Lineage Induction Medium

Fig. 1. Generation of megakaryocytes and platelets from adipose tissues in vitro. A schematic outline and images of the adherent cells (see Note 1) derived from adipose tissues and at different stages of differentiation into adipocytes and then megakaryocyte (MK) lineages. To obtain MKs and platelets from adipose tissues, the adherent cells from adipose tissues initially differentiated into adipocytes. The cell population comprising approximately 80% of cells is positive for Oil Red-O-staining positive (right panel of day 12) and is used to further differentiate into MK lineages. In the lower panel, loosely attached cells on +day 12 and floating cells on +days 12 and 17 morphologically resemble MKs.

well of a six-well plate) in 2 mL/well MK lineage induction medium. Of the adipose tissue-derived cells cultured in MK lineage induction medium for 8–12 days, the floating and loosely adherent cells will morphologically resemble MKs. Under electron microscopic observation (see Subheading 3.6), adipose tissue-derived large sized cells have organelles typical of native MKs, such as granules, demarcation membranes, and lobulated nuclei, and adipose tissue-derived small-sized cells show features typical of platelets, such as granules, mitochondria, and an open canalicular system (5).

5. As described below (Subheading 3.4), the surface marker expressions of the MKs and platelets and the DNA ploidy of the MKs are analyzed by flow cytometry on days 8–12 after differentiation into the MK lineage. Also, platelet functional assays (Subheading 3.5) and electron microscopy study (Subheading 3.6) (Fig. 2) are performed.

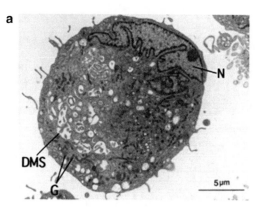

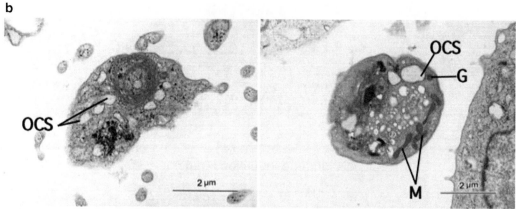

Fig. 2. (**a**, **b**) The morphology of (**a**) megakaryocytes and (**b**) platelets derived from adipose tissues, analyzed using transmission electron microscopy. (**a**) In adipose tissue-derived megakaryocytes, typical organelles such as granules (G), demarcation membrane system (DMS), and lobulated nuclei (N) were evident. (**b**) Adipose tissue-derived platelets contained granules (G), mitochondria (M), and open canalicular system (OCS).

3.2. Differentiation from 3T3L1 Cell Line into MK Lineages In Vitro

1. Maintain 3T3L1 cells in DMEM with 10% fetal bovine serum.

2. To differentiate the 3T3L1 cells into adipocytes, culture the cells in adipocyte induction medium (DMEM with 10% fetal bovine serum, supplemented with 1.6 μM insulin, 1 μM dexamethasone, 0.5 mM IBMX, and 0.2 mM indomethacin) for 10 days.

3. Apply the Oil Red-O-staining protocol (Subheading 3.3) on these cells.

4. To differentiate the 3T3L1 cells into the MK lineage, the confluent cells in a six-well plate are trypsinized and detached from the plate, and transferred to a six-well plate. The cells are cultured in MK lineage induction medium for 12–14 days. By day 3, loosely adherent and floating cells will form large spheres (see Note 5). Cells within floating spheres will begin to separate into clusters of individual cells during culture.

3.3. Oil Red-O-Staining

1. Wash adherent cells twice with PBS, and fix in 10% formalin for 15 min at 37°C.

2. Wash the cells twice with PBS and soak in 60% isopropanol for 1 min at room temperature.

3. Immerse in the Oil Red-O- staining solution for 15 min at 37°C, wash with 60% isopropanol, and place in PBS. Oil Red-O-positive cells (adipocytes) are determined by microscopic examination (see Note 6).

3.4. Flow Cytometry Analysis

1. Cells are collected and centrifuged at $700 \times g$ for 5 min at room temperature. The cell pellets are resuspended in 5 mM EDTA in PBS and incubated with antibodies for CD41 (diluted 1:10), a surface marker that is present throughout MK differentiation (12), or CD42b (diluted 1:10), a surface marker for the late stage of MK differentiation (12), for 60 min at room temperature and diluted using PBS without divalent cations. An isotype control is used in flow cytometry analysis. These cell samples are used for flow cytometry analysis (5) (see Note 7).

2. DNA ploidy (13) in adipose tissue- and 3T3L1-derived cells on days 8–10 is also examined by flow cytometry as follows. After staining with FITC conjugated anti-CD41 antibody incubate with a 50-fold volume of methanol for 30 min on ice. Wash the cells twice with PBS, resuspend in PBS containing 20 μg/mL RNAase, and incubate for 60 min at 37°C. Incubate the cell samples with PI (50 μg/mL) for 30 min on ice and dilute with PBS for analysis by flow cytometry (5) (see Note 7).

3.4.1. Fibrinogen Binding and P-Selectin Surface Expression

Platelets are unique in that they are activated by certain agonists resulting in interaction between fibrinogen and its receptor, CD41/CD61 (also known as glycoprotein IIb/IIIa or the integrin $\alpha_{IIb}\beta_3$) (14).

1. Collect cells and centrifuge at $150 \times g$ for 15 min at room temperature. The supernatant containing platelet-sized cells is then further centrifuged at $900 \times g$ for 7 min at room temperature.

2. Resuspend the pellet from the supernatant in divalent cation-free HEPES–Tyrode's buffer. Proceed to functional assays in step 3 (fibrinogen binding) or step 4 (P-selectin binding).

3. Stimulate the adipose tissue-derived small-sized cells with 0.5 U/mL thrombin or 10 μM ADP, 10 μM epinephrine, and 50 μM PAR1-activating peptide in divalent cation-free HEPES–Tyrode's buffer plus 1.5 mM $CaCl_2$ and 2 mM $MgCl_2$ in the presence of 100 μg/mL Alexa Fluor 488-labeled fibrinogen for 5 min at room temperature. Alexa Fluor 488-labeled fibrinogen (100 μg/mL) binding to platelet-sized small cells derived from 3T3L1 cells is also performed upon stimulation with 0.5 U/mL thrombin or 10 μM ADP and 1 mM

PAR4-activating peptide in divalent cation-free HEPES–Tyrode's buffer plus 1.5 mM $CaCl_2$ and 2 mM $MgCl_2$ for 5 min at room temperature. The cell samples are diluted with divalent cation-free HEPES–Tyrode's buffer plus 1.5 mM $CaCl_2$ and 2 mM $MgCl_2$ and analyzed by flow cytometry (5).

4. Incubate with anti-P-selectin antibody in the presence of 0.5 U/mL thrombin or 10 µM ADP, 10 µM epinephrine, and 50 µM PAR1-activating peptide in divalent cation-free HEPES–Tyrode's buffer plus 1.5 mM $CaCl_2$ and 2 mM $MgCl_2$ for 30 min at room temperature. Dilute with divalent cation-free HEPES–Tyrode's buffer plus 1.5 mM $CaCl_2$ and 2 mM $MgCl_2$ and analyze by flow cytometry (5).

3.4.2. Electron Microscopy

1. Fix the cultured cells by the addition of an equal volume of 2% glutaraldehyde in 0.1 M phosphate buffer (pH 7.4), and fix at 4°C for 60 min.

2. Centrifuge the cells at $700 \times g$ for 5 min at 4°C. Dissect the pellets into blocks of 1-mm diameter cubes, wash with the phosphate buffer five times, and then postfix with 1% osmium tetroxide in the same buffer at 4°C for 60 min.

3. Dehydrate the fixed cells with a graded ethanol series, embed in Epon, and then polymerize at 60°C for 48 h.

4. Prepare ultrathin sections using an ultramicrotome (Ultracut, Leica Microsystems, Wetzlar, Germany) and mount on copper grids.

5. Sections are electron-stained with 2% uranyl acetate for 5 min, washed with distilled water and then stained with 0.1% lead citrate for 5 min.

6. The electron-stained sections are examined with a JEM 1200EX transmission electron microscope (JEOL, Tokyo, Japan) at an accelerating voltage of 80 kV.

4. Notes

1. The data sheet of the primary human subcutaneous preadipocytes (Lonza) indicates that this product is isolated from adipose tissues by enzymatic digestion and selective culturing techniques. Adipose tissues contain various cell types, most of which are preadipocytes and mesenchymal stem cells. In the culture system described, the adherent cells within the Lonza product are used, although whether these cells are preadipocytes, mesenchymal stem cells, or adipocyte-derived stem cells is controversial.

2. If mouse subcutaneous adipose tissues are used as the starting material in this culture system, these tissues are digested with collagenase type I (Wako, final concentration, 1 mg/mL) for 60 min at 37°C. The digested cells are washed twice using PBS without divalent cations (Sigma) and resuspended in DMEM with 10% fetal bovine serum (item 2, Subheading 2.1). To obtain the MK lineage cells from these cells, low (2–4) passage number cells are trypsinized and then cultured in the MK lineage induction medium (item 1, Subheading 2.2) for 8–14 days.

3. Primary human subcutaneous preadipocytes (Lonza) show substantial lot-to-lot variation in their ability to differentiate into mature adipocytes. The timing of medium change from adipocyte induction medium to MK lineage induction medium is determined based on the morphologic features, such as the lipid droplets.

4. The experimental conditions described allow for the production of a large number of MKs and platelets, as assessed by flow cytometric analysis with anti-CD41 antibody: adipocyte precursor cells are initially cultured in adipocyte induction medium for 12 days to differentiate into mature adipocytes, followed by another 12 days of culture in MK lineage induction medium to differentiate into MK lineage. In a pilot study using a small amount of adipocyte precursor cells (approximately 10^5 cells), they were first cultured in adipocyte induction medium to differentiate into mature adipocytes. After the designated number of days of culture, i.e., days 0, 6, 9, and 12, subsequent culture in MK lineage induction media for 12 days was performed, and the number of MKs that were CD41-positive large-sized cells was assessed. The number of MKs, i.e., CD41-positive cells – isotype-reactive cells, was negligible, 3500, 7900, and 9500, in the day 0, 6, 9, and 12 samples, respectively (5).

5. The morphological features were more easily observed during differentiation from the 3T3L1 cell line into MK lineages than during differentiation from primary adipose tissues.

6. Oil Red-O-staining stains triglycerides and lipids and is therefore used as a marker of mature adipocytes. After 12 days in adipocyte induction medium, approximately 80% of cells should be Oil-red-O positive.

7. On days 8–10, approximately 70% of the loosely adherent and approximately 30% of the floating cells morphologically resemble MKs and express CD41 and CD41/CD42b. Among the cells expressing CD41, approximately 20% were dead cells, assessed by PI-staining. The DNA ploidy of 3T3L1-derived CD41+ cells ranges from 2 to 16 N.

Acknowledgments

The authors thank Dr. Hidenori Suzuki, Center for Electron Microscopy, Tokyo Metropolitan Institute of Medical Science, for providing details of the protocol used in the electron microscopy studies and Mr. Akira Sonoda, Central Research Laboratory, School of Medicine, Keio University, for providing technical advice.

References

1. Gomillion CT, Burg KJ (2006) Stem cells and adipose tissue engineering. Biomaterials 27: 6052–6063
2. Casteilla L, Dani C (2006) Adipose tissue-derived cells: from physiology to regenerative medicine. Diabetes Metab 32: 393–401
3. Gimble JM, Katz AJ, Bunnell BA (2007) Adipose-derived stem cells for regenerative medicine. Circ Res 100: 1249–1260
4. Battiwalla M, Hematti P (2009) Mesenchymal stem cells in hematopoietic stem cell transplantation. Cytotherapy 11: 503–515
5. Matsubara Y, Saito E, Suzuki H, Watanabe N, Murata M, Ikeda Y (2009) Generation of megakaryocytes and platelets from human subcutaneous adipose tissues. Biochem Biophys Res Commun 378: 716–720
6. Trayhurn P (2005). Endocrine and signalling role of adipose tissue: new perspectives on fat. Acta Physiol Scand 184: 285–293
7. Matsubara Y, Suzuki H, Ikeda Y, Murata M (2010) Generation of megakaryocytes and platelets from preadipocyte cell line 3T3-L1, but not the parent cell line 3T3, in vitro. Biochem Biophys Res Commun 402: 796–800
8. Mackall JC, Student AK, Polakis SE, Lane MD (1976) Induction of lipogenesis during differ-entiation in a preadipocyte cell line. J Biol Chem 251: 6462–6464
9. Cowherd RM, Lyle RE, McGehee RE Jr (1999) Molecular regulation of adipocyte differentiation. Semin Cell Dev Biol 1: 3–10
10. Zauli G, Bassini A, Vitale M, Gibellini D, Celeghini C, Caramelli E, Pierpaoli S, Guidotti L, Capitani S (1997) Thrombopoietin enhances the alpha IIb beta 3-dependent adhesion of megakaryocytic cells to fibrinogen or fibronectin through PI 3 kinase. Blood 89: 883–895
11. Kerrigan SW, Gaur M, Murphy RP, Shattil SJ, Leavitt AD (2004) Caspase-12: a developmental link between G-protein-coupled receptors and integrin alphaIIbbeta3 activation. Blood 104: 1327–1334
12. Tomer A (2004) Human marrow megakaryocyte differentiation: multiparameter correlative analysis identifies von Willebrand factor as a sensitive and distinctive marker for early (2 N and 4 N) megakaryocytes. Blood 104: 2722–2727
13. Chang Y, Bluteau D, Debili N, Vainchenker W (2007) From hematopoietic stem cells to platelets. J Thromb Haemost 5: 318–327
14. Coller BS, Shattil SJ (2008) The GPIIb/IIIa (integrin alpha IIb beta3) odyssey: a technology-driven saga of a receptor with twists, turns, and even a bend. Blood 112: 3011–3025

Chapter 18

Purification of Native Bone Marrow Megakaryocytes for Studies of Gene Expression

Gwen Tolhurst, Richard N. Carter, Nigel Miller, and Martyn P. Mahaut-Smith

Abstract

Megakaryocytes constitute less than 1% of all marrow cells, therefore purification of these giant platelet precursor cells represents a challenge. We describe two methods to ultra-purify mature megakaryocytes from murine marrow for the purpose of extracting RNA suitable for studies of gene expression. In the first approach, unit velocity gradients are used to enrich for megakaryocytes, which are then selected by fluorescence-activated cell sorting based upon size and high surface expression of CD41. In the second method, individual megakaryocytes, identified by their distinct morphology, are extracted using glass suction pipettes. Despite the small numbers of cells that can be isolated via the latter technique, recent studies have demonstrated how this pure population can be used to detect mRNA transcripts encoding ion channels and other proteins in the native megakaryocyte.

Key words: Megakaryocyte RNA, Purification, PCR, Single-cell PCR, Ion channel

1. Introduction

The megakaryocyte is a highly specialised cell designed to manufacture blood platelets (1–3). It has a number of unique features, including a highly invaginated plasma membrane (the demarcation membrane system, DMS) that serves to increase the amount of surface membrane, and a polyploidic nucleus that supports increased gene expression in these large cells. In addition, the megakaryocyte develops secretory granules, surface receptors, and signalling mechanisms that are required for function in the anuclear platelet. Although culture systems have been described that yield large cells

Jonathan M. Gibbins and Martyn P. Mahaut-Smith (eds.), *Platelets and Megakaryocytes:*
Volume 3, Additional Protocols and Perspectives, Methods in Molecular Biology, vol. 788,
DOI 10.1007/978-1-61779-307-3_18, © Springer Science+Business Media, LLC 2012

capable of generating platelet-like bodies, the megakaryocytes that develop in vitro may differ from those that grow in vivo within the complex microenvironment of the marrow. Thus, the ability to screen molecular messages in megakaryocytes from marrow (or spleen in the case of some animals which retain this as an adult organ of thrombopoiesis) represents a useful tool in the arsenal of techniques to study megakaryopoiesis. In addition, the megakaryocyte is often used as a surrogate for investigating platelet signalling since the precursor cell must express most, if not all proteins of its anuclear progeny (4, 5). In fact the megakaryocyte provides distinct advantages over the platelet in this regard; for example, the large size of the megakaryocyte facilitates application of the patch clamp technique for studies of ion channels and membrane potential, and for introduction of large molecules through the whole-cell configuration (6–9). Furthermore, there are concerns over the integrity of mRNA in the anuclear platelet, and therefore screening by RT-PCR (reverse transcription-polymerase chain reaction) in the late stage megakaryocyte may be more indicative of the proteins manufactured for platelet function.

Mature megakaryocytes constitute <1% of cells within the marrow, therefore their purification represents a substantial challenge. RT-PCR of material extracted from single cells during whole-cell patch clamp has been reported for several cell types (10–12). In addition, laser capture microdissection can be used to extract identified cells from whole tissues (13, 14). However, since marrow is easily dispersed into a suspension of cells without enzymatic treatment and megakaryocytes can be easily identified by their large size or unique surface markers, patch clamp/laser microdissection offer no great advantage. This chapter describes two approaches we have developed to purify intact megakaryocytes from murine marrow. Firstly, the combination of a unit velocity density gradient and fluorescence-activated cell sorting (FACS) of CD41+ cells allows separation of a highly purified population of thousands of megakaryocytes from the femoral/tibial marrow of the mouse. Secondly, glass pipettes with openings slightly larger than a conventional patch pipette can be used to isolate individual megakaryocytes. In the latter approach, some patience and dedication can extract >50 megakaryocytes on a good day. This may not seem a large number of cells, however, when we first started this approach, we were amazed that standard RT-PCR could detect mRNA transcripts for ion channels using material from only a few megakaryocytes (15). This is probably a consequence of the polyploidic nucleus, which greatly increases the level of gene expression (16) to allow megakaryocytes to fulfil their role of replacing approximately one-tenth of the entire platelet population every day.

2. Materials

2.1. Purification of Megakaryocytes by Unit Velocity Albumin Gradients and Fluorescence-Activated Cell Sorting

The protocol using unit velocity albumin gradients is based upon the methods of Leven (17), whose chapter should be consulted for a detailed treatise of megakaryocyte separation based upon density and size.

2.1.1. Extraction of Marrow

1. 10× calcium- and magnesium-free Hank's salt solution (CMFH): (in mM): 53.7 KCl, 4.4 KH_2PO_4, 1370 NaCl, 41.7 Na_2CO_3, 3.38 Na_2HPO_4, and 55.6 glucose, pH 7.4 (NaOH) (see Note 1).
2. CATCH buffer: 1× CMFH with 12.9 mM sodium citrate, 1.4 mM adenosine, 2.7 mM theophylline, pH 7.4 (NaOH).
3. Bovine serum albumin (BSA, standard lyophilised powder).
4. Apyrase (type VII): stock 320 U/ml in sterile water.
5. PGE_1: stock 1 mg/ml in ethanol.
6. Total CATCH medium: CATCH buffer with 35 mg/ml BSA, 1 μl/ml apyrase stock and 1 μl/ml, PGE_1 stock (see Note 2).
7. Mice (two animals are normally sufficient for the procedure); typically C57/Bl6 mice aged between 14 and 16 weeks (see Note 3).
8. Sterile dissection instruments.
9. Sterile microfuge tubes and 15 ml centrifuge tubes.
10. 5 ml plastic sterile syringes.
11. 2 mm bore silicone tubing, cut into short (approximately 2 cm) lengths; the bore may have to be different depending upon the diameter of bones used. Sterilise by flushing with 0.1 M NaOH, 1 mM EDTA followed by DEPC-treated water (see Note 4) and then sterile RNase- and DNase-free PBS.
12. Sterile 35 mm petri dishes.
13. Cell rotator capable of gently mixing cells in tubes at a rate of approximately 0.2 Hz.
14. 100 μm pore nylon mesh (Cadisch Precision Meshes Ltd, London, UK).
15. A low-speed centrifuge capable of holding 15 ml tubes.

2.1.2. Unit Velocity Gradient Centrifugation to Enrich for Megakaryocytes

1. BSA stock solution: 200 ml water, 15.5 ml 10× CMFH, 58 g BSA, 97.2 mg adenosine, and 128.4 mg theophylline, pH 7.4 with NaOH (see Note 2).
2. CATCH buffer (see Subheading 2.1.1, item 2).

3. 0.22 μm filters.

4. Sterile 15 ml centrifuge tubes.

2.1.3. Purification of Megakaryocytes by Fluorescence-Activated Cell Sorting

1. CATCH buffer (see Subheading 2.1.1, item 2) containing 2 mM Na_4EGTA, 0.32 U/ml apyrase VII, and 1 μg/ml PGE_1.

2. FITC-conjugated rat anti-mouse CD41 (BD Pharmingen).

3. Pre-prepared RNA lysis buffer containing 4 ng/μl Poly-A-carrier RNA (see Note 5).

4. Sterile 1.5 ml eppendorf tubes.

5. Tin foil.

6. 75 μm pore filter (Dako UK Ltd, UK).

7. A Beckman Coulter 3 laser MoFlo flow cytometer (Dako UK Ltd, Ely, UK) or equivalent, capable of sorting large diameter cells at low pressure and equipped with a laser capable of exciting FITC at 488 nm with an appropriate emission filter, e.g., 530/30 nm.

8. Fluorescence microscope with appropriate lamp/laser and filters to detect FITC-labelled cells (used to validate the cell sorting by FACS).

9. Cell rotator capable of gently mixing cells at approximately 0.2 Hz.

2.2. Extraction of Individual Megakaryocytes Using Glass Pipettes

1. Dissection and tissue preparation items (7–13, Subheading 2.1.1).

2. Apyrase (type VII, Sigma-Aldrich): stock 320 U/ml in sterile water.

3. Pre-prepared RNA lysis buffer containing poly-A-carrier RNA, kept on ice (see Note 5).

4. RNase- and DNase-free phosphate-buffered saline (PBS).

5. Solutions for cleaning the perfusion system and chamber: 0.1 M NaOH, 1 mM EDTA, and DEPC-treated water (see Note 4).

6. Sterile eppendorf tubes, 0.5 and 1.5 ml.

7. Microscope (inverted configuration) with 200–400× magnification, mounted on an anti-vibration table.

8. Thin wall borosilicate glass capillaries with filament, 1.5 mm o.d., 1.17 mm i.d., 100 mm length (Harvard Apparatus Ltd, Edenbridge, Kent UK) (see Note 6).

9. Suitable micromanipulator allowing micron-level precision movements of glass micropipettes, for example, a Narishige hydraulic manipulator (Narishige International) or Luigs and Neumann piezo-electric manipulator (Luigs & Neumann, Ratingen, Germany).

10. Perspex patch pipette holder for 1.5 mm o.d. glass with suction side-port. The holder should be firmly mounted to the micromanipulator, which in our system is achieved via the headstage of a patch clamp amplifier as we normally use a patch clamp rig to carry out the separation.

11. 1 mm bore silicone tubing (approximately 1–2 m in length) attached at one end to the suction side-port of the Perspex patch pipette holder, and at the other to a 1 ml syringe. This is used to apply positive or negative pressure. Some users prefer to apply pressure by mouth. A three-way plastic tap between the syringe and tubing allows the applied pressure to be easily maintained.

12. Suitable perfusion chamber for mounting on the inverted microscope (e.g., Warner Instruments, Hamden, CT, USA, sell a massive range). Our chamber is manufactured in-house from a 5 mm thick Perspex plate with a central 20×10 mm hole and an access hole drilled in the side for perfusion inflow. The base is formed by item 13, sealed with vacuum grease. The chamber volume is approximately 500 µl.

13. Borosilicate coverglass, thickness no. 1 or 1.5, suitable for forming the base of the chamber; e.g., 22×40 mm coverslips (Warner Instruments) (see Note 6).

14. Perfusion system. We use a gravity-driven inflow from a reservoir of solution held in a 50 ml sterile plastic syringe tube. Solution inflow is controlled by a three-way sterile plastic tap connected to the end of the syringe tube. A suitable gauge hypodermic needle with its final bevelled section removed is used to attach the plastic tap to portex tubing that delivers solution to the perfusion chamber. A 5 ml sterile plastic syringe attached to the side port of the tap enables flushing of air bubbles from the system. Excess solution is extracted from the chamber via a bevelled tube glued to the top of the chamber. A suction pump connected to a waste container can be used to drive the outflow. Alternatively, peristaltic pumps can also be used for the perfusion system. The flow rate is approximately 2–3 ml/min.

15. Patch pipette puller capable of pulling pipettes with wide (5–25 µm) tip openings, e.g., DMZ Universal puller (Zeitz Instruments, GmbH Ltd, Munich, Germany).

16. Fire-polishing rig for reducing the glass pipette tip diameter. Commercial systems are available (e.g., Narishige model MF-830 microforge, Narishige Instruments). Ours is a homemade system consisting of a 0.5 mm diameter platinum iridium wire (Goodfellow Metals, UK) mounted on an inverted microscope and coupled to a variable power supply (0–6 V, 10 A, 50 Hz).

2.3. RNA Extraction, cDNA Generation, and PCR

The main focus of this methods chapter is the purification of native marrow megakaryocytes (Subheadings 3.1 and 3.2) into a lysis buffer that allows stabilisation and extraction of RNA. Following extraction, the mRNA can be reverse transcribed and analysed by end-point or quantitative PCR. Nowadays, these molecular approaches are mostly carried out with off-the-shelf kits, which include detailed instructions from the manufacturer. Therefore, this final section (and accompanying methods in Subheading 3.3) provides a general description of the approach we used to screen megakaryocyte-derived RNA for expression of a specific gene family, e.g., the transient receptor potential (TRP) family of ion channels (15).

1. Qiagen RNeasy Micro (Qiagen) or similar kit designed for extraction of high-quality RNA from a small sample.

2. Sensiscript RT kit (Qiagen) or similar kit designed for reverse transcription.

3. RNase inhibitor (e.g., 20–40 U/μl RNasin (Promega)).

4. Oligo(dT) and random primers (Promega).

5. PCR: *Taq* polymerase (Qiagen) and mastermix buffer with 1.5 mM $MgCl_2$ and 0.2 mM dNTPs and primer pairs (1 μM) for the gene(s) of interest.

6. Standard gel electrophoresis equipment and reagents for analysis of PCR products. Alternatively, real-time (quantitative) PCR may be used (see Chapter 12, this volume for further discussion).

3. Methods

3.1. Purification of Megakaryocytes by Unit Velocity Albumin Gradients and Fluorescence Activated Cell Sorting

3.1.1. Extraction of Marrow

1. Mice are sacrificed immediately prior to the experiment in accordance with local and national permissions and guidelines (see Note 3). Dissect out the femoral and tibial bones, keeping the bones intact (see Note 7). Remove as much attached muscle and other tissue as possible from the outside of the bone.

2. Immerse each bone in total CATCH medium while dissecting the remaining bones.

3. Fill a 5 ml syringe with total CATCH medium and attach to a short length of silicone tubing. This can be achieved using a hypodermic needle or a trimmed-down pipettor tip (20–200 μl volume).

4. Cut off both the ends of the bone (which contain little marrow) to expose the marrow.

5. Attach the free end of the silicone tubing to one end of the bone and flush out the marrow into a petri dish.

6. Transfer all marrow from two animals to a 15 ml centrifuge tube and place on a rotator at 16°C (see Note 8) for 2 h. The sample will be in about 10 ml total volume.

7. Gently triturate to break up any remaining clumps of marrow.

8. Pass the sample through 100 μm pore nylon mesh.

9. Spin at approximately $150 \times g$ for 5 min and resuspend in 2 ml total CATCH medium.

10. Split the sample into two (see Note 9).

3.1.2. Enriching Megakaryocytes Using BSA Unit Velocity Gradients

Megakaryocytes are enriched by passing through two successive BSA unit velocity ($1g$) gradients. Pass the two samples, prepared above (Subheading 3.1.1), through separate BSA unit velocity gradients.

1. The day before the separation, prepare the BSA stock as described in Subheading 2.1.2, item 1.

2. On the day of experimentation, prepare a 2:1 (e.g., 8 ml BSA and 4 ml CATCH buffer) and a 1:2 (e.g., 4 ml BSA and 8 ml CATCH buffer) (v/v) stock BSA: CATCH buffer solution (see also ref. 17). All solutions to be filtered using a 0.22 μm filter.

3. Prepare the enrichment gradient as follows: in a 15 ml centrifuge tube, add 3.6 ml stock BSA solution. Carefully overlay 1.8 ml of 2:1 (v/v) stock BSA: CATCH solution followed by 1.8 ml 1:2 (v/v) stock BSA: CATCH solution. Lastly, add 1 ml of marrow suspension prepared in Subheading 3.1.1. Place in a rack and leave at $1 \times g$ (i.e., no centrifugation) for 30 min at room temperature.

4. Discard the top 4 ml of the gradient and centrifuge the remaining solution at approximately $150 \times g$ for 5 min.

5. Resuspend the pellet in 1 ml CATCH buffer.

6. Apply the cell suspension to a freshly prepared gradient, thereby repeating steps 3 and 4.

7. Resuspend the final pellet in 1 ml CATCH buffer containing 2 mM EGTA, 0.32 U/ml apyrase VII and 1 μg/ml PGE$_1$.

3.1.3. Separation of Megakaryocytes by Fluorescence-Activated Cell Sorting

1. For each of the two samples processed in Subheading 3.1.2, add 4 μl FITC-conjugated anti-CD41 antibody to 10 μl CATCH buffer in a 1.5 ml microfuge tube and add the 1 ml cell suspension obtained at the end of Subheading 3.1.2. Protect the sample from light and gently rotate at room temperature for 1 h.

2. Pass through a 75 μm filter (see Note 10).

3. Pool the two samples and analyse on a suitable cell sorter at low pressure and wide bore nozzle to avoid activation or rejection of the megakaryocytes. On the Dako MoFlo™ MLS flow cytometer

we found that a pressure setting of 10 p.s.i. and a 200 μm nozzle were suitable. A 100 mW argon laser (or equivalent) is used to generate an excitation wavelength of 488 nm and fluorescence emission recorded using a 530/30 nm band pass filter.

4. Given the high levels of CD41 on megakaryocytes (see fluorescence image in Fig. 1a), these cells can be sorted based upon high forward scatter (indicative of large size) and high CD41-FITC fluorescence. Adjust the settings of the sorting gate (R1, Fig. 1a) to ensure a pure collection of healthy, intact megakaryocytes. Cells reading high CD41 fluorescence, but low in forward scatter consist mainly of megakaryocytes with a transparent appearance (T, Figs. 1a and 2a), and are to be avoided (see Note 8). Region S (Fig. 1a) contains smaller diameter megakaryocytes, however, this area is contaminated with non-megakaryocytic cells.

5. Initially, collect a sample of gated cells and check the validity of the sort using a fluorescence microscope. Once confirmed, a high purity sort mode can be used to separate cells into a 96-well plate, or appropriate vessel for imaging, or into a sterile microfuge tube containing 500 μl RNA lysis buffer (Subheading 2.1.3, item 3) for RNA extraction. Samples collected into lysis buffer can be stored at −80°C until processed.

6. Samples collected for imaging can be viewed using a confocal fluorescence microscope or an equivalent imaging system. Transfer the sample into a suitable chamber containing the CATCH solution described in Subheading 2.1.3, item 1. CD41-FITC-labelled cells are visualised by excitation with an argon laser (or equivalent), at a wavelength of 488 nm and with emission collected at >505 nm (Fig. 1a).

3.2. Extraction of Individual Megakaryocytes Using Glass Pipettes

1. Follow steps 1–6 of Subheading 3.1, but use PBS with apyrase type VII in place of total CATCH medium. A single mouse is sufficient for this form of megakaryocyte extraction.

2. Aliquot the marrow suspension into 1.5 ml microfuge tubes using a sterile pasteur pipette, ensuring a small clump of marrow is transferred to each tube (see Note 8).

Fig. 1. Purification of murine megakaryocytes by unit velocity gradients and fluorescence-activated cell sorting of CD41+ cells. (**a**) Appearance of cells dispersed from murine marrow (initial) and following enrichment for megakaryocytes using unit velocity gradients (cells passed consecutively through two gradients 1 and 2). Cells were then labelled with a FITC-conjugated CD41 antibody (before sorting, CD41+), and separated by FACS (after sorting, CD41+). Scale bars 50 μm. R1 in the scatter plot shows the gate used to collect cells with high forward scatter and high fluorescence. Cells in region T, with high CD41 fluorescence and low forward scatter consist mainly of megakaryocytes with a transparent appearance (see also Fig. 2a and Note 8). Region S contains smaller diameter megakaryocytes and non-megakaryocytic cells. (**b**) Gel electrophoresis analysis of TRPC transcripts in purified murine marrow CD41+ cells compared with murine brain. PCR products are shown for each TRPC primer pair from brain (B, 35 cycles) or CD41+ cells (M, two rounds of PCR of 35 and 25 cycles). Only TRPC6 and TRPC1 were detected in CD41+ cells, i.e., megakaryocytes. Of the transcripts obtained for TRPC1, the strongest signal was at 213 bp, as expected for the full length sequence and previously reported variants. Reproduced from Carter et al. 2006 (15) with permission from John Wiley & Sons.

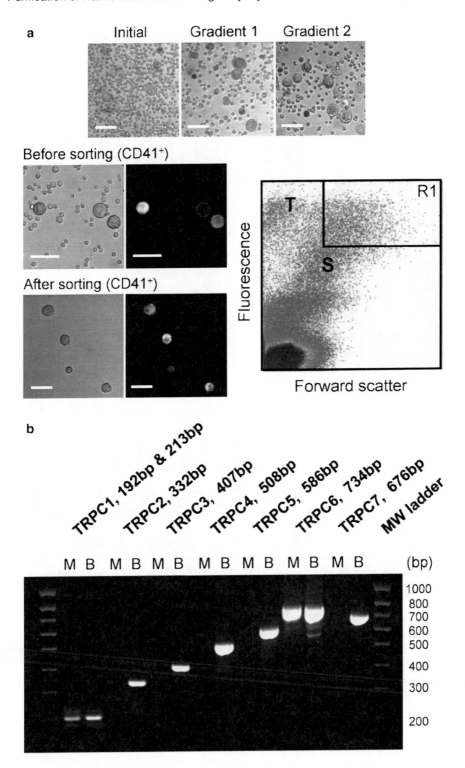

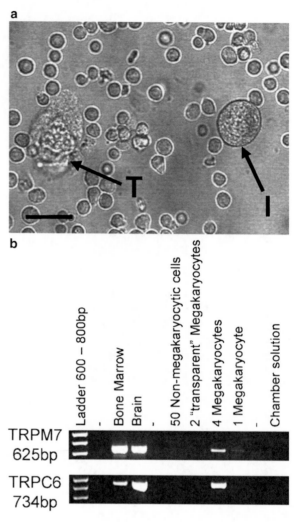

Fig. 2. RT-PCR detection of transcripts in individually extracted primary murine mega-karyocytes. (**a**) Image of marrow cells showing a normal intact megakaryocyte (I) and a transparent megakaryocyte (T). Scale bar 20 μm. (**b**) Gel electrophoresis analysis of PCR products with gene-specific primers for either TRPM7 or TRPC6. mRNA was extracted from individually collected megakaryocytes or small non-megakaryocytic cells, whole brain or whole marrow, reverse transcribed, and PCR performed (35 rounds for brain and marrow and 45 rounds for all other samples). TRPM7 was just detectable after 45 PCR cycles in material from a single intact megakaryocyte, whereas a strong signal was detected for both TRPC6 and TRPM7 when four intact megakaryocytes were combined. No signal was detected for ≈50 non-megakaryocytic cells, two transparent megakaryo-cytes or the bath saline control. In other experiments no PCR signal was detected for B-actin in samples of up to ten transparent megakaryocytes (Richard Carter, unpublished observations). Reproduced from Carter et al. 2006 (15) with permission from John Wiley & Sons.

3. Leave for 2 h on a rotator at 16°C (see Note 8).

4. Prepare the perfusion chamber by attachment of the glass coverslip with silicone grease and connect the inflow and outflow perfusion lines.

5. Flush the chamber and perfusion system with 0.1 M NaOH, 1 mM EDTA followed by DEPC water then sterile RNase- and DNase-free PBS.

6. Fill the chamber with sterile RNase- and and DNase-free PBS and add approximately 10 μl of the cell preparation. Wait for cells to appear at the base of the chamber (approximately 1 or 2 min) and then perfuse approximately 40 ml PBS. The perfusion clears many smaller cells and cellular debri from the bath, whilst leaving the megakaryocytes in the chamber. If the megakaryocytes are also swept away, allow a slightly longer settling time or reduce the perfusion rate by lowering the height of the reservoir.

7. Scan the chamber for clearly visible, individual intact megakaryocytes, free from attached and emperipolesed smaller cells.

8. Pull a glass pipette with a tip approximately one-third to one-half the diameter of a selected megakaryocyte. If required, use the fire-polishing rig to reduce the tip to a suitable diameter.

9. Attach the pipette to the holder and apply positive pressure through the silicone tubing. Insert the pipette tip into the chamber solution, manoeuvring it with the micromanipulator so the tip is positioned above the megakaryocyte. Ensure that no bath solution has entered the pipette tip and that no cells or cellular debri have attached to the pipette tip.

10. Move the pipette tip down to gently touch the megakaryocyte. Apply suction such that the megakaryocyte is either attached to the end of the tip or is partially pulled into the pipette (with as little solution as possible). To ensure that the megakaryocyte will not fall off, move the tip across the air–water interface a couple of times, refocusing on the tip each time to check the megakaryocyte is still on the end of, or inside the pipette.

11. Move the pipette out of the chamber and place the tip in the 75 μl prepared RNA lysis buffer (Subheading 2.2, item 3) and break the tip on the side of the eppendorf. Close the lid, vortex the tube after addition of each megakaryocyte and store on ice.

12. Collect further megakaryocytes, pooling approximately 20 per tube. It is possible to carry out the mRNA extraction and RT-PCR for material from a single megakaryocyte, however, using the protocol described below we found that only the most highly expressed TRP channels were detected in material from a single megakaryocyte (Subheading 3.3, Fig. 2b). In addition to megakaryocytes, it is important to process several

negative controls. Examples include a cell-free sample of PBS transferred from the perfusion chamber in the same way used for the megakaryocytes. Other useful controls include lysis buffer alone and glass pipette tips without saline crushed in the lysis buffer. These should all generate negative PCR results, but if positive will allow one to determine the source of contamination. A number of small cells (i.e., non-megakaryocytes) can also be collected as a further comparison.

13. After collection of all megakaryocytes, disrupt and homogenise the sample by vortexing for 1 min. At this stage the sample can either be processed accordingly, e.g., through RNA microcolumns (Subheading 3.3) or stored at −80°C for up to a couple of months.

3.3. RNA Extraction, Reverse Transcription, and Analysis by PCR

As explained above, this section provides only a general description of the molecular approaches we have used (15) to investigate the presence of transcripts within RNA from purified populations of megakaryocytes.

1. RNA extraction: We found that the Qiagen RNeasy Micro kit provides a reliable, quick and easy to follow protocol for extracting total RNA. Disrupt the cells in the supplied lysis buffer (see Note 5) then add ethanol to the lysate to provide ideal binding conditions to the RNeasy spin column. Add the lysate to the spin column and treat with DNase 1 to remove any contaminating genomic DNA (see Note 11). Wash away contaminates and DNase 1 and elute concentrated RNA in RNase- and DNase-free water. To maximise the concentration, elute the RNA in the smallest volume possible, e.g., 14 μl using the Qiagen RNeasy Micro kit.

2. Reverse transcription: A number of companies supply kits suitable for generating complementary DNA (cDNA) from the extracted RNA by a reverse transcription protocol. Our studies of TRP channels (15) (Figs. 1b and 2b) used Qiagen's Sensiscript RT kit. 8 μl of eluted RNA was used per 20 μl reverse transcription reaction, with a combination of oligo(dT) (10 μM) and random (1 μM) primers. A RNase inhibitor (RNasin, 20–40 U/reaction) was present throughout the reaction to prevent RNA degradation. Please refer to the manufacturers' instructions regarding enzyme concentration and the temperature protocol to be used, as this varies between suppliers.

3. PCR: Use 4 μl cDNA sample per 20 μl endpoint PCR reaction. Again, there are a number of *Taq* polymerase enzymes supplied by a number of companies that one could use and are supplied with advice on conditions for use; we have had good success with the *Taq* polymerase from Qiagen at 0.5 U/20 μl reaction. Reagent concentrations for each PCR reaction should

be optimised for the specific target, primers and enzyme used. Generally, 1.5 mM $MgCl_2$, 0.2 mM dNTPs, and 1 µM of each forward and reverse primer is a good starting point. The reaction involves an initial denaturation step (95°C, 3 min) followed by 35 cycles of denaturation (95°C, 30 s), annealing (5°C below T_m of primers, 30 s), and extension (72°C, time will depend on amplicon size); the reaction is then completed following a final extension step for 10 min at 72°C.

4. To analyse the products of the PCR, run on a 1–2% agarose gel containing 0.5 µg/ml ethidium bromide and visualise with a UV illumination system alongside an appropriate-sized DNA ladder.

4. Notes

1. All solutions are sterilised by filtration through 0.22 µm pore filters. Sterile conditions and standard approaches to reduce contamination of samples should be established, e.g., wear gloves, do not breathe excessively over samples, clean all areas with ethanol, use molecular grade chemicals, and restrict use of chemicals to these experiments. We use filter-tips when using pipettors, and purchase a set of pipettors specifically for these molecular experiments.

2. Add BSA powder to the buffer the day before megakaryocyte isolation. Do not stir the solution; just leave overnight at 4°C to allow the BSA to dissolve. If apyrase type VII and PGE_1 are used, add immediately before use.

3. Researchers should obtain necessary licenses and ethical permissions as required by their Institution and country of residence to allow work to be conducted on animal tissue.

4. To make diethylpyrocarbonate (DEPC)-treated water, add 1 ml of 0.1% DEPC to 1 L distilled water. Shake vigorously for a couple of minutes and leave overnight at room temperature to mix. Autoclave and cool prior to use. Alternatively, purchase RNase-free water from a commercial supplier such as Sigma.

5. All RNA extraction kits supply lysis buffer. Follow the manufacturers' instructions regarding preparation of this solution prior to use. We recommend the addition of Poly-A-carrier RNA or equivalent to reduce the loss of sample RNA. The carrier RNA will not interfere with down stream reactions.

6. The borosilicate glass tubing and the glass coverslips are sterilised by wrapping in aluminium foil and baking in an oven for 4 h at 220°C. We use thin-walled borosilicate glass as this facilitates the pulling of pipettes with large diameter tips on our puller.

However, any patch glass can be used if the puller characteristics are suitable.

7. This can be performed in a flow hood, however in practise, we generally just clean an area of bench with alcohol and perform in the open, attempting to keep conditions as sterile as possible.

8. Immediately after isolating fresh marrow, many of the megakaryocytes are in a transparent state (Fig. 2a), in which the integrity of the plasma membrane appears to be compromised and one can clearly see the membranes of the multilobular nucleus. Compared to normal intact megakaryocytes, transparent megakaryocytes provide little or no mRNA (15), presumably because this has been lost into the extracellular environment. During the ensuing hours, a greater number of normal, intact megakaryocytes emerge (Fig. 2a). We do not fully understand why this occurs, but it may be a consequence of the migration of megakaryocytes from undissociated marrow, as shown in explant cultures (see Eckly et al., Chapter 13, this volume). It is therefore advisable to leave small clumps of tissue in each tube when preparing the marrow. We have also detected an overall improved appearance of intact megakaryocytes when the marrow is kept slightly below standard ambient temperatures; we routinely use 16°C.

9. We find splitting the material in two and passing through separate BSA unit velocity gradients improves megakaryocyte enrichment and reduces "clumping" of material. The samples are then pooled immediately prior to FACS.

10. The 75 μM filter is required to remove large particles/cellular material that would otherwise clog the 200 μm nozzle of the cell sorter. Analysis of the sample after passage through the filter clarified that this filter size does not remove many of the megakaryocytes.

11. We strongly recommend treating your RNA samples with DNase 1 to remove any contaminating genomic DNA, which may give false-positive results in your PCR reactions. If DNase 1 is not supplied with the kit and part of the protocol, an additional step can be added at the end using Ambion® DNAse 1 (Applied Biosystems).

Acknowledgements

We thank Dr Stefan Amisten for comments on the manuscript. The techniques described in this chapter were developed as part of investigations of platelet and megakaryocyte ion channels, funded by the British Heart Foundation.

References

1. Wright, J. H. (1910) The histogenesis of the blood platelets *Journal of Morphology* **21**, 263–278.

2. Wright, J. H. (1906) The origin and nature of the blood plates *Boston Medical and Surgical Journal* **154**, 643–645.

3. Italiano, J. E., Jr. and Shivdasani, R. A. (2003) Megakaryocytes and beyond: the birth of platelets *J Thromb. Haemost.* **1**, 1174–1182.

4. Shattil, S. J. and Leavitt, A. D. (2001) All in the family: primary megakaryocytes for studies of platelet $\alpha_{IIb}\beta_3$ signaling *Thromb. Haemost.* **86**, 259–265.

5. Tolhurst, G., Vial, C., Leon, C., Gachet, C., Evans, R. J., and Mahaut-Smith, M. P. (2005) Interplay between $P2Y_1$, $P2Y_{12}$, and $P2X_1$ receptors in the activation of megakaryocyte cation influx currents by ADP: evidence that the primary megakaryocyte represents a fully functional model of platelet P2 receptor signaling *Blood* **106**, 1644–1651.

6. Mason, M. J. and Mahaut-Smith, M. P. (2004) Measurement and manipulation of intracellular Ca^{2+} in single platelets and megakaryocytes *Methods Mol. Biol.* **273**, 251–276.

7. Mahaut-Smith, M. P. (2004) Patch-clamp recordings of electrophysiological events in the platelet and megakaryocyte *Methods Mol. Biol.* **273**, 277–300.

8. Mason, M. J. and Mahaut-Smith, M. P. (2001) Voltage-dependent Ca^{2+} release in rat megakaryocytes requires functional IP_3 receptors *J Physiol* **533**, 175–183.

9. Tertyshnikova, S. and Fein, A. (1998) Inhibition of inositol 1,4,5-trisphosphate-induced Ca^{2+} release by cAMP-dependent protein kinase in a living cell *Proc Natl Acad Sci USA* **95**, 1613–1617.

10. Monyer, H. and Jonas, P. (2010) in *Single Channel Recording* (Sakmann-B and Neher-H, Eds.).

11. Jonas, P., Racca, C., Sakmann, B., Seeburg, P. H., and Monyer, H. (1994) Differences in Ca^{2+} permeability of AMPA-type glutamate receptor channels in neocortical neurons caused by differential GluR-B subunit expression *Neuron* **12**, 1281–1289.

12. Lambolez, B., Audinat, E., Bochet, P., Crepel, F., and Rossier, J. (1992) AMPA receptor subunits expressed by single Purkinje cells *Neuron* **9**, 247–258.

13. de Preter, K., Vandesompele, J., Heimann, P., Kockx, M. M., Van Gele, M., Hoebeeck, J., De Smet, E., Demarche, M., Laureys, G., Van Roy, N., De Paepe, A., and Speleman, F. (2003) Application of laser capture microdissection in genetic analysis of neuroblastoma and neuroblastoma precursor cells *Cancer Lett.* **197**, 53–61.

14. Iyer, E. P. and Cox, D. N. (2010) Laser capture microdissection of Drosophila peripheral neurons *J. Vis. Exp.*

15. Carter, R. N., Tolhurst, G., Walmsley, G., Vizuete-Forster, M., Miller, N., and Mahaut-Smith, M. P. (2006) Molecular and electrophysiological characterization of transient receptor potential ion channels in the primary murine megakaryocyte *J. Physiol* **576**, 151–162.

16. Raslova, H., Roy, L., Vourc'h, C., Le Couedic, J. P., Brison, O., Metivier, D., Feunteun, J., Kroemer, G., Debili, N., and Vainchenker, W. (2003) Megakaryocyte polyploidization is associated with a functional gene amplification *Blood* **101**, 541–544.

17. Leven, R. M. (2004) Isolation of primary megakaryocytes and studies of proplatelet formation *Methods Mol. Biol.* **272**, 281–292.

Chapter 19

Assessment of Megakaryocyte Migration and Chemotaxis

Alexandra Mazharian

Abstract

Cell migration is a highly integrated multistep process that plays an essential role during development and disease. Megakaryocytes (MKs) are specialized precursor cells that produce platelets and release them into the circulation. MK migration from the proliferative osteoblastic niche within the bone marrow (BM) environment to the capillary-rich vascular niche is an essential step for platelet production. Among the chemokines that may play a central role in cell migration, the stromal cell-derived factor 1α (SDF1α) also known as CXCL12 has been described to act as a potent chemoattractant for MKs (1, 2). This biological effect is mediated by the SDF1α receptor CXCR4 (Fusin), which is expressed on haematopoietic stem cells, MKs and platelets. The Dunn chemotaxis chamber in conjunction with the time-lapse microscopy is a powerful tool that enables the user to observe directly the morphological response of cells to chemoattractant in real time. This chapter describes the Dunn chemotaxis chamber to study the migration of primary BM-derived MKs in response to a gradient of SDF1α. In combination with genetically modified mice, this provides a powerful approach to directly investigate the role of specific proteins in MK migration and chemotaxis.

Key words: Primary bone marrow-derived megakaryocytes, Migration, Dunn chemotaxis chamber, Chemotactic gradient formation, Time-lapse microscopy

1. Introduction

Directed cell migration (chemotaxis) is fundamental during embryogenesis (3), wound healing (4), and inflammatory responses (5). MK migration and its regulation by the chemokine SDF1α and its receptor CXCR4 play an important role in platelet formation (6). The technique described in this chapter has been developed to observe and study the intracellular signalling pathways and morphological changes that are required for MKs to initiate a migratory response towards the chemoattractant SDF1α. Chemotaxis can be studied using a Boyden chamber or transwell

Jonathan M. Gibbins and Martyn P. Mahaut-Smith (eds.), *Platelets and Megakaryocytes: Volume 3, Additional Protocols and Perspectives*, Methods in Molecular Biology, vol. 788, DOI 10.1007/978-1-61779-307-3_19, © Springer Science+Business Media, LLC 2012

assay (7). However, both these methods are based on scoring cells that have migrated into or through a filter membrane towards a source of chemotactic factor. In these assays, the local concentration gradients of chemotactic factor in and around the pores of the filter membrane are variable and unknown. Furthermore, the migratory behaviour of the cells cannot be observed and is therefore deduced from the final distribution of the cell population. To overcome these experimental issues, the Dunn chemotaxis chamber has been developed (Fig. 1) (8). In this chamber, the migratory behaviour of MKs can be directly observed on a matrix in response to chemoattractant. The Dunn chemotaxis chamber is a modification of the Zigmond chamber (9), which allows the direct observation of slowly moving cells in a concentration gradient that is stable over time (8, 10). The chamber consists of a glass microscope slide with two concentric annular wells ground into the centre of one face of the glass slide, referred to as the inner and the outer wells. An annular bridge separates the two wells. The bridge is optically polished to lay precisely 20 μm below the surrounding glass slide. When the inner well is filled with the control medium and the outer well filled with a medium containing the chemoattractant, a directed linear diffusion gradient becomes established across the bridge between these two wells. MKs are seeded onto a glass-coated coverslip which is then inverted over the chamber. The MKs that lie directly above the bridge are viewed during the assay.

The Dunn chemotaxis chamber has been used to characterize the migratory response of human macrophages to colony-stimulating

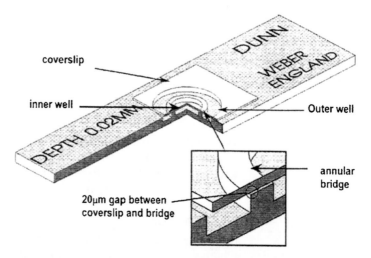

Fig. 1. View of the Dunn chemotaxis chamber with the correct position of the coverslip to allow access to the outer well. The chamber consists of a glass microscope slide with two annular wells separated by a bridge 1 mm wide. If the inner well is filled with control medium and the outer well filled with medium containing the chemoattractant, a linear gradient is established between the two wells. Adapted from a figure kindly provided by Professor Gareth Jones, King's College London.

factor-1 (CSF-1), fibroblasts to platelet-derived growth factor (PDGF) (8) and thrombospondin (11), and neutrophils to interleukin-8 (12). We have recently used the Dunn chemotaxis chamber to elucidate the roles for platelet endothelial cell adhesion molecule-1 (PECAM-1) (13), the protein tyrosine phosphatase CD148 (14), Src Family Kinases (SFKs) and PLCγ2 in MK migration on fibronectin matrix and towards a SDF1α gradient (15). In this chapter, we provide description of the use of this chamber to study the response of primary BM-derived MKs to SDF1α, a protocol that can be extended to study the migratory behaviour of MKs and other cell types derived from genetically modified mice and in response to various chemoattractants.

2. Materials

1. Murine bone marrow: bone marrow MKs are isolated from C57BL6 wild-type mice purchased from Harlan (Oxon, UK). Young adult mice should be used as a decrease in haematopoietic tissue is observed in older animals.

2. Dissection tools: Scissors and forceps are cleaned with ethanol before and after use.

3. Sterile tissue culture plastics and accessories: 100 mm diameter tissue culture dishes, 15 and 50 ml polypropylene tubes (BD Falcon, Oxford, UK), 70 μm nylon filter (BD Falcon, Oxford, UK), 1 and 20 ml syringes, 25-gauge needles. For BSA density gradient, 15 ml tubes must be used.

4. Complete DMEM medium: DMEM supplemented with 1% L-glutamine, 1% penicillin–streptomycin and 10% foetal bovine serum (PAA, Somerset, UK).

5. ACK buffer: 0.15 M NH_4Cl; 1 mM $KHCO_3$; 0.1 mM Na_2EDTA in distilled water. Adjust the pH to 7.3 with NaOH. This solution is autoclaved, sterilized, and stored at 4°C for up to 2 months.

6. Rat anti-mouse antibodies for cell depletion: anti-Ly-6G, anti-CD11b (eBioscience, Hatfield, UK), anti-CD45R/B220, anti-CD16/32 (BD Pharmingen, Oxford, UK).

7. Anti-rat Dynabeads (Invitrogen, Paisley, UK).

8. Magnetic bead separator: DynaMag magnetic bead separator: (Dynal Biotech Invitrogen, Paisley, UK).

9. 22×22 mm glass coverslips (VWR, Teddington, UK). These are washed with detergent, rinsed with distilled water, soaked for 10 min in concentrated 1 M HCl, then washed extensively with distilled water and stored in 70% ethanol until use.

The ethanol is removed by washing several times with distilled water (see Note 1).

10. Dunn chemotaxis chambers are supplied by Weber Scientific International (Teddington, UK). Sealing the coverslip onto the chamber requires dental wax, which can be purchased from Agar Scientific (Stansted, UK). It is recommended to clean the chamber immediately after use. The most important principle to observe when cleaning the Dunn chemotaxis chamber is to avoid touching the bridge. The chamber is cleaned with ethanol 70% then rinsed extensively with distilled water.

11. Coverslips were coated with either bovine plasma fibronectin 20 μg/ml (Calbiochem, Nottingham, UK) diluted in PBS, human plasma fibrinogen 100 μg/ml (Enzyme Research Laboratories, South Bend, IN) diluted in PBS or fibrillar-type I collagen 100 μg/ml diluted in collagen diluent (Nycomed, Zurich, Switzerland) 2 h at room temperature or overnight at 4°C. Fresh solutions of adhesive ligands are made on the day of use. On the day of the experiment, the coverslips are washed with PBS, blocked with denatured BSA (5 mg/ml) (First Link Ltd, Birmingham, UK) for 1 h at room temperature, and washed with PBS before use. To denature the BSA, boil for 10 min and allow to cool before use.

12. Complete Stempro medium: Stempro-34 Serum-Free Medium (Invitrogen, Paisley, UK) supplemented with 1% L-glutamine, 1% penicillin–streptomycin (PAA, Somerset, UK) and 2.6% serum replacement factor (Invitrogen, Paisley, UK). This sterile solution is stored at 4°C.

13. For migration experiments, a solution of 1:20 dilution of complete Stempro medium is prepared, using Stempro-34 Serum-Free Medium (supplemented with 1% L-glutamine, 1% penicillin-streptomycin) as the diluent. This sterile solution is stored at 4°C.

14. SDF1α (Peprotech, London, UK). A 100 μg/ml stock solution is prepared with sterile distilled water. Aliquots are stored at –20°C. To provide reproducible and accurate concentration gradients during the chemotaxis assay, SDFα is used at a final concentration of 300 ng/ml, diluted in complete Stempro medium. The MKs are maintained in a humidified incubator at 37°C in the presence of 5% CO_2.

15. Stem cell factor (SCF) (Peprotech, London, UK). A 20 μg/ml stock solution is prepared with sterile distilled water. Aliquots are stored at –20°C. SCF is used at a final concentration of 20 ng/ml, diluted in complete Stempro medium.

16. Thrombopoietin (TPO) (Peprotech, London, UK). A 50 μg/ml stock solution is prepared with sterile distilled water. Aliquots are stored at –20°C. TPO is used at a final concentration of 50 ng/ml, diluted in complete Stempro medium.

17. Bovine serum albumin (BSA) (Fraction V) (First Link Ltd, Birmingham, UK): 3 and 1.5% solutions of BSA diluted in sterile PBS (Invitrogen, Paisley, UK) are made for use in a unit velocity gradient. All BSA solutions are filtered, stored, and kept sterile at 4°C for up to 2 weeks.

18. A 1 ml syringe with fine needle (25 gauge) is used for refilling the outer well of the Dunn chemotaxis chamber with the complete Stempro medium containing the chemoattractant.

19. Migration is observed in the Dunn chemotaxis chamber using a microscope adapted for time-lapse recording. The Dunn chemotaxis chamber is placed in a humidified chamber on a heated stage at 37°C within an Axiovert 200 inverted high-end microscope (Zeiss, Welwyn Garden City, UK) equipped with a Hamamatsu Orca 285 cooled digital camera. Time-lapse images are digitally captured once a every minute for 3 h. Slidebook (3I, http:// www.intelligent-imaging.com/) and Image J (http:// www.rsb.info.nih.gov/ij/) software are used to acquire and process images (see Subheading 3.3). To determine the chemotactic potential of MKs requires the analysis of a large number of cells. Therefore, a low-magnification objective lens (10× or 20×) is used whereby the full width of the Dunn chemotaxis chamber bridge can be visualized.

20. Transwell, 24-well cell clusters, with 8 μm pore filters were obtained from Costar (Cambridge, MA).

21. FITC-conjugated anti-mouse GPIIb antibody was purchased from BD Pharmingen (Oxford, UK).

3. Methods

3.1. Isolation and Culture of Murine Primary Bone Marrow-Derived Megakaryocytes

1. After sacrifice of the mice, isolate the femurs and tibiae. Remove all attached muscle and connective tissue. It is important that all of the tissue surrounding the bones is scraped away to prevent contamination of the bone marrow cells. Immerse the intact bones in 70% ethanol on ice for up to 15 min. Provided the bones are intact, the ethanol will not affect the marrow cells, and will help reduce bacterial growth in subsequent cultures.

2. Under the hood, carefully cut the ends off the bones and flush out the bone marrow into a 50 ml Falcon tube with complete DMEM medium using a 20 ml syringe and attached to a 25-gauge needle to pierce the knee end of the bones. Move the needle up and down with a constant medium stream to remove most of the marrow. Use approximately 10 ml of medium per mouse, reusing flushed medium as required.

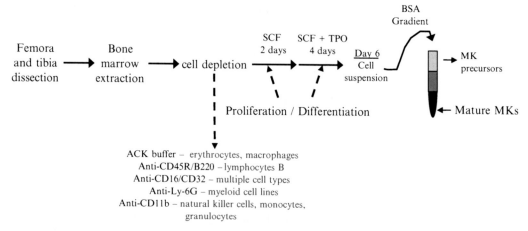

Fig. 2. Purification and culture of murine primary bone marrow-derived megakaryocytes. Bone marrow cells were obtained from femora and tibiae of C57BL6 mice by flushing. Erythrocytes were lysed using ACK buffer and the CD16/CD32+ Ly-6 G+CD45R/B220+ CD11b+cells were depleted using immunomagnetic beads (sheep anti-rat IgG Dynabeads). The remaining population was cultured with 20 ng/ml murine SCF at 37°C under 5% CO_2 for 2 days. Cells were then cultured for a further 4 days in the presence of 20 ng/ml murine SCF and 50 ng/ml murine TPO. After 4 days of culture in the presence of TPO, the cell population was enriched for mature MKs using a 1.5/3% BSA gradient under gravity (1 × g) for 45 min at room temperature.

3. Centrifuge the cell suspension for 5 min at $200 \times g$ at 4°C and resuspend the pellet in ACK buffer in order to lyse the red blood cells and macrophages. Use 6 ml per mouse. Incubate at room temperature for 5 min (Fig. 2).

4. Filter the cells through a 70 μm nylon filter into a 50 ml Falcon tube and centrifuge the cell suspension for 5 min at $200 \times g$ at 4°C.

5. Resuspend the cells in 1 ml of complete DMEM medium containing the following antibodies for depletion of cells: anti-Ly-6G, anti-CD11b, anti-CD45R/B220, and anti-CD16/32 (3 μl of each antibody (0.5 mg/ml) per mouse). Incubate on ice for 30 min.

6. Whilst cells are incubating, prepare anti-rat Dynabeads. Use 50 μl of Dynabeads per mouse and wash three times with 1 ml of complete DMEM medium. Resuspend the beads in 50 μl of complete DMEM medium per mouse.

7. Add a further 1 ml of complete DMEM medium to the cells and centrifuge for 5 min at $200 \times g$ at 4°C.

8. Resuspend the cells in 1 ml of complete DMEM medium and add 50 μl washed beads per mouse.

9. Centrifuge the cells for 5 min at $200 \times g$ and resuspend in the same medium by pipetting up and down. Then use a magnet to pull down the beads.

10. Transfer the supernatant to a fresh tube, add 1 ml of complete DMEM medium to the beads and mix. Use the magnet a second

time to further deplete unwanted cell types and pool the supernatants from both depletion steps. Discard the beads.

11. Centrifuge for 5 min at $200 \times g$ and resuspend the cells in 1 ml of complete Stempro medium per mouse with the addition of 20 ng/ml of SCF. Incubate the cells at 37°C under 5% CO_2 for 2 days.

12. After 48 h, transfer the cells to a 15 ml Falcon tube, centrifuge for 5 min at $200 \times g$. Discard the supernatant and resuspend the pellet in 1 ml complete Stempro medium supplemented with 20 ng/ml SCF and 50 ng/ml TPO. Incubate the cells at 37°C under 5% CO_2 for 4 days to allow MK differentiation.

13. After 4 days of culture in the presence of SCF and TPO, enrich the cell population for mature MKs using a 1.5/3% BSA gradient under gravity ($1 \times g$). The gradient consists of 4 ml of 3% BSA on the bottom of a 15 ml tube and 4 ml of 1.5% BSA on top. The cell suspension (1 or 2 ml per mouse) is layered on top of the BSA and left for 45 min at room temperature.

14. Take the upper 5 or 6 ml of the gradient/cell mixture and centrifuge for 5 min at $200 \times g$ at room temperature. Resuspend the pellet in 1 ml complete Stempro medium supplemented with 20 ng/ml SCF and 50 ng/ml TPO for a further 4 days to allow additional MK differentiation.

15. Take the bottom 4 ml and centrifuge for 5 min at $200 \times g$ at room temperature. Resuspend the cells in complete Stempro medium. This is the MK-enriched population of cells and can be used immediately for migration experiments.

3.2. Assembly of the Dunn Chemotaxis Chamber

1. Ensure the humidified time-lapse stage is heated to 37°C in the presence of 5% CO_2 and that the time-lapse recording can begin immediately after the chamber is assembled.

2. Seed the previously cultured MKs (see Subheading 3.1, step 15) onto the 22 mm² coverslip coated with fibronectin, fibrinogen or collagen. Allow the MKs to attach to the coverslip for 1 h at 37°C and then gently wash the coverslip with complete Stempro medium.

3. Place the Dunn chemotaxis chamber on tissue paper. The chamber should be completely dry before processing. Do not attempt to wipe the Dunn chemotaxis chamber as contact with the bridge can result in damage that may affect the gradient.

4. Fill both wells of the Dunn chemotaxis chamber with 1:20 dilution of complete Stempro medium.

5. Carefully invert the coverslip over the two wells in the centre of the Dunn chemotaxis chamber. It is important that inverting the coverslip over the chamber does not incorporate any bubbles into the chamber as these can affect the gradient

(see Note 2). The coverslip should be placed slightly off centre as a small gap remains in the outer well. However, the coverslip should completely cover the inner well so that the outer well can be drained without any loss of medium from the inner well (Fig. 1).

6. Lightly press down the coverslip around the edges and mop up excess medium with Whatman paper.

7. Tear a piece of Whatman paper and place a small corner of the torn edge just inside the outer well gap until it starts to absorb the medium. Leave the filter paper to absorb the entire medium in the outer well. It is important not to move or lift the paper as this can introduce air into the Dunn chemotaxis chamber.

8. Gently fill the outer well with 1:20 dilution of complete Stempro medium supplemented with the chemoattractant SDF1α (300 ng/ml) using a 1 ml syringe with a sterile 25-gauge needle. Avoid any air bubbles (see Notes 2 and 3 for testing the gradient formation).

9. Wax the sides of the coverslip ensuring that the chamber is completely sealed. Once the wax has set, wash the remaining coverslip surface with distilled water and dry with Whatman paper.

10. Immediately place the Dunn chemotaxis chamber on the microscope stage and start imaging. See Note 4 for details on fixing cells at the end of the recording.

3.3. Time-Lapse Microscopy

1. The live image from the microscope is viewed on the computer monitor using Slidebook software and a region of the bridge is selected for recording. Ideally, there should be between 2 and 5 cells in a field of view (MKs are large cells of 50 μm in diameter) and the cells should be evenly spaced.

2. The high concentration end of the chemotactic gradient (outer well) is then aligned to the top of the monitor screen. This is achieved by rotating the camera. This is extremely useful for tracking and mathematical analysis of chemotactic behaviour.

3. To analyze MK migration in response to a gradient of SDF1α, we use a time-lapse interval of 1 min and film the MKs for 3 h. The majority of large mature MKs respond slowly to a SDF1α gradient and start to move after 1 h (Fig. 3). MK migration has been previously shown to be divided into three distinct phases, namely adhesion, polarization and migration (13).

4. Once the assay is finished, the recorded sequence of images can be analyzed using Image J software (see Subheading 2, item 18). Each cell must be individually tracked. In our laboratory, only MKs present in the first frame are tracked. We select the centre of each MK and track for the 3 h recording period. The net result of tracking is a set of cell trajectories, each consisting of a sequence of (x, y) position coordinates obtained from a single cell.

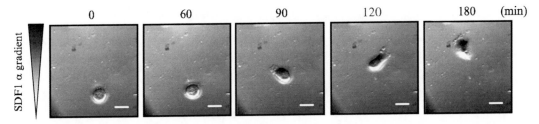

Fig. 3. Differential interference contrast (DIC) images of megakaryocyte migration in response to SDF1α gradient within the Dunn chemotaxis chamber. The outer well contained complete Stempro medium supplemented with SDF1α 300 ng/ml and time-lapse recording taken at 1 min intervals for a 3-h period (scale bar = 20 μm). Figure from Mazharian et al. (15).

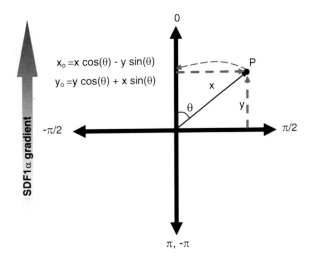

Fig. 4. The formula. All coordinates (*x*, *y*) collected from Slidebook are taken relative to the starting position and are corrected to an angle (*θ*) relative to the position of the SDF1α gradient (*θ*). All angles are measured in radians. The formula rotates any position vector in a specified radian-measured angle.

5. To analyze the direction and migratory behaviour of the MKs, all the coordinates of the migratory MKs are collected over time using Image J tracking software.

6. The MK trajectories can be plotted in Microsoft Excel. Each MK has its starting position set to the zero point of the scatter plot. Therefore, the relative starting position of any cell is always $(0, 0)$. Then, all the relative coordinates are corrected to an angle (*θ*) measured using Image J software at the start of the experiment, using the Formula (Fig. 4) (see Note 5). This formula rotates any position vector in a specified radian-measured angle. Serially plot the subsequent sequence of position coordinates for each timed interval for every MK until the end of the recording time. Figure 5 demonstrates the results obtained from migration of primary BM-derived MKs in response to a linear gradient of SDF1α (15).

Outer well

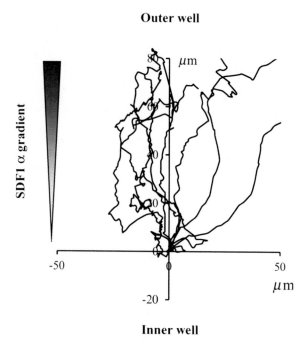

Inner well

Fig. 5. Vector plot of megakaryocyte migration in response to SDF1α gradient in the Dunn chemotaxis chamber. The outer well contained complete Stempro medium supplemented with SDF1α 300 ng/ml and time-lapse recordings were taken at 1 min intervals for a 3-h period. The intersection of the *x*- and *y*-axis was taken to be the starting point of each cell path and the source of SDF1α is at the top of the figure. Figure from Mazharian et al. (15).

7. These data can be used to calculate the distance and the velocity of MK movements.

3.4. Transwell Assay

The transwell migration assay can also be used to examine MK migration using the Boyden chamber. The Boyden chamber originally introduced by Boyden for analysis of leukocyte chemotaxis (7) is ideally suited for quantitative analysis of different migratory responses of cells. The Boyden chamber consists of two medium-filled compartments separated by microporous membrane. The cells are placed in the upper compartment and are allowed to migrate through the pores of the membrane into the lower compartment, in which the chemoattractant agent is present. After a specific time period, the membrane between the two compartments can be fixed and stained, and the number of cells that have migrated to the lower compartment is determined (Fig. 6). The method described in this section is intended specifically for measuring the migration of primary murine BM-derived MKs.

1. Place 500 μl of serum-free medium containing 300 ng/ml SDF1α in the lower compartment of the Boyden chamber. For controls, add serum-free medium without agonist (see Note 6).

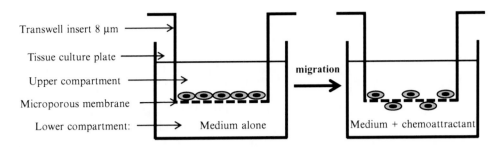

$$\text{Percentage migration} = 100 \times \frac{\text{number of migrated cells}}{\text{number of input cells}}$$

Fig. 6. Schematic representation of the transwell migration assay. A transwell insert is placed in a well of a multiwell cell culture plate forming a migration chamber. The migration chamber consists of an upper and lower compartment with a microporous membrane in between. If the lower compartment contains the chemoattractant, cells may actively migrate from the upper to the lower compartment.

2. Add 100 μl of MKs at a concentration of 1×10^6/ml in complete Stempro medium to the upper compartment on an 8 μm pore transwell insert (see Note 7).

3. Incubate for 1–3 h at 37°C in a humidified atmosphere containing 5% CO_2 (see Note 8).

4. Carefully remove the upper compartment and recover in the same volume the MKs from the upper and bottom compartments for counting using a haemocytometer. Label both cell fractions with FITC-conjugated anti-mouse GPIIb antibody to determine the percentage of migrated GPIIb positive cells by flow cytometry.

5. All assays should be performed in triplicate. Data are presented either as a chemotaxis index calculated as the ratio of number of migrating cells in response to SDF1α/number of migrating cells to medium, or as the percentage of migrated cells calculated as follows: 100× number of migrated cells/number of input cells (Fig. 6). For morphological studies, the different cell fractions obtained can be put on coverslips, fixed, and stained for specific markers (e.g., actin).

6. This transwell assay can also be performed after coating the transwell insert membrane with an extracellular matrix protein similar to endothelial basement membrane such as collagen, fibronectin or with endothelial cells (1, 16–20). Coating with BSA is used as the control. For migration inhibition studies, MKs were preincubated with an inhibitor of choice at an adequate concentration for 30 min at 37°C, washed once with complete Stempro medium and then placed in transwells as described above.

4. Notes

1. The coverslip should be as thick as the microscope optics permits.

2. Successful assembly of the Dunn chemotaxis chamber requires patience and practice. It is advisable for a new user to practice assembling the chamber initially using a wet coverslip with no cells attached. It is essential that the chamber contains no bubbles even after 4 h of incubation. The presence of bubbles particularly in the inner well seriously interferes with the formation of the gradient and produces unreliable results. Bubbles can result from incorrect sealing of the chamber but are most commonly the result of the coverslip being lifted away from the chamber during the draining of the outer well.

3. The time taken for a gradient to form between the two wells of the Dunn chemotaxis chamber depends on the molecular weight of the chemotactic factor. Factors with a molecular weight between 350 and 370 Da will form a gradient in 10 min, whereas factors with a molecular weight between 10 and 20,000 Da will take 30 min (8).

4. The coverslip is removed from the chamber at the end of the recording and the cells are fixed. To fix the cells, carefully lift the coverslip away from the chamber and immediately place the coverslip in fixative (e.g. 4% paraformaldhehyde). The coverslip can be stained for different proteins such as actin fibres following normal protocols.

5. The starting position of each cell trajectory should be ideally taken at the start of the experiment. Only cells whose starting position lies more than 30 μm from the nearest edge of the recording field should be analyzed.

6. It is important to dedicate a number of transwells for the detection of non-specific cell migration, which occurs in the absence of any chemotactic agents.

7. When choosing the transwell pore size, is it important to take into account that cells should not diffuse through the filter pores, but must actively transmigrate through them by changing their shape. Filters with a pore size of 5 or 8 μm have been reported in the literature for murine BM-derived MKs (1, 16–20).

8. It is also necessary to determine the optimal time over which to measure the migration of the cells, since it is important to establish a sufficient window to allow the accurate measurement of the migration. Three hours incubation in transwell or 24 h in matrix endothelial-coated transwells at 37°C under 5% CO_2 has been routinely used to study MK migration (1, 16–20).

Acknowledgements

The author would like to thank Professor Steve Watson at the University of Birmingham who has contributed significantly to the ideas discussed in this chapter, for providing constructive discussion and critical reading of this chapter. In addition, the author would like to thank Dr. Tarvinder Dhanjal, who provided guidance and helpful discussion during the development of the migration assay using the Dunn chemotaxis chamber. The author's work was supported by the British Heart Foundation.

References

1. Hamada, T., Mohle, R., Hesselgesser, J., Hoxie, J., Nachman, R. L., Moore, M. A., and Rafii, S. (1998) Transendothelial migration of megakaryocytes in response to stromal cell-derived factor 1 (SDF-1) enhances platelet formation. *J Exp Med 188*, 539–48.

2. Wang, J. F., Liu, Z. Y., and Groopman, J. E. (1998) The alpha-chemokine receptor CXCR4 is expressed on the megakaryocytic lineage from progenitor to platelets and modulates migration and adhesion. *Blood 92*, 756–64.

3. Dormann, D., and Weijer, C. J. (2003) Chemotactic cell movement during development. *Curr Opin Genet Dev 13*, 358–64.

4. Seppa, H., Grotendorst, G., Seppa, S., Schiffmann, E., and Martin, G. R. (1982) Platelet-derived growth factor in chemotactic for fibroblasts. *J Cell Biol 92*, 584–8.

5. Jones, G. E. (2000) Cellular signaling in macrophage migration and chemotaxis. *J Leukoc Biol 68*, 593–602.

6. Avecilla, S. T., Hattori, K., Heissig, B., Tejada, R., Liao, F., Shido, K., Jin, D. K., Dias, S., Zhang, F., Hartman, T. E., Hackett, N. R., Crystal, R. G., Witte, L., Hicklin, D. J., Bohlen, P., Eaton, D., Lyden, D., de Sauvage, F., and Rafii, S. (2004) Chemokine-mediated interaction of hematopoietic progenitors with the bone marrow vascular niche is required for thrombopoiesis. *Nat Med 10*, 64–71.

7. Boyden, S. (1962) The chemotactic effect of mixtures of antibody and antigen on polymorphonuclear leucocytes. *J Exp Med 115*, 453–66.

8. Zicha, D., Dunn, G. A., and Brown, A. F. (1991) A new direct-viewing chemotaxis chamber. *J Cell Sci 99 (Pt 4)*, 769–75.

9. Zigmond, S. H., and Hirsch, J. G. (1973) Leukocyte locomotion and chemotaxis. New methods for evaluation, and demonstration of a cell-derived chemotactic factor. *J Exp Med 137*, 387–410.

10. Dunn, G. A., and Zicha, D. (1993) Long-term chemotaxis of neutrophils in stable gradients: preliminary evidence of periodic behavior. *Blood Cells 19*, 25–39; discussion 39–41.

11. Orr, A. W., Elzie, C. A., Kucik, D. F., and Murphy-Ullrich, J. E. (2003) Thrombospondin signaling through the calreticulin/LDL receptor-related protein co-complex stimulates random and directed cell migration. *J Cell Sci 116*, 2917–27.

12. Zicha, D., Allen, W. E., Brickell, P. M., Kinnon, C., Dunn, G. A., Jones, G. E., and Thrasher, A. J. (1998) Chemotaxis of macrophages is abolished in the Wiskott-Aldrich syndrome. *Br J Haematol 101*, 659–65.

13. Dhanjal, T. S., Pendaries, C., Ross, E. A., Larson, M. K., Protty, M. B., Buckley, C. D., and Watson, S. P. (2007) A novel role for PECAM-1 in megakaryocytokinesis and recovery of platelet counts in thrombocytopenic mice. *Blood 109*, 4237–44.

14. Senis, Y. A., Tomlinson, M. G., Ellison, S., Mazharian, A., Lim, J., Zhao, Y., Kornerup, K. N., Auger, J. M., Thomas, S. G., Dhanjal, T., Kalia, N., Zhu, J. W., Weiss, A., and Watson, S. P. (2009) The tyrosine phosphatase CD148 is an essential positive regulator of platelet activation and thrombosis. *Blood 113*, 4942–54.

15. Mazharian, A., Thomas, S. G., Dhanjal, T. S., Buckley, C. D., and Watson, S. P. (2010) Critical role of Src-Syk-PLC{gamma}2 signaling in megakaryocyte migration and thrombopoiesis. *Blood 116*, 793–800.

16. Giet, O., Van Bockstaele, D. R., Di Stefano, I., Huygen, S., Greimers, R., Beguin, Y., and Gothot, A. (2002) Increased binding and

defective migration across fibronectin of cycling hematopoietic progenitor cells. *Blood 99*, 2023–31.

17. Aiuti, A., Webb, I. J., Bleul, C., Springer, T., and Gutierrez-Ramos, J. C. (1997) The chemokine SDF-1 is a chemoattractant for human CD34+ hematopoietic progenitor cells and provides a new mechanism to explain the mobilization of CD34+ progenitors to peripheral blood. *J Exp Med 185*, 111–20.

18. Lane, W. J., Dias, S., Hattori, K., Heissig, B., Choy, M., Rabbany, S. Y., Wood, J., Moore, M. A., and Rafii, S. (2000) Stromal-derived factor 1-induced megakaryocyte migration and platelet production is dependent on matrix metalloproteinases. *Blood 96*, 4152–9.

19. Gilles, L., Bluteau, D., Boukour, S., Chang, Y., Zhang, Y., Robert, T., Dessen, P., Debili, N., Bernard, O. A., Vainchenker, W., and Raslova, H. (2009) MAL/SRF complex is involved in platelet formation and megakaryocyte migration by regulating MYL9 (MLC2) and MMP9. *Blood 114*, 4221–32.

20. Zou, Z., Schmaier, A. A., Cheng, L., Mericko, P., Dickeson, S. K., Stricker, T. P., Santoro, S. A., and Kahn, M. L. (2009) Negative regulation of activated alpha-2 integrins during thrombopoiesis. *Blood 113*, 6428–39.

Part III

**Perspectives on the Study of Platelet
and Megakaryocyte Function**

Chapter 20

Megakaryopoiesis and Thrombopoiesis: An Update on Cytokines and Lineage Surface Markers

Ming Yu and Alan B. Cantor

Abstract

Megakaryopoiesis is the process by which mature megakaryocytes (MKs) develop from hematopoietic stem cells (HSCs). The biological function of MKs is to produce platelets, which play critical roles in hemostasis and contribute to angiogenesis and wound healing. The generation of platelets from MKs is termed thrombopoiesis. The cytokine thrombopoietin (TPO) is the major regulator of megakaryopoiesis and thrombopoiesis. It binds to its surface receptor, c-Mpl, and acts through multiple downstream signaling pathways, including the PI-3 kinase-Akt, MAPK, and ERK1/ERK2 pathways. However, non-TPO pathways, such as the SDF1/CXCR4 axis, Notch signaling, src family kinases, integrin signaling, and Platelet Factor 4/low-density lipoprotein receptor-related protein 1, have more recently been recognized to influence megakaryopoiesis and thrombopoiesis in vitro and in vivo. In this chapter, we review megakaryopoiesis and thrombopoiesis with emphasis on cell surface marker changes during their differentiation from HSCs, and the classical cytokines that affect these developmental stages. We also discuss non-TPO regulators and their effects on in vitro culture systems.

Key words: Megakaryopoiesis, Thrombopoietin, Hematopoietic stem cells, Thrombopoiesis, c-Mpl, Megakaryocyte, Megakaryocyte progenitor, Cytokine, Surface marker, CD41

1. Introduction

Megakaryocytes (MKs) are named for their large nucleus, i.e., mega (large) karyo (nucleus) cyte (cell), and are large polyploid blood cells with diameters ranging from 20 to 100 μm (1). The large size of MKs is thought to be the result of endomitosis, a physiological abortive mitosis characterized by DNA replication without nuclear and cellular division resulting in DNA ploidy up to 128 N. The biological function of MKs is to produce platelets, which play critical roles in hemostasis, wound healing, and angiogenesis.

Jonathan M. Gibbins and Martyn P. Mahaut-Smith (eds.), *Platelets and Megakaryocytes: Volume 3, Additional Protocols and Perspectives*, Methods in Molecular Biology, vol. 788, DOI 10.1007/978-1-61779-307-3_20, © Springer Science+Business Media, LLC 2012

Thrombopoiesis is the process by which MKs produce thrombocytes. In mammals, thrombocytes are anucleate cell fragments called platelets. In humans, around 10^{11} platelets are produced each day (2).

2. Hierarchical Generation of MKs from Hematopoietic Stem Cells

Like all blood cells, MKs are ultimately derived from hematopoietic stem cells (HSCs). Using a combination of flow cytometry and semisolid medium cell culture, it has been possible to dissect the hierarchical relationships between different cell populations during hematopoiesis (for review, see ref. 3). Murine systems provided the initial and most detailed mapping information, but equivalent immunophenotypic populations in humans have been recently characterized. A depiction of the overall hierarchical relationships between defined cell populations, with emphasis on derivation of the MK lineage, is shown in Fig. 1.

2.1. Hematopoietic Stem Cells

Hematopoiesis initiates in two distinct waves during the development of most vertebrates: a short-lived "primitive" stage that develops in extraembryonic structures and is characterized by embryonic globin gene expression and a later "definitive" stage that develops intraembryonically in the aorto-gonadal-mesonephros (AGM) region of the developing embryo (for review, see ref. 4). Definitive HSCs are derived from ventral mesoderm. Patterning of the mesoderm in preparation of subsequent hematopoietic development is mainly controlled by the bone morphogenetic protein (BMP) signaling. Hedgehog signaling amplifies stem cells and affects their differentiation through modulation of the BMP pathway (5).

HSCs are small, quiescent cells that constitute less than 0.1% of nucleated bone marrow cells. They are functionally characterized by the dual attributes of self-renewal and pluripotency (the ability to differentiate into all of the blood lineages). HSCs are highly enriched in a population of $Lin^-Sca1^+c-Kit^+$ (LSK) cells. A subpopulation capable of long-term hematopoietic reconstitution is called long-term HSCs (LT-HSCs). By definition, these cells can reconstitute a lethally irradiated animal and provide multilineage hematopoiesis for the lifetime of the animal. The immediate progeny of LT-HSCs are called short-term HSCs (ST-HSCs), a subpopulation with reduced self-renewal capacity and potential for only short-term reconstitution. Human and mouse LT-HSCs express different surface markers. Mouse LT-HSCs are $Lin^-c-Kit^+Sca1^+Flt3^-CD34^-CD150^+Thy1.1^{+/low}$ and human LT-HSCs are $Lin^-CD34^+CD38^-CD90^+CD45RA^-$ (3, 6) (Table 1). Mouse ST-HSCs are defined as $Lin^-c-Kit^+Sca1^+Flt3^-CD34^+CD150^+Thy1.1^{+/low}$ Mac-1^{low}.

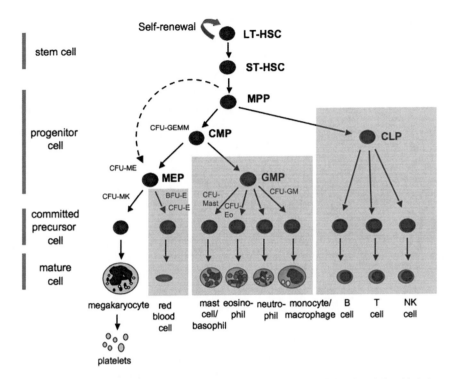

Fig. 1. Hierarchical pathways in hematopoietic development. Schematic diagram depicting the relationship between distinct identifiable progenitor populations during hematopoiesis. Nonmegakaryocytic differentiation pathways are shaded. The dashed line represents an alternate pathway of MEP derivation (9, 10). *LT-HSC* long-term hematopoietic stem cell, *ST-HSC* short-term hematopoietic stem cell, *MPP* multipotent progenitor, *CMP* common myeloid progenitor, *GMP* granulocyte-macrophage progenitor, *CLP* common lymphoid progenitor, *MEP* megakaryocyte–erythroid progenitor, *MkP* megakaryocyte progenitor, *CFU* colony-forming unit.

Table 1
Flow cytometric markers for distinct hematopoietic progenitor populations during megakaryopoiesis (3, 6, 8, 11, 15)

Cell population	Mouse	Human
LT-HSC	Lin⁻c-Kit⁺Sca1⁺Flt3⁻CD34⁻CD150⁺Thy1.1$^{+/low}$	Lin⁻CD34⁺CD38⁻CD90⁺CD45RA⁻
ST-HSC	Lin⁻c-Kit⁺Sca1⁺Flt3⁻CD34⁺CD150⁺Thy1.1$^{+/low}$ Mac-1^{low}	
MPP	Lin⁻c-Kit⁺Sca1⁺CD34⁺Flt3⁺CD150⁻Thy1.1⁻Mac-1^{low}CD4^{low} (early) Lin⁻c-Kit⁺Sca1⁺CD34⁺Flt3^{high}, CD150⁻ Thy1.1⁻Mac-1^{low}CD4^{low} (late)	Lin⁻CD34⁺CD38⁻CD90⁻CD45RA⁻
CMP	Lin⁻c-Kit⁺Sca1$^{-/low}$CD34⁺FcγR^{low}IL7R$_\alpha$⁻	Lin⁻CD34⁺CD38⁺IL3Ra^{low}CD45RA⁻
MEP	Lin⁻c-Kit⁺Sca1⁻CD34⁻FcγR^{low}IL7R$_\alpha$⁻	Lin⁻CD34⁺CD38⁺IL3Ra⁻CD45RA⁻
MkP	Lin⁻c-Kit⁺Sca1⁻CD150⁺CD41⁺ or CD9⁺CD41⁺FcRγ^{low}c-kit⁺Sca-1⁻IL7R$_\alpha$⁻Thy1.1⁻Lin⁻	

LT-HSC long-term hematopoietic stem cell, *ST-HSC* short-term hematopoietic stem cell, *MPP* multipotent progenitor, *CMP* common myeloid progenitor, *MEP* megakaryocyte–erythroid progenitor, *MkP* megakaryocyte progenitor

2.2. Multipotent Progenitors

Multipotent progenitors (MPPs) are multipotent cells. However, in contrast to HSCs, MPPs do not have substantial self-renewal capacity. These cells are highly heterogeneous. Early mouse MPP cells are characterized by the immunophenotype Lin⁻c-Kit⁺Sca1⁺CD34⁺ Flt3⁺ CD150⁻Thy1.1⁻Mac-1ˡᵒʷCD4ˡᵒʷ. Late mouse MPPs are Lin⁻c-Kit⁺Sca1⁺CD34⁺Flt3ʰⁱᵍʰ CD150⁻ Thy1.1⁻Mac-1ˡᵒʷCD4ˡᵒʷ. Both MPP populations also express c-Mpl. Human MPPs are characterized by the immunophenotype Lin⁻CD34⁺CD38⁻CD90⁻CD45RA⁻. They mainly reside in bone marrow in adults, but can also be detected in peripheral blood (7).

2.3. Common Myeloid Progenitors

Beyond MPPs, hematopoiesis bifurcates into cells that are restricted to myeloid or lymphoid cell fates (8). Common myeloid progenitors (CMPs) lack lymphoid potential, and are marked by IL-7R$_\alpha$⁻Lin⁻c-Kit⁺Sca-1⁻ᐟˡᵒʷCD34⁺FcγRˡᵒʷ for mouse (8) and Lin⁻CD34⁺CD38⁺IL3R$_\alpha$ˡᵒʷCD45RA⁻ for humans (6).

2.4. Megakaryocyte/ Erythroid Progenitors

Megakaryocyte/erythroid progenitors (MEPs) are the bipotential progenitors of megakaryocyte and erythroid cells derived from CMPs. Mouse MEPs are IL-7R$_\alpha$⁻Lin⁻c-Kit⁺Sca1⁻CD34⁻FcγR⁻ (8) and human MEPs are Lin⁻CD34⁺CD38⁺IL3R$_\alpha$⁻CD45RA⁻. Recent studies indicate that MEPs can also directly arise from early MPPs (9, 10) (see Fig. 1).

MEPs express both c-Mpl and erythropoietin receptor (EpoR) (11). Interestingly, thrombopoietin (TPO)- or c-Mpl-deficient mice have a lower number of erythroid and myeloid progenitors, although the levels of circulating white and erythoid cells are not affected in these mice (12). The mechanisms that influence the cell fate decision of MEPs for the erythroid and megakaryocyte lineages remain incompletely understood. Recent work suggests that functional cross-antagonism between the major MK transcription factors Fli1 and the major erythroid transcription factor EKLF plays a role (13, 14). How cytokines may regulate this process is not known.

2.5. Megakaryocytes

The burst-forming unit-megakaryocyte (BFU-MK) is the most primitive, committed MK progenitor cell, and is capable of producing numerous colony-forming unit-megakaryocyte (CFU-MK), a later-stage, more differentiated MK progenitor. The immunophenotype of clonogenic murine MK progenitors has recently been reported by two groups as CD9⁺CD41⁺Fc$_\gamma$Rˡᵒʷc-Kit⁺Sca-1⁻ IL7R$_\alpha$⁻Thy1.1⁻Lin⁻ (15) or Lin⁻c-Kit⁺Sca1⁻CD150⁺CD41⁺ cells (11). The former represents ~0.01% of the total nucleated bone marrow cells. The immunophenotype for human MK progenitors has not been reported as of yet. For bone marrow-derived human MKs, expression of the myeloid CD45 and immunoglobulin G (IgG)-Fc$_\gamma$RII receptor (CDw32) increases with maturation, whereas expression of HLA-DR declines (16).

CD41 (GPIIb), a key platelet surface glycoprotein, was initially thought to be a specific marker for MKs. However, expression of CD41 was found to mark the initiation of definitive hematopoiesis in the mouse embryo, and most yolk sac (YS) and AGM hematopoietic progenitor cells are CD41+ (17). Only a minority of bone marrow and fetal liver hematopoietic progenitors express this antigen (18). CD41 is also expressed on the surface of human and mouse mast cells (19, 20). CD42b (GPIb), which is expressed later than CD41 in MKs, is not expressed on mast cells.

3. Cytokines and MK Development

3.1. Thrombopoietin Signaling in Megakaryopoiesis

TPO is the primary regulator of megakaryopoiesis and thrombopoiesis. It was purified and cloned by five independent groups in 1994 (21–27). Binding of TPO to c-Mpl results in dimerization of the receptor and activation of a number of downstream signaling pathways, including the PI-3 kinase-Akt, MAPK, and ERK1/ERK2 pathways (reviewed by Kaushansky (28)). Unlike a number of other cytokine receptors, c-Mpl does not have intrinsic tyrosine kinase activity, but recruits cytosolic tyrosine kinases, such as JAK-2, to its cytoplasmic tail. This leads to tyrosine phosphorylation of c-Mpl itself and recruitment of signal transducers and activator of transcription (STAT) factors, such as STAT5 and STAT3 (29). The adaptor protein Lnk negatively regulates TPO signaling (30).

Other classical cytokines synergize with TPO in promoting megakaryopoiesis. Combinations of IL-3, IL-6, erythropoietin (EPO), and especially stem cell factor (SCF) enhance MK colony number and size when cultured from murine spleen or bone marrow in soft agar (31). SCF enhances TPO-induced STAT5 signaling in human megakaryocyte progenitors through JAK2 and Src kinase in the human megakaryocytic leukemic cell line MO7e and in primary human MK progenitors (32). Other cytokines, including IL-9 and IL-11, have been used in addition to the above cytokines to expand mouse MKs and CD34+CD41+ human MK progenitors in vitro (33, 34). G-CSF inhibits mouse MK colony formation stimulated by EPO, TPO, or IL-6, but not that stimulated by IL-3 (31).

3.2. Non-TPO Signaling in Megakaryopoiesis

Both c-Mpl and TPO knockout mice are severely thrombocytopenic, having ~15% of normal platelet levels and reduced MK numbers (35–37). Importantly, these animals show no abnormalities in MK and platelet structure, suggesting a primary role for TPO signaling in quantitative aspects of megakaryopoiesis and thrombopoiesis. The fact that TPO$^{-/-}$ and c-Mpl$^{-/-}$ mice have some residual platelet production indicates that receptors and ligands other than c-Mpl/TPO must also be able to support megakaryopoiesis. A number of such pathways, including some negative regulatory pathways, have been described over the past several years.

Both human and mouse MKs express CXCR4, the chemokine receptor of SDF-1 (38–40). SDF-1 and fibroblast growth factor-4 (FGF-4) enhance MK development and promote thrombopoiesis in TPO$^{-/-}$ and c-Mpl$^{-/-}$ mice (41). These effects involve enhancing MK progenitor interactions with bone marrow vascular sinusoidal structures and are mediated in part through vascular cell adhesion molecule-1 (VCAM-1) and very late antigen-4 (VLA-4). In humans, dysregulation of SDF-1/CXCR4 axis in the megakaryocytic lineage leads to thrombocythemia (42). Adenovirus-mediated transfer of FGF-4 gene induces increased levels of platelet count in vivo (43).

Gilliland and colleagues recently showed that Notch signaling markedly enhances megakaryopoiesis both in vitro and in vivo (44). These effects could be abrogated by inhibition of Notch signaling either with gamma-secretase inhibitors or by expression of the dominant-negative molecule Mastermind-like 1.

Src family tyrosine kinases (SFKs) play key roles in cell proliferative signaling (45). They typically act downstream of cytokine receptors, integrins, and focal adhesion complexes. A number of SFKs, including Fyn, Lyn, Fgr, Hck, Src, and Yes, are expressed in primary MKs, and Fyn and Lyn are activated upon TPO receptor signaling (46). Drachman and colleagues have shown that deficiency of Lyn or treatment with the SFK inhibitor SU6656 markedly enhances MK maturation, polyploidization, and platelet release (47–49). In fact, src inhibitors are now included in clinical research protocols for human in vitro platelet production (50).

Focal adhesion kinase (FAK) is a nonreceptor tyrosine kinase that is stimulated by integrin engagement, but also signals downstream of cytokine receptors. Kaushanksy and colleagues showed that megakaryocyte-specific deletion of FAK results in marked enhancement of megakaryopoiesis in vivo (51). Interactions between MKs and integrins or stroma have previously been shown to influence TPO-dependent megakaryocyte proliferation and maturation (52).

Poncz and colleagues recently demonstrated that platelet factor 4 (PF4) negatively regulates megakaryopoiesis in an autocrine fashion through low-density lipoprotein receptor-related protein 1 (LRP1) expressed on the MK surface (53).

Thrombospondin-1 (TSP-1), an angiogenesis inhibitor, also impairs human and mouse megakaryopoiesis in vitro via CD36 signaling (54).

3.3. Common Features Between MKs and HSCs

MKs share a number of specific features with HSCs, including their response to cytokines and other signaling pathways (for review, see ref. 55). It is worth noting that c-Mpl, the primary regulator of megakaryopoiesis and thrombopoiesis, is also expressed on HSCs and multipotent progenitors (11, 56). TPO or c-Mpl knockout mice show a reduced HSC number (57), and humans with congenital amegakaryocytic thrombocytopenia (CAMT) due to inactivating c-Mpl mutations often develop complete bone marrow failure (58). c-Mpl does not control the number of HSCs, but rather facilitates the early expansion of differentiating clones (59). Both human and mouse CD34+ HSCs express CXCR4 and its ligand SDF-1α, and CXCR4- or SDF-1α-deficient mice fail to establish bone marrow hematopoiesis (60, 61). Like megakaryocyte progenitors (MKPs), synergism between TPO and other cytokines has been observed for early multipotent progenitor cells. MPPs from yolk sac and the embryo proper (EP) proliferate in the presence of TPO and SCF, whereas their numbers are maintained by TPO alone (62).

4. Thrombopoiesis

MKs undergo endomitotic cell cycles during their terminal maturation, leading to polyploidy and markedly enlarged cell size. Endomitosis is ultimately used to direct the protein and lipid synthesis to create an extensive internal membranous labyrinth called the demarcation membrane system (DMS) (63, 64). This structure is contiguous with the plasma membrane (65). At the terminal stage of MK maturation, the cytoplasm forms a complex network of pseudopodal processes termed proplatelets (66–68). These are intermediates in platelet biogenesis, and appear as long tubular structures with bead-like protrusions linked by thin cytoplasmic bridges. Recent work by Shivdasani and colleagues indicates that the DMS serves as an extensive membrane reservoir during proplatelet formation (69). Microtubules are essential for proplatelet formation, and defects in their assembly lead to an absence of proplatelet generation and thrombocytopenia (70). Inhibition of actin polymerization has no effect on the extension of proplatelets, but impedes branching necessary for the amplification of platelet production. The driving force for proplatelet elongation is microtubule sliding (68).

4.1. Thrombopoiesis and the Bone Marrow Vascular Niche

Although in vitro studies demonstrate that proplatelet and platelet formation can occur in the absence of other cell types, this process occurs at vascular sinusoidal structures within the bone marrow space in vivo. Elegant in vivo imaging studies of terminal MK maturation have recently been performed using the murine calvarium model (71). These demonstrate attachment of MKs to the bone marrow side of vascular endothelial sinusoidal structures. Proplatelet processes are then extended in or around the vascular sinus endothelial cells and into the circulation, where they are then shed into the circulation. The resolution in this study was not great enough to identify individual platelets, but these were presumably also shed. These studies suggest that shear forces in the circulation may contribute to platelet and proplatelet shedding. Physical interactions between MK precursors and vascular sinusoidal structures, in combination with the earlier work on chemokines and adhesion molecules, point to the likely importance of cell–cell and cell–extracellular matrix contacts in terminal MK maturation and thrombopoiesis in vivo.

4.2. Platelet Surface Receptors

Platelet maturation and development involve the expression of specialized and highly regulated transmembrane receptors on their surface, prominent among which are the integrin family glycoproteins (GPs). These receptors facilitate platelet adhesion, spreading, and aggregation and promote thrombus development through receptor–ligand interactions. Surface markers expressed on the surface of platelets are summarized in Table 2. Interestingly, many of these platelet surface markers are also expressed on the MK surface. The potential functional role of these glycoproteins on megakaryopoiesis has not been fully explored.

Table 2
Platelet cell surface glycoproteins

Receptor	Ligand	Function
GPIa/IIa ($\alpha_2\beta_1$)	Collagen	Adhesion
GPIb/IX/V	Von Willebrand Factor (vWF)	Adhesion
GPIc/IIa ($\alpha_5\beta_1$)	Fibronectin	Adhesion, stabilizing GPIa/IIa
GPIIb/IIIa ($\alpha_{IIb}\beta_3$)	Thrombospondin Fibrinogen Vitronectin vWF	Aggregation
GPIV	Thrombospondin	Adhesion
GPVI	Collagen	Signal transduction Activation
Vitronectin ($\alpha_v\beta_3$)	Thrombospondin Vitronectin	Adhesion
VLA-6 ($\alpha_6\beta_1$)	Laminin	Adhesion

5. Culture Systems

The purification and cloning of TPO has enabled efficient in vitro culture of MKs. A number of different protocols have been described for mouse and human primary cells, as well as from embryonic stem cells (see Chapters 14–17, this volume, for more details). The collagen-based system from Stem Cell Technologies (Mega-Cult™) is widely used for semisolid medium culture of murine and human CFU-Mk. This is a collagen-containing Isocove's Modified Dulbecco's Medium (IMDM)-based semisolid medium that includes bovine serum albumin, recombinant human insulin, iron-saturated human transferrin, L-glutamine, and 2-mercaptoethanol. Recombinant TPO (50 ng/ml), IL-3 (10 ng/ml), and IL-6 (10 ng/ml) are recommended. The addition of SCF, in combination with TPO and IL-3, enhances the number of MK colonies produced from mouse bone marrow and spleen (31). Bone marrow cells are typically cultured for 10–12 days at 37°C, 5% CO_2, in a humidified environment. The cultures are then dried and stained for acetylcholinesterase activity or GPIIb/IIIa expression to identify MK colonies.

Liquid MK culture systems for single cell suspensions derived from yolk sac and fetal liver typically utilize IMDM with 15% heat-inactivated fetal calf serum and 5–10 ng/ml recombinant TPO. The dispersed fetal liver cells generate an adherent stromal layer when cultured on plastic. The MKs grow in suspension, loosely adhered to the stromal layer. Use of stromal cell lines that constitutively express Notch ligand markedly enhance MK generation, as described earlier (44). Fetal liver-derived MK progenitors produce significantly more MKs than adult bone marrow-derived cells. The molecular basis for this is not known, but likely relates to broad developmental differences between fetal, neonatal, and adult MKs (72, 73).

Well-characterized liquid culture systems have been developed for the in vitro differentiation of human CD34$^+$ cells into MKs. In one such protocol, human cord blood- or bone marrow-derived CD34$^+$ cells are differentiated into MKs using a two-phase culture system (74). This involves an initial culturing of the cells in serum-free medium consisting of IMDM, 20% BIT serum substitute (albumin/insulin/transferrin), low-density lipoproteins, 2-mercaptoethanol, TPO, SCF, FL, and IL-6. After 7 days, the culture medium is changed to include the optimal cytokine cocktail, TPO, SCF, IL-6, and IL-9, and cultured for a total of up to 14 additional days. Methods for culturing MKs from murine and human embryonic stem cells have been published (75, 76) and are continuing to be refined.

6. Conclusions

Because of the extreme rarity of MKs in the bone marrow, characterization of their surface markers and cytokine pathways initially lagged behind other blood lineages. However, the cloning and purification of TPO facilitated the efficient culturing of MKs, allowing for considerable progress in these areas. The ability to prospectively isolate committed MK progenitors (at least in the murine system) represents a significant new advance and should further enable MK developmental studies. The recognition and identification of TPO-independent pathways that also influence megakaryopoiesis has provided new insights into MK biology and platelet biogenesis. Further understanding of the pivotal molecular events that occur when MK precursors interact with bone marrow vascular sinusoidal structures during physiological MK terminal maturation should contribute to further improvements in MK in vitro culture systems.

References

1. Sun, L., Hwang, W. Y., and Aw, S. E. (2006) Biological characteristics of megakaryocytes: specific lineage commitment and associated disorders. *Int J Biochem Cell Biol* **38**, 1821–6.

2. Chang, Y., Bluteau, D., Debili, N., and Vainchenker, W. (2007) From hematopoietic stem cells to platelets. *J Thromb Haemost* **5** Suppl 1, 318–27.

3. Chao, M. P., Seita, J., and Weissman, I. L. (2008) Establishment of a normal hematopoietic and leukemia stem cell hierarchy. *Cold Spring Harb Symp Quant Biol* **73**, 439–49.

4. Orkin, S. H., and Zon, L. I. (2008) Hematopoiesis: an evolving paradigm for stem cell biology. *Cell* **132**, 631–44.

5. Bhardwaj, G., Murdoch, B., Wu, D., Baker, D. P., Williams, K. P., Chadwick, K., Ling, L. E., Karanu, F. N., and Bhatia, M. (2001) Sonic hedgehog induces the proliferation of primitive human hematopoietic cells via BMP regulation. *Nat Immunol* **2**, 172–80.

6. Manz, M. G., Miyamoto, T., Akashi, K., and Weissman, I. L. (2002) Prospective isolation of human clonogenic common myeloid progenitors. *Proc Natl Acad Sci USA* **99**, 11872–7.

7. Cesselli, D., Beltrami, A. P., Rigo, S., Bergamin, N., D'Aurizio, F., Verardo, R., Piazza, S., Klaric, E., Fanin, R., Toffoletto, B., Marzinotto, S., Mariuzzi, L., Finato, N., Pandolfi, M., Leri, A., Schneider, C., Beltrami, C. A., and Anversa, P. (2009) Multipotent progenitor cells are present in human peripheral blood. *Circ Res* **104**, 1225–34.

8. Akashi, K., Traver, D., Miyamoto, T., and Weissman, I. L. (2000) A clonogenic common myeloid progenitor that gives rise to all myeloid lineages. *Nature* **404**, 193–7.

9. Adolfsson, J., Mansson, R., Buza-Vidas, N., Hultquist, A., Liuba, K., Jensen, C. T., Bryder, D., Yang, L., Borge, O. J., Thoren, L. A., Anderson, K., Sitnicka, E., Sasaki, Y., Sigvardsson, M., and Jacobsen, S. E. (2005) Identification of Flt3+ lympho-myeloid stem cells lacking erythro-megakaryocytic potential a revised road map for adult blood lineage commitment. *Cell* **121**, 295–306.

10. Forsberg, E. C., Serwold, T., Kogan, S., Weissman, I. L., and Passegue, E. (2006) New evidence supporting megakaryocyte-erythrocyte potential of flk2/flt3+ multipotent hematopoietic progenitors. *Cell* **126**, 415–26.

11. Pronk, C. J., Rossi, D. J., Mansson, R., Attema, J. L., Norddahl, G. L., Chan, C. K., Sigvardsson, M., Weissman, I. L., and Bryder, D. (2007) Elucidation of the phenotypic, functional, and molecular topography of a myeloerythroid progenitor cell hierarchy. *Cell Stem Cell* **1**, 428–42.

12. Carver-Moore, K., Broxmeyer, H. E., Luoh, S. M., Cooper, S., Peng, J., Burstein, S. A., Moore, M. W., and de Sauvage, F. J. (1996) Low levels of erythroid and myeloid progenitors in thrombopoietin-and c-mpl-deficient mice. *Blood* **88**, 803–8.

13. Starck, J., Cohet, N., Gonnet, C., Sarrazin, S., Doubeikovskaia, Z., Doubeikovski, A., Verger, A., Duterque-Coquillaud, M., and Morle, F. (2003) Functional cross-antagonism between transcription factors FLI-1 and EKLF. *Mol Cell Biol* 23, 1390–402.

14. Frontelo, P., Manwani, D., Galdass, M., Karsunky, H., Lohmann, F., Gallagher, P. G., and Bieker, J. J. (2007) Novel role for EKLF in megakaryocyte lineage commitment. *Blood* 110, 3871–80.

15. Nakorn, T. N., Miyamoto, T., and Weissman, I. L. (2003) Characterization of mouse clonogenic megakaryocyte progenitors. *Proc Natl Acad Sci USA* 100, 205–10.

16. Tomer, A. (2004) Human marrow megakaryocyte differentiation: multiparameter correlative analysis identifies von Willebrand factor as a sensitive and distinctive marker for early (2 N and 4 N) megakaryocytes. *Blood* 104, 2722–7.

17. Mikkola, H. K., Fujiwara, Y., Schlaeger, T. M., Traver, D., and Orkin, S. H. (2003) Expression of CD41 marks the initiation of definitive hematopoiesis in the mouse embryo. *Blood* 101, 508–16.

18. Mitjavila-Garcia, M. T., Cailleret, M., Godin, I., Nogueira, M. M., Cohen-Solal, K., Schiavon, V., Lecluse, Y., Le Pesteur, F., Lagrue, A. H., and Vainchenker, W. (2002) Expression of CD41 on hematopoietic progenitors derived from embryonic hematopoietic cells. *Development* 129, 2003–13.

19. Berlanga, O., Emambokus, N., and Frampton, J. (2005) GPIIb (CD41) integrin is expressed on mast cells and influences their adhesion properties. *Exp Hematol* 33, 403–12.

20. Oki, T., Kitaura, J., Eto, K., Lu, Y., Maeda-Yamamoto, M., Inagaki, N., Nagai, H., Yamanishi, Y., Nakajima, H., Kumagai, H., and Kitamura, T. (2006) Integrin alphaIIbbeta3 induces the adhesion and activation of mast cells through interaction with fibrinogen. *J Immunol* 176, 52–60.

21. Bartley, T. D., Bogenberger, J., Hunt, P., Li, Y. S., Lu, H. S., Martin, F., Chang, M. S., Samal, B., Nichol, J. L., Swift, S., and et al. (1994) Identification and cloning of a megakaryocyte growth and development factor that is a ligand for the cytokine receptor Mpl. *Cell* 77, 1117–24.

22. de Sauvage, F. J., Hass, P. E., Spencer, S. D., Malloy, B. E., Gurney, A. L., Spencer, S. A., Darbonne, W. C., Henzel, W. J., Wong, S. C., Kuang, W. J., and et al. (1994) Stimulation of megakaryocytopoiesis and thrombopoiesis by the c-Mpl ligand. *Nature* 369, 533–8.

23. Kaushansky, K., Lok, S., Holly, R. D., Broudy, V. C., Lin, N., Bailey, M. C., Forstrom, J. W.,

Buddle, M. M., Oort, P. J., Hagen, F. S., and et al. (1994) Promotion of megakaryocyte progenitor expansion and differentiation by the c-Mpl ligand thrombopoietin. *Nature* 369, 568–71.

24. Kuter, D. J., Beeler, D. L., and Rosenberg, R. D. (1994) The purification of megapoietin: a physiological regulator of megakaryocyte growth and platelet production. *Proc Natl Acad Sci USA* 91, 11104–8.

25. Lok, S., Kaushansky, K., Holly, R. D., Kuijper, J. L., Lofton-Day, C. E., Oort, P. J., Grant, F. J., Heipel, M. D., Burkhead, S. K., Kramer, J. M., and et al. (1994) Cloning and expression of murine thrombopoietin cDNA and stimulation of platelet production in vivo. *Nature* 369, 565–8.

26. Wendling, F., Maraskovsky, E., Debili, N., Florindo, C., Teepe, M., Titeux, M., Methia, N., Breton-Gorius, J., Cosman, D., and Vainchenker, W. (1994) cMpl ligand is a humoral regulator of megakaryocytopoiesis. *Nature* 369, 571–4.

27. Sohma, Y., Akahori, H., Seki, N., Hori, T., Ogami, K., Kato, T., Shimada, Y., Kawamura, K., and Miyazaki, H. (1994) Molecular cloning and chromosomal localization of the human thrombopoietin gene. *FEBS Lett* 353, 57–61.

28. Kaushansky, K. (2005) The molecular mechanisms that control thrombopoiesis. *J Clin Invest* 115, 3339–47.

29. Bacon, C. M., Tortolani, P. J., Shimosaka, A., Rees, R. C., Longo, D. L., and O'Shea, J. J. (1995) Thrombopoietin (TPO) induces tyrosine phosphorylation and activation of STAT5 and STAT3. *FEBS Lett* 370, 63–8.

30. Tong, W., and Lodish, H. F. (2004) Lnk inhibits Tpo-mpl signaling and Tpo-mediated megakaryocytopoiesis. *J Exp Med* 200, 569–80.

31. Metcalf, D., Di Rago, L., and Mifsud, S. (2002) Synergistic and inhibitory interactions in the in vitro control of murine megakaryocyte colony formation. *Stem Cells* 20, 552–60.

32. Drayer, A. L., Boer, A. K., Los, E. L., Esselink, M. T., and Vellenga, E. (2005) Stem cell factor synergistically enhances thrombopoietin-induced STAT5 signaling in megakaryocyte progenitors through JAK2 and Src kinase. *Stem Cells* 23, 240–51.

33. Ahmed, N., Khokher, M. A., and Hassan, H. T. (1999) Cytokine-induced expansion of human CD34+ stem/progenitor and CD34+CD41+ early megakaryocytic marrow cells cultured on normal osteoblasts. *Stem Cells* 17, 92–9.

34. Lazzari, L., Henschler, R., Lecchi, L., Rebulla, P., Mertelsmann, R., and Sirchia, G. (2000) Interleukin-6 and interleukin-11 act synergistically

with thrombopoietin and stem cell factor to modulate ex vivo expansion of human CD41+ and CD61+ megakaryocytic cells. *Haematologica* **85**, 25–30.

35. Alexander, W. S., Roberts, A. W., Nicola, N. A., Li, R., and Metcalf, D. (1996) Deficiencies in progenitor cells of multiple hematopoietic lineages and defective megakaryocytopoiesis in mice lacking the thrombopoietic receptor c-Mpl. *Blood* **87**, 2162–70.

36. Gurney, A. L., Carver-Moore, K., de Sauvage, F. J., and Moore, M. W. (1994) Thrombocytopenia in c-mpl-deficient mice. *Science* **265**, 1445–7.

37. Bunding, S., Widmer R, Lipari, T, Rangell, L, Steinmetz, H, Carver-Moore, K, Moore, MW, Keller, GA, de Sauvage, FJ. (1997) Normal platelets and megakaryocytes are produced in vivo in the absence of thrombopoietin. *Blood* **90**, 3423–3429.

38. Wang, J. F., Liu, Z. Y., and Groopman, J. E. (1998) The alpha-chemokine receptor CXCR4 is expressed on the megakaryocytic lineage from progenitor to platelets and modulates migration and adhesion. *Blood* **92**, 756–64.

39. Riviere, C., Subra, F., Cohen-Solal, K., Cordette-Lagarde, V., Letestu, R., Auclair, C., Vainchenker, W., and Louache, F. (1999) Phenotypic and functional evidence for the expression of CXCR4 receptor during megakaryocytopoiesis. *Blood* **93**, 1511–23.

40. Perez, L. E., Desponts, C., Parquet, N., and Kerr, W. G. (2008) SH2-inositol phosphatase 1 negatively influences early megakaryocyte progenitors. *PLoS One* **3**, e3565.

41. Avecilla, S. T., Hattori, K., Heissig, B., Tejada, R., Liao, F., Shido, K., Jin, D. K., Dias, S., Zhang, F., Hartman, T. E., Hackett, N. R., Crystal, R. G., Witte, L., Hicklin, D. J., Bohlen, P., Eaton, D., Lyden, D., de Sauvage, F., and Rafii, S. (2004) Chemokine-mediated interaction of hematopoietic progenitors with the bone marrow vascular niche is required for thrombopoiesis. *Nat Med* **10**, 64–71.

42. Salim, J. P., Goette, N. P., Lev, P. R., Chazarreta, C. D., Heller, P. G., Alvarez, C., Molinas, F. C., and Marta, R. F. (2009) Dysregulation of stromal derived factor 1/CXCR4 axis in the megakaryocytic lineage in essential thrombocythemia. *Br J Haematol* **144**, 69–77.

43. Sakamoto, H., Ochiya, T., Sato, Y., Tsukamoto, M., Konishi, H., Saito, I., Sugimura, T., and Terada, M. (1994) Adenovirus-mediated transfer of the HST-1 (FGF4) gene induces increased levels of platelet count in vivo. *Proc Natl Acad Sci USA* **91**, 12368–72.

44. Mercher, T., Cornejo, M. G., Sears, C., Kindler, T., Moore, S. A., Maillard, I., Pear, W. S., Aster, J. C., and Gilliland, D. G. (2008) Notch signaling specifies megakaryocyte development from hematopoietic stem cells. *Cell Stem Cell* **3**, 314–26.

45. Yeatman, T. J. (2004) A renaissance for SRC. *Nat Rev Cancer* **4**, 470–80.

46. Lannutti, B. J., Shim, M. H., Blake, N., Reems, J. A., and Drachman, J. G. (2003) Identification and activation of Src family kinases in primary megakaryocytes. *Exp Hematol* **31**, 1268–74.

47. Lannutti, B. J., Blake, N., Gandhi, M. J., Reems, J. A., and Drachman, J. G. (2005) Induction of polyploidization in leukemic cell lines and primary bone marrow by Src kinase inhibitor SU6656. *Blood* **105**, 3875–8.

48. Lannutti, B. J., and Drachman, J. G. (2004) Lyn tyrosine kinase regulates thrombopoietin-induced proliferation of hematopoietic cell lines and primary megakaryocytic progenitors. *Blood* **103**, 3736–43.

49. Lannutti, B. J., Minear, J., Blake, N., and Drachman, J. G. (2006) Increased megakaryocytopoiesis in Lyn-deficient mice. *Oncogene* **25**, 3316–24.

50. Sullenbarger, B., Bahng, J. H., Gruner, R., Kotov, N., and Lasky, L. C. (2009) Prolonged continuous in vitro human platelet production using three-dimensional scaffolds. *Exp Hematol* **37**, 101–10.

51. Hitchcock, I. S., Fox, N. E., Prevost, N., Sear, K., Shattil, S. J., and Kaushansky, K. (2008) Roles of focal adhesion kinase (FAK) in megakaryopoiesis and platelet function: studies using a megakaryocyte lineage specific FAK knockout. *Blood* **111**, 596–604.

52. Fox, N. E., and Kaushansky, K. (2005) Engagement of integrin alpha4beta1 enhances thrombopoietin-induced megakaryopoiesis. *Exp Hematol* **33**, 94–9.

53. Lambert, M. P., Wang, Y., Bdeir, K. H., Nguyen, Y., Kowalska, M. A., and Poncz, M. (2009) Platelet factor 4 regulates megakaryopoiesis through low-density lipoprotein receptor-related protein 1 (LRP1) on megakaryocytes. *Blood* **114**, 2290–8.

54. Yang, M., Li, K., Ng, M. H., Yuen, P. M., Fok, T. F., Li, C. K., Hogg, P. J., and Chong, B. H. (2003) Thrombospondin-1 inhibits in vitro megakaryocytopoiesis via CD36. *Thromb Res* **109**, 47–54.

55. Huang, H., and Cantor, A. B. (2009) Common features of megakaryocytes and hematopoietic stem cells: what's the connection? *J Cell Biochem* **107**, 857–64.

56. Satoh, Y., Matsumura, I., Tanaka, H., Ezoe, S., Fukushima, K., Tokunaga, M., Yasumi, M., Shibayama, H., Mizuki, M., Era, T., Okuda, T.,

and Kanakura, Y. (2008) AML1/RUNX1 works as a negative regulator of c-Mpl in hematopoietic stem cells. *J Biol Chem* **283**, 30045–56.

57. Solar, G. P., Kerr, W. G., Zeigler, F. C., Hess, D., Donahue, C., de Sauvage, F. J., and Eaton, D. L. (1998) Role of c-mpl in early hematopoiesis. *Blood* **92**, 4–10.

58. van den Oudenrijn, S., Bruin, M., Folman, C. C., Peters, M., Faulkner, L. B., de Haas, M., and von dem Borne, A. E. (2000) Mutations in the thrombopoietin receptor, Mpl, in children with congenital amegakaryocytic thrombocytopenia. *Br J Haematol* **110**, 441–8.

59. Abkowitz, J. L., and Chen, J. (2007) Studies of c-Mpl function distinguish the replication of hematopoietic stem cells from the expansion of differentiating clones. *Blood* **109**, 5186–90.

60. Zou, Y. R., Kottmann, A. H., Kuroda, M., Taniuchi, I., and Littman, D. R. (1998) Function of the chemokine receptor CXCR4 in haematopoiesis and in cerebellar development. *Nature* **393**, 595–9.

61. Mohle, R., Bautz, F., Rafii, S., Moore, M. A., Brugger, W., and Kanz, L. (1998) The chemokine receptor CXCR-4 is expressed on CD34+ hematopoietic progenitors and leukemic cells and mediates transendothelial migration induced by stromal cell-derived factor-1. *Blood* **91**, 4523–30.

62. Huang, X., Sakamoto, H., and Ogawa, M. (2009) Thrombopoietin controls proliferation of embryonic multipotent hematopoietic progenitors. *Genes Cells* **14**, 851–60.

63. Yamada, E. (1957) The fine structure of the megakaryocyte in the mouse spleen. *Acta Anat (Basel)* **29**, 267–90.

64. Behnke, O. (1968) An electron microscope study of the megacaryocyte of the rat bone marrow. I. The development of the demarcation membrane system and the platelet surface coat. *J Ultrastruct Res* **24**, 412–33.

65. Mahaut-Smith, M. P., Thomas, D., Higham, A. B., Usher-Smith, J. A., Hussain, J. F., Martinez-Pinna, J., Skepper, J. N., and Mason, M. J. (2003) Properties of the demarcation membrane system in living rat megakaryocytes. *Biophys J* **84**, 2646–54.

66. Italiano, J. E., Jr., Lecine, P., Shivdasani, R. A., and Hartwig, J. H. (1999) Blood platelets are assembled principally at the ends of proplatelet processes produced by differentiated megakaryocytes. *J Cell Biol* **147**, 1299–312.

67. Radley, J. M., and Haller, C. J. (1982) The demarcation membrane system of the megakaryocyte: a misnomer? *Blood* **60**, 213–9.

68. Patel, S. R., Richardson, J. L., Schulze, H., Kahle, E., Galjart, N., Drabek, K., Shivdasani, R. A., Hartwig, J. H., and Italiano, J. E., Jr. (2005) Differential roles of microtubule assembly and sliding in proplatelet formation by megakaryocytes. *Blood* **106**, 4076–85.

69. Schulze, H., Korpal, M., Hurov, J., Kim, S. W., Zhang, J., Cantley, L. C., Graf, T., and Shivdasani, R. A. (2006) Characterization of the megakaryocyte demarcation membrane system and its role in thrombopoiesis. *Blood* **107**, 3868–75.

70. Cramer, E. M., Norol, F., Guichard, J., Breton-Gorius, J., Vainchenker, W., Masse, J. M., and Debili, N. (1997) Ultrastructure of platelet formation by human megakaryocytes cultured with the Mpl ligand. *Blood* **89**, 2336–46.

71. Junt, T., Schulze, H., Chen, Z., Massberg, S., Goerge, T., Krueger, A., Wagner, D. D., Graf, T., Italiano, J. E., Jr., Shivdasani, R. A., and von Andrian, U. H. (2007) Dynamic visualization of thrombopoiesis within bone marrow. *Science* **317**, 1767–70.

72. Pastos, K. M., Slayton, W. B., Rimsza, L. M., Young, L., and Sola-Visner, M. C. (2006) Differential effects of recombinant thrombopoietin and bone marrow stromal-conditioned media on neonatal versus adult megakaryocytes. *Blood* **108**, 3360–2.

73. Hu, Z., Slayton, W. B., Rimsza, L. M., Bailey, M., Sallmon, H., and Sola-Visner, M. C. Differences between newborn and adult mice in their response to immune thrombocytopenia. *Neonatology* **98**, 100–8.

74. Cortin, V., Garnier, A., Pineault, N., Lemieux, R., Boyer, L., and Proulx, C. (2005) Efficient in vitro megakaryocyte maturation using cytokine cocktails optimized by statistical experimental design. *Exp Hematol* **33**, 1182–91.

75. Fujimoto, T. T., Kohata, S., Suzuki, H., Miyazaki, H., and Fujimura, K. (2003) Production of functional platelets by differentiated embryonic stem (ES) cells in vitro. *Blood* **102**, 4044–51.

76. Gaur, M., Kamata, T., Wang, S., Moran, B., Shattil, S. J., and Leavitt, A. D. (2006) Megakaryocytes derived from human embryonic stem cells: a genetically tractable system to study megakaryocytopoiesis and integrin function. *J Thromb Haemost* **4**, 436–42.

Chapter 21

Using Zebrafish (*Danio rerio*) to Assess Gene Function in Thrombus Formation

Christopher M. Williams and Alastair W. Poole

Abstract

Cardiovascular and cerebrovascular disease is the major cause of death in the developed world, with a high burden of disease and substantial pharmaceutical investment to manage it (WHO, Global Burden of Disease, 2004 Update, W.H. Organisation, Editor. 2008). Platelets, as the principal mediators of thrombus formation, are a primary pharmaceutical target, with attenuation of platelet function and thrombus formation significantly reducing the incidence of myocardial infarction and stroke. Haemostasis, however, may also be affected by antithrombotics, leading to spontaneous and/or prolonged bleeding as a potentially severe side effect. Developing a comprehensive understanding of the mechanisms involved in platelet function and thrombus formation is anticipated to identify drug targets that may effectively manage vascular disease without an impact on haemostasis. Despite the progress in characterising individual genes in platelet function and thrombosis, using gene knockout and transgenic mice over the past decade or so, there is still much to be uncovered.

Investigating gene function using mouse models is a substantial investment and a considerable amount of work, with a relevant phenotype not guaranteed. As such, a new model is needed for the effective screening of novel genes that have been identified as having potential roles in platelet function or cardiovascular disease by genomic association and comparative expression studies (Nature, 447(7145): 661–678, 2007; Nat Genet, 41(11): 1182–1190, 2009; N Engl J Med, 357(5): 443–453, 2007; Blood, 109(8): 3260–3269, 2007). Here, we highlight and discuss the relevance of the zebrafish (Danio rerio) as a model for studying thrombosis, the current techniques that are employed to assess gene function in a zebrafish model of thrombosis, and how an effective genetic screen may be constructed.

Key words: Thrombosis, Thrombus, Zebrafish, Thrombocytes, Laser, Morpholino, Knock-down, Mutant

1. Introduction

The pathological role of platelets in cardio- and cerebrovascular disease and thrombosis has led to the generation of many pharmaceutical agents aiming to attenuate platelet function and prevent the

Jonathan M. Gibbins and Martyn P. Mahaut-Smith (eds.), *Platelets and Megakaryocytes: Volume 3, Additional Protocols and Perspectives*, Methods in Molecular Biology, vol. 788, DOI 10.1007/978-1-61779-307-3_21, © Springer Science+Business Media, LLC 2012

formation of occlusive thrombi that lead to myocardial infarction and stroke (1). These drugs have been targeted at platelet adhesion (e.g. tirofiban, abciximab, eptifibatide (6)) or targeted to prevent/inhibit the amplification signals ADP or thromboxane A_2 (e.g. aspirin, clopidogrel, prasugrel (7, 8)). Despite the effectiveness of these drugs, side effects can be severe. Aspirin, though cost-effective, can cause gastrointestinal (GI) bleeding after prolonged usage (9, 10). Patients treated with the $P2Y_{12}$ receptor antagonist clopidogrel can also show bleeding in the GI tract, which can be increased when co-administered with aspirin, though a subset of patients has been reported to be unresponsive to clopidogrel treatment (7, 11–14). Inhibitors of the primary platelet adhesion receptor, the $\alpha_{IIb}\beta_3$ integrin (e.g. tirofiban, abciximab), also have undesirable side effects. As such potent inhibitors of platelet function, patients on $\alpha_{IIb}\beta_3$ antagonists can also suffer increased bleeding or in some instances can result in increased mortality due to a paradoxical increase in platelet activation (15–17). To this end, the field of thrombosis and haemostasis is continually striving to achieve a greater understanding of the mechanisms involved in platelet function and thrombus formation. This greater understanding may allow the development of more effective pharmacological agents to manage myocardial infarction and stroke while maintaining effective haemostasis. Recent genome-wide association and comparative expression studies have identified numerous novel genes with potential roles in cardiovascular disease and platelet function (2–5). Murine gene knockout models have greatly assisted in understanding the molecular basis of platelet function and thrombus formation since their development in the early 1990s; however, with the wealth of genomic data presented in recent years, the need for a model that can quickly screen gene function in thrombosis, prior to a more in-depth characterisation, is becoming apparent. The zebrafish (Danio rerio) is one such model.

Zebrafish are small, tropical, freshwater fish originating from the River Ganges (18, 19). Maturing within 3–4 months, adult zebrafish have high fecundity and are capable of breeding every 1–2 weeks, with a single pair able to lay up to 200 eggs (19, 20). Mating at first light, egg/embryo production can be managed by controlling the lighting conditions, providing a distinct advantage to the investigator with respect to the development of methodical, experimental procedures. The fertilised eggs are relatively large and translucent and are amenable to the introduction of soluble compounds, such as antisense morpholinos (described later) via microinjection approaches (21, 22). The translucency of the developing embryos/larvae is another key reason for the emergence of zebrafish as a thrombosis model as the vasculature can be clearly visualised by bright-field microscopy, with thrombus development easily identified and followed.

The zebrafish is a well-characterised model and has an open research community that supports valuable resources, such as the

Zebrafish Information Network, ZFIN (http://www.zfin.org/cgi-bin/webdriver?MIval=aa-ZDB_home.apg), a database that brings together information on zebrafish genes, mutant and transgenic lines, antibodies, anatomy and gene expression, publications, genomic databases, and community links and the zebrafish Book – a guide for the laboratory use of zebrafish (20). Much of the zebrafish genome is mapped, with many zebrafish genes showing high sequence homology with human orthologues (23, 24). Researchers should be aware, however, that although the zebrafish genome is now fully sequenced, its level of annotation is currently poorer than that for human or mouse. When searching for genes in genome browsers, such as Ensembl, it is therefore necessary to consider the potential for incorrect or incompletely annotated sequences. To this end, careful and rigorous searching and validation is required to confirm that the correct fish orthologue is being studied. Also, teleost-specific gene duplications mean that some zebrafish genes have paralogues when only a single orthologue is present in humans (18, 24, 25). Such duplication events need to be considered when studying novel genes in zebrafish, as duplicate genes may provide redundancy of gene function, and so yield false negatives. Despite this, the conservation of thrombosis and haemostasis in zebrafish, and its ease of use as a model organism, place zebrafish in an ideal position for a rapid screen of in vivo gene function in thrombosis.

2. Zebrafish Thrombocytes and Haemostasis

Diverging from humans, where anucleate platelets are "born" from precursor cells (megakaryocytes), zebrafish possess nucleated thrombocytes, which instead mature from thrombocyte precursors in the zebrafish vascular niche (26). The thrombocyte shares the same functions and regulatory mechanisms as human platelets, with the coagulation pathways, vasculogenesis, and haematopoiesis all conserved in zebrafish (24, 26–33), though the mechanisms that drive zebrafish thrombopoiesis are not fully described. Demonstrating the conservation of the coagulation pathways, adult zebrafish swimming in water treated with warfarin showed reduced factor X and protein C activity, delayed clotting times, and spontaneous bleeding (29). Likewise, knock-down of factor VIII (34) and prothrombin (35) activity in zebrafish embryos using antisense morpholinos (described later) reduced thrombus size and increased time to occlusion, respectively, in a laser injury model of thrombosis, reproducing their expected mammalian phenotypes. Injection of human FITC-fibrinogen results in its accumulation at the site of thrombus formation, clearly demonstrating the highly conserved nature of the coagulation system (36). With regards to thrombopoiesis, despite the absence of a megakaryocyte stage in thrombocyte development, thrombopoietin and its receptor c-Mpl are conserved and remain essential for thrombocyte development, as morpholino knock-down of c-Mpl results in a reduction in circulating thrombocytes (33). Likewise, two transcription factors that regulate megakaryocyte and haematopoietic progenitor differentiation, GATA1 and FLI1, also regulate thrombocyte maturation in zebrafish (37).

As with coagulation and thrombopoiesis, the functional and regulatory machinery of human platelets has also been shown to be conserved, with the $\alpha_{IIb}\beta_3$ integrin (CD41/CD61), GPIb-V-IX, PAR, and P2Y receptors all present, though the collagen receptor GPVI is still to be identified (24). Through the pioneering efforts of Dr Jagadeeswaran and his laboratory, functional responses, such as aggregation, calcium release, P-selectin exposure, annexin V binding, dense-granule secretion, and thrombus formation, have all been demonstrated in zebrafish and thrombocytes, establishing zebrafish as an appropriate organism to model mammalian thrombosis (18, 26, 30–32, 36, 38).

To aid in the analysis of thrombocyte development, Lin et al. (2006) created a CD41-GFP transgenic zebrafish line (33). Using CD41-GFP embryos, CD41 expression was detected within 2 days post fertilisation (dpf), together with the observation of CD41-GFP+ cells in the haematopoietic tissue between the caudal artery and caudal vein (the zebrafish equivalent of fetal liver) (33, 39). In these embryos, a dual population of CD41-GFP+ cells was also observed: CD41-GFP^low and CD41-GFP^high. Here, CD41-GFP^low cells are considered haematopoietic progenitor cells with CD41-GFP^high cells being thrombocytes/prothrombocytes (Fig. 1). CD41-GFP+ thrombocytes can be detected in the circulation between 2 and 3 dpf, indicating that thrombosis can be studied in

Fig. 1. CD41-GFP+ cells in CD41-GFP zebrafish embryos. A representative photomicrograph of the haematopoietic tissue between the caudal artery and caudal vein in a 4dpf CD41-GFP zebrafish embryo. A CD41-GFP^high cell identified as a thrombocyte/prothrombocyte is indicated by the *dashed white arrow* with a cluster of CD41-GFP^low cells, likely to be haematopoietic progenitors, indicated by the *solid white arrows*.

any embryo older than this (33, 39). Indeed, in laser injury models of thrombosis in 4 dpf zebrafish, thrombus formation is driven by CD41-GFP$^+$ cells (34).

3. Zebrafish as a Model for Thrombosis

Several in vivo models of thrombosis have been developed in mice over the years and include models based upon mechanical, chemical (and photochemical), or laser-induced damage to the vascular endothelium (reviewed in ref. 40, 41). The most commonly used are the laser injury and ferric chloride (FeCl$_3$) models. The laser injury model uses a pulsed nitrogen dye laser integrated into a microscope to cause discrete damage to an artery in a tissue with transparent vasculature, such as the mesentery or cremaster muscle (42, 43). This can be used in conjunction with fluorescent labelling of platelets and followed by intravital time-lapse microscopy, allowing quantification of thrombus dynamics. The topical application of FeCl$_3$ is widely used in chemical-induced thrombosis, where it results in the denudation of the endothelium and pronounced thrombus formation (44). Models of FeCl$_3$-induced thrombosis include the application of FeCl$_3$ to either the microvasculature of the mesentery or the more physiologically relevant carotid/femoral arteries, with thrombus formation followed by intravital microscopy (44) or a Döppler flow probe (45), respectively. There is some contention as to which model is the most physiologically relevant, with each having its own advantages and disadvantages; however, all the models require surgical training and substantial experience to generate reproducible data.

With many of the key components of thrombosis and haemostasis shown to be conserved in zebrafish, they present as a physiologically relevant organism with which to model thrombosis. There are, however, several practical advantages that make them a highly suitable model to rapidly screen gene function. The main practical advantage, as previously stated, is that developing zebrafish embryos/larvae are translucent. This allows for the straightforward visualisation of the zebrafish vasculature by standard bright-field microscopy without any manipulation by the investigator, therefore requiring less training and time to become suitably experienced. This is in contrast to murine models of microvascular thrombosis which require immunofluorescent labelling of platelets for reliable visualisation and quantification, as well as significant surgical manipulation to allow such imaging (42, 46). In zebrafish, both the FeCl$_3$ (36) and laser injury (34, 36, 38) models have been successfully applied, with the laser injury model proving to result in faster, reversible thrombus formation compared to the irreversible thrombus formation seen with FeCl$_3$ (30, 36). Another advantage

Table 1
Summary of parameters that can be measured in the zebrafish model of thrombosis

Parameter	Descriptor	Summary of parameter
TTA	Time to attachment of initial cell	Readout of thrombus initiation, including thrombocyte activation and thrombocyte–subendothelium interactions
TTO	Time to vessel occlusion	Readout of thrombus growth, including thrombocyte activation and aggregation
TTD	Time to thrombus dissolution	Readout of thrombus stability and retraction
TSA	Thrombus surface area	Readout of thrombus growth, including thrombocyte activation and aggregation

of the laser injury model is that the intensity of the laser, and its rate and duration of fire, can be varied. This adds an element of flexibility to thrombus induction, allowing the control of vascular damage and so peak thrombus size. The extent of thrombus formation can, therefore, be managed depending on the parameters to be measured, which can include measurements of thrombus kinetics (34, 38) and thrombus growth (34) (outlined in Table 1 and Fig. 2). The use of time-lapse microscopy with good temporal resolution is vital for the accurate measurement of the parameters used for the analysis of thrombus kinetics. A successful laser injury of the vessel wall results in a characteristic visual depression or twitch (Fig. 2b). This indication of a wound can be used as a marker from which to measure two parameters: the time to adhesion (TTA) of the initial cell (Fig. 2c) and the time to occlusion (TTO) of the vessel (Fig. 2d).

Time to attachment measurements are used as a readout of platelet–endothelial interactions (typically occurring within 30 s of wounding) and are possible due to the larger nature of zebrafish thrombocytes compared to human platelets (34, 38, 47). TTO is a common parameter for in vivo models of thrombosis as occlusive thrombi are the primary cause of myocardial infarction and stroke. In zebrafish, the formation of occlusive thrombi typically occurs within 90 s and can be considered as a readout of thrombus growth, including the extent of thrombocyte activation and aggregation (34, 38). However, a substantial wound is required to consistently induce the formation of occlusive thrombi (38), with O'Connor et al. (2009) reporting that only 10% of thrombi in their laser injury model of thrombosis became occlusive (34). When occlusive

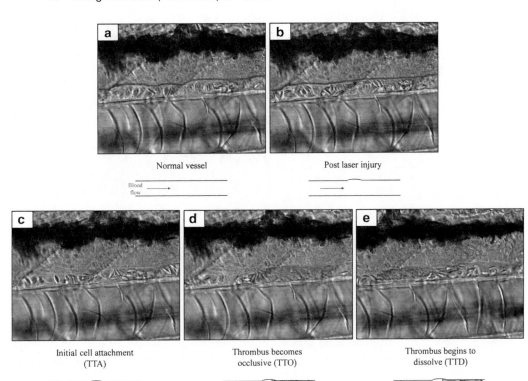

Fig. 2. Time course for laser-induced thrombus formation in a 3 dpf zebrafish embryo presenting measurable parameters. Wild-type zebrafish embryos (3 dpf) were wounded by laser at a section of caudal artery, above the cloaca, with the resulting thrombus formation followed over 10 min by DIC microscopy. Parts (**a**) and (**b**) show the vessel wall before and after injury, with the characteristic depression seen in the wall post laser damage, as indicated by the *grey* outline of the vessel wall and asterisk. Part (**c**) demonstrates the initial cell adhesion event, with (**d**) showing the growth of the thrombus leading to the total occlusion of the vessel, before its eventual clearance (**e**). Parts (**c–e**) also demonstrate the measurement of thrombus surface area (TSA; hue).

thrombi do form, they typically begin to clear within 2–10 min of becoming occlusive (34, 38), allowing the measurement of the time to dissolution (TTD) of the thrombus, which can give a readout of thrombus stability or retraction (Fig. 2e). Measurements of thrombus surface area (TSA) have also been used as a readout for thrombus growth (34), but unlike TTO measurements they require the size of the thrombus to be defined and measured using image analysis software, such as NIH ImageJ (e.g. Fig. 2c–e). To this end, the CD41-GFP transgenic zebrafish line presents itself as an ideal tool to use to this end as the analysis of GFP-thrombus formation, captured by fluorescence time-lapse microscopy, is straightforward and yields a wealth of data with high temporal resolution (Fig. 3a). Unfortunately, not all thrombi, especially large thrombi, are entirely composed of CD41-GFP⁺ cells (Fig. 3b), making analysis of these thrombi by fluorescence inaccurate. The manual measurement of TSA is considerably more time intensive; so to use

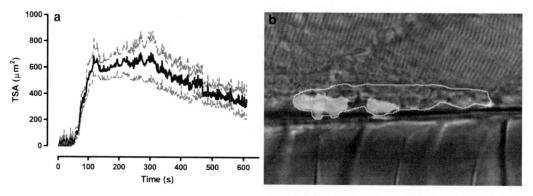

Fig. 3. Analysis of thrombus surface area using CD41-GFP zebrafish embryos. CD41-GFP zebrafish embryos (4 dpf) were wounded by laser at a section of caudal artery, above the cloaca, with the resulting thrombus formation followed over 10 min by infrared-DIC and fluorescence time-lapse microscopy. The developing thrombus was defined as a region of interest with the thrombus surface area (TSA) measured every second by Image J 1.42 using threshold, greyscale images from the fluorescence channel. (**a**) The TSA as measured by threshold GFP fluorescence (*n*=6). Data are shown as mean (*black*) ± SEM (*grey*). (**b**) Representative image of a thrombus (*white outline*) from a CD41-GFP zebrafish embryo containing both GFP-thrombocytes (*white*) and non-GFP cells.

the zebrafish as a model to screen for the functional role of novel genes in thrombosis, a compromise is required between the return of quantitative data and the time expended to analyse it. In their recent work, O'Connor et al. (34) calculated TSA for every minute of their time courses, before using a 2-min time point to compare the effects of morpholino administration. Since thrombus formation is a dynamic process, such a snapshot comparison may not be ideal, but as O'Connor et al. have shown it can be effectively used for the functional analysis of novel platelet genes in thrombus formation.

4. Assessing Gene Function: Mutants and Morpholinos

In zebrafish, gene function can be analysed through the generation of mutant lines or by transient knock-down using antisense morpholinos. Targeting-induced local lesions in genomes (TILLING) is a reverse genetic strategy for the production of mutant lines (48). The approach identifies heterozygous mutations in specific genes of interest in a population of mutagenised embryos. These mutagenised embryos are typically generated by breeding male fish with chemically mutagenised germ cells (after treatment with mutagens, such as ethylnitrosourea (ENU), which yield multiple, random point mutations) with wild-type females (49–52). Mutagenised populations can also be generated by insertional mutagenesis approaches, such as injection of retrovirus into blastula-stage embryos (50, 51), though these are considered less efficient at generating mutants than chemical mutagenesis (50). The screened

F1 generation of heterozygous mutants can then be used to generate mutant lines homozygous for loss of function mutations in the gene of interest within a couple of generations (50, 51). The production of a stable mutant line is advantageous in terms of flexibility, with the ability to perform in vivo thrombosis analysis at the embryonic stage and additional functional assays using adult blood. The production of mutant lines in this manner does, however, have its disadvantages. The derivation of homozygous mutants requires several generations and requires large-scale facilities (50, 52). This requires substantial input in terms of time and cost and would, therefore, usually be prohibitive for a single lab; however, several resources are available whereby the generation of mutants can be requested. These include the zebrafish TILLING Project (https:// www.webapps.fhcrc.org/science/tilling/index.php) and the zebrafish Mutation Resource (http://www.sanger.ac.uk/Projects/ D_rerio/mutres/), though these can still take up to a year to generate a suitable mutant. Homozygous mutants for the target gene may also contain many heterozygous mutations from the mutagenesis, so it is possible that any phenotype observed may be the result of one of these, necessitating the need for phenotypic rescue experiments with the wild-type gene to confirm the cause of the phenotype (48).

Despite the versatility that mutant lines can offer, the timescale for the generation of mutant lines does not necessarily facilitate the screening of a large number of genes. To this end, antisense morpholinos present themselves as a suitable tool. Morpholinos are analogues of phosphodiester DNA and are used to rapidly knock down gene function by targeting the RNA transcript – ablating translation or severely compromising protein function/expression by causing aberrant RNA splicing (22, 53). Structurally, they contain hexameric morpholino rings instead of pentameric ribose sugars and non-ionic phosphorodiamidate linkages instead of ionic phosphodiester bonds while still using the same bases (Fig. 4). These result in morpholinos being resistant to cellular nucleases and having few protein-binding interactions while binding nucleic acids with comparable affinity (21, 22, 53–56). Gene Tools LLC (USA) is currently the only commercial producer of antisense morpholinos, designing morpholinos based on appropriate sequences provided by the end-user (57). The morpholino sequences Gene Tools return are purportedly free of secondary structure, but the responsibility is on the end user to confirm sequence specificity for the target transcript via BLAST database searches. The standard experimental morpholinos are 25 base pairs long to allow specific binding to the target sequence and with high affinity (21, 22, 53, 58). In zebrafish, microinjection of morpholinos into the yolk sacs of 1–4 cell embryos is the most effective mode of delivery (58). Microinjection into the yolk allows for morpholino dispersion throughout the developing embryo, so achieving whole embryo

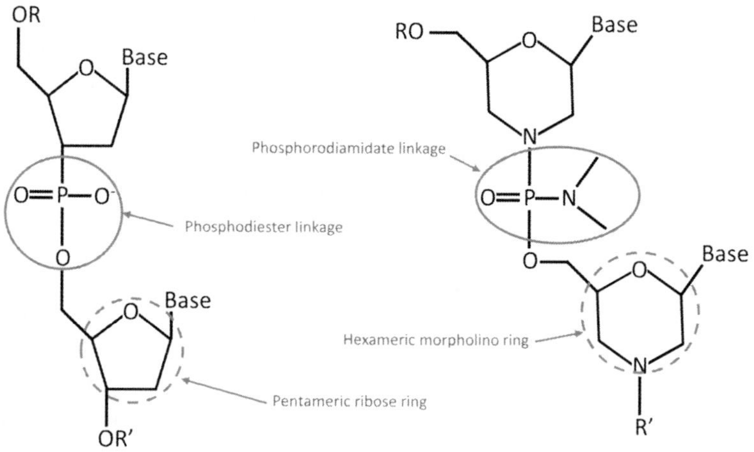

Fig. 4. A structural comparison of the sequence of DNA and morpholino oligonucleotides. The structure of phosphodiester DNA is represented on the left and morpholino structure is represented on the right. Dashed grey circles indicate the cyclic components, whereas solid grey circles indicate linkages. R and R′ denote the continuation of the 5′ and 3′ nucleotide sequence, respectively.

knock-down of gene function (22, 30). In zebrafish embryos, morpholinos are stable for several days, typically maintaining protein knock-down for up to 5 days, though their effects can wane as the embryo grows (22). Morpholinos can also have potential off-target effects, such as an increase in p53 expression and apoptosis, that can impede the identification of a phenotype or lead to the misidentification of a non-specific phenotype (21, 22, 58). Controlling for off-target effects can include the use of dual, sequence-independent morpholinos, mismatch control morpholinos, or phenotypic rescue by co-injection of a modified copy of the target mRNA that resists morpholino binding (21, 22). The concept behind using dual, sequence-independent morpholinos is that if both give the same phenotype then it is likely a specific effect. Likewise, if a mismatch control morpholino or rescued morpholino has no phenotype, then that observed using the experimental morpholino is likely to be a specific effect.

Two types of morpholinos are typically designed and used. The first are translation-blocking morpholinos. These are designed to target the translation start sequence of a mature mRNA. The binding of a morpholino at this site prevents ribosomal progression and so cessation of protein synthesis (Fig. 5a) (21, 54). However, the effectiveness of such morpholinos can be hard to quantify in the absence of commercial antibodies against zebrafish homologues or novel targets, a drawback of the zebrafish system. The other type of morpholino that can be used is designed against the splice junctions of a pre-mRNA. Typically, splice-acceptor and splice-donor sites of internal exons are used as these result in the "skipping"

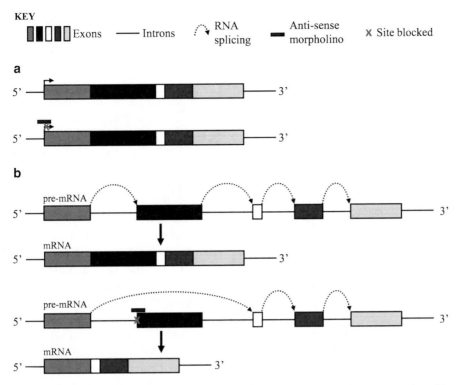

Fig. 5. Design of translation-blocking and splice-blocking morpholinos. (**a**) *Top*: An mRNA under normal conditions, which allows translation (*arrow*). *Bottom*: A translation-blocking morpholino binds across the translation start site, inhibiting ribosome progression. (**b**) *Top*: A pre-mRNA with normal exon–intron splicing. *Bottom*: A splice-blocking morpholino targeted at a splice-acceptor site, hides the site and results in the excision of the proceeding exon from the final mRNA.

of the targeted exon and its absence from the mature mRNA (Fig. 5b), though other sites can be used with varying effects (reviewed ref. 59). The excision of an exon results in the translation of a truncated protein. If the excised exon contains an incomplete codon, a frame shift will occur that will either result in the introduction of a nonsense/mis-sense mutation or a premature stop codon (TAG, TAA, TGA) (21, 59). The latter, if introduced early in the mRNA, results in the formation of a highly truncated, and likely non-functional, peptide sequence. Despite this, splice-blocking morpholinos can give more variable success than translation-blocking morpholinos. Cryptic splice sites can become uncovered and substitute for the blocked site, and indeed the final splice variant may have no deleterious effect on protein function (21, 59). However, unlike translation-blocking morpholinos, the action of splice-blocking morpholinos can be readily quantified by reverse transcriptase PCR (rtPCR), where the destruction of the transcript or a shift in the molecular weight is a readout of morpholino effectiveness (34, 59).

As stated earlier, the time period over which morpholinos produce an effective knock-down of gene expression is limited to around 5 days. However, this is sufficient for a functional screen based upon in vivo thrombosis in embryos less than 5 days of age and has been successfully used by O'Connor et al. to identify several novel genes that regulate thrombus formation (34). As Jagadeeswaran et al. have previously demonstrated, zebrafish can also be used for other functional assays, such as micro-aggregation, calcium signalling, P-selectin exposure, and dense-granule secretion (18, 26, 30–32, 36, 38). Unfortunately, these assays require blood samples from adult fish, by which time any morpholinos microinjected into early embryos have long worn off. However, recently, Kim et al. have shown that Vivo-Morpholinos, morpholinos that are covalently linked to octa-guanidine dendrimers to allow penetration of cell membranes (again produced by Gene Tools LLC), can be successfully used in live adult zebrafish (60, 61). Here, the authors used a Vivo-Morpholino designed against zebrafish CD41 to knock down CD41 expression in thrombocytes. By evaluating the effects using the techniques previously developed in the Jagadeeswaran lab, the anti-CD41 Vivo-Morpholino was shown to reduce aggregation and increase bleeding times in a gill-bleeding assay – mirroring what is seen with $\alpha_{IIb}\beta_3$ antagonists or in patients with bleeding disorders associated with abnormal $\alpha_{IIb}\beta_3$ function or expression (6, 17, 62). Together, standard morpholinos and Vivo-Morpholinos can cover what makes mutant lines more versatile, such as being able to utilise the translucency of the embryos for thrombosis assays and the capability to do more in-depth functional studies using adults, but without the substantial investment in time.

5. Conclusions

With the emerging need for a rapid and effective screen to identify novel regulators of platelet function and thrombus formation, zebrafish represents a physiologically relevant model with practical advantages over traditional murine knockout and transgenic models. The use of antisense morpholinos to rapidly knock down gene expression in developing zebrafish embryos/larvae, coupled with their translucency, allows for the rapid assessment of gene function in an in vivo model of thrombosis that requires considerably less time and training than comparable murine models. The generation of mutant lines allows for a more in-depth characterisation of thrombocyte function using blood from adult fish in functional assays (aggregation, P-selectin exposure, calcium release, annexin V binding, etc.), though for the purposes of a rapid functional screen, the time frame for the generation of homozygous mutants is not favourable. However, the recent demonstration of the effectiveness of Vivo-Morpholinos in adult zebrafish opens up these functional assays to rapid characterisation. While not a replacement for murine knockout models, zebrafish have the potential to be used as an inexpensive high-throughput screen to identify novel regulators of thrombus formation that merit further characterisation in a murine model, so streamlining research.

References

1. WHO, Global Burden of Disease, 2004 Update, W.H. Organisation, Editor. 2008.

2. Genome-wide association study of 14,000 cases of seven common diseases and 3,000 shared controls. Nature, 2007. 447(7145): p. 661–78.

3. Soranzo, N., et al., A genome-wide meta-analysis identifies 22 loci associated with eight hematological parameters in the HaemGen consortium. Nat Genet, 2009. 41(11): p. 1182–1190.

4. Samani, N.J., et al., Genomewide association analysis of coronary artery disease. N Engl J Med, 2007. 357(5): p. 443–53.

5. Macaulay, I.C., et al., Comparative gene expression profiling of in vitro differentiated megakaryocytes and erythroblasts identifies novel activatory and inhibitory platelet membrane proteins. Blood, 2007. 109(8): p. 3260–9.

6. Boersma, E., et al., Platelet glycoprotein IIb/IIIa inhibitors in acute coronary syndromes: a meta-analysis of all major randomised clinical trials. The Lancet, 2002. 359(9302): p. 189–198.

7. Wiviott, S.D., et al., Prasugrel versus clopidogrel in patients with acute coronary syndromes. N Engl J Med, 2007. 357(20): p. 2001–15.

8. Baigent, C., et al., ISIS-2: 10 year survival among patients with suspected acute myocardial infarction in randomised comparison of intravenous streptokinase, oral aspirin, both, or neither. BMJ, 1998. 316(7141): p. 1337–1343.

9. Sørensen, H.T., et al., Risk of upper gastrointestinal bleeding associated with use of low-dose aspirin. American Journal of Gastroenterology, 2000. 95(9): p. 2218–2224.

10. Faulkner, G., et al., Aspirin and bleeding peptic ulcers in the elderly. BMJ, 1988. 297(6659): p. 1311–3.

11. Fabre, J.E. and M.E. Gurney, Limitations of current therapies to prevent thrombosis: a need for novel strategies. Mol Biosyst, 2010. 6(2): p. 305–15.

12. Jaremo, P., et al., Individual variations of platelet inhibition after loading doses of clopidogrel. J Intern Med, 2002. 252(3): p. 233–8.

13. Gachet, C., P2 receptors, platelet function and pharmacological implications. Thromb Haemost, 2008. 99(3): p. 466–72.

14. Serebruany, V.L., et al., Variability in platelet responsiveness to clopidogrel among 544 individuals. Journal of the American College of Cardiology, 2005. 45(2): p. 246–251.

15. Jennings, L.K., Mechanisms of platelet activation: need for new strategies to protect against platelet-mediated atherothrombosis. Thromb Haemost, 2009. 102(2): p. 248–57.

16. Chew, D.P., et al., Increased Mortality With Oral Platelet Glycoprotein IIb/IIIa Antagonists : A Meta-Analysis of Phase III Multicenter Randomized Trials. Circulation, 2001. 103(2): p. 201–206.

17. Jones, M.L., et al., RGD-ligand mimetic antagonists of integrin alphaIIb beta3 paradoxically enhance GPVI-induced human platelet activation. Journal of Thrombosis and Haemostasis, 2010. 8(3): p. 567–576.

18. Jagadeeswaran, P., et al., zebrafish: from hematology to hydrology. Journal of Thrombosis and Haemostasis, 2007. 5(s1): p. 300–304.

19. Streisinger, G., et al., Production of clones of homozygous diploid zebrafish (Brachydanio rerio). Nature, 1981. 291(5813): p. 293–6.

20. Westerfield, M., The zebrafish Book. A guide for the laboratory use of zebrafish (Danio rerio). 2000, University of Oregon Press: Eugene.

21. Eisen, J.S. and J.C. Smith, Controlling morpholino experiments: don't stop making antisense. Development, 2008. 135(10): p. 1735–43.

22. Bill, B.R., et al., A Primer for Morpholino Use in zebrafish. zebrafish, 2009. 6(1): p. 69–77.

23. EMBL-EMI and WTSI. Ensembl Genome Database. Available from: http://www.ensembl.org/index.html.

24. Lang, M.R., et al., Haemostasis in *Danio rerio* – Is the zebrafish a useful model for platelet research? Journal of Thrombosis and Haemostasis, 2010. 9999(999A).

25. Cvejic, A., et al., Analysis of WASp function during the wound inflammatory response - live-imaging studies in zebrafish larvae. J Cell Sci, 2008. 121(19): p. 3196–3206.

26. Jagadeeswaran, P., et al., Identification and characterization of zebrafish thrombocytes. British Journal of Haematology, 1999. 107(4): p. 731–738.

27. Jagadeeswaran, P. and Y.C. Liu, A hemophilia model in zebrafish: analysis of hemostasis. Blood Cells Mol Dis, 1997. 23(1): p. 52–7.

28. Gottgens, B., et al., Transcriptional regulation of the stem cell leukemia gene (SCL)—comparative analysis of five vertebrate SCL loci. Genome Res, 2002. 12(5): p. 749–59.

29. Jagadeeswaran, P. and J.P. Sheehan, Analysis of Blood Coagulation in the zebrafish. Blood Cells, Molecules, and Diseases, 1999. 25(4): p. 239–249.

30. Jagadeeswaran, P., et al., Zebrafish: a genetic model for hemostasis and thrombosis. Journal of Thrombosis and Haemostasis, 2005. 3(1): p. 46–53.

31. Thattaliyath, B., M. Cykowski, and P. Jagadeeswaran, Young thrombocytes initiate the formation of arterial thrombi in zebrafish. Blood, 2005. 106(1): p. 118–124.

32. Gregory, M. and P. Jagadeeswaran, Selective Labeling of zebrafish Thrombocytes: Quantitation of Thrombocyte Function and Detection during Development. Blood Cells, Molecules, and Diseases, 2002. 28(3): p. 418–427.

33. Lin, H.-F., et al., Analysis of thrombocyte development in CD41-GFP transgenic zebrafish. Blood, 2005. 106(12): p. 3803–3810.

34. O'Connor, M.N., et al., Functional genomics in zebrafish permits rapid characterization of novel platelet membrane proteins. Blood, 2009. 113(19): p. 4754–62.

35. Day, K., N. Krishnegowda, and P. Jagadeeswaran, Knockdown of prothrombin in zebrafish. Blood Cells, Molecules, and Diseases, 2003. 32(1): p. 191–198.

36. Gregory, M., R. Hanumanthaiah, and P. Jagadeeswaran, Genetic analysis of hemostasis and thrombosis using vascular occlusion. Blood Cells Molecules and Diseases, 2002. 29(3): p. 286–295.

37. Jagadeeswaran, P., et al., Loss of GATA1 and gain of FLI1 expression during thrombocyte maturation. Blood Cells, Molecules, and Diseases, 2010. 44(3): p. 175–180.

38. Jagadeeswaran, P., R. Paris, and P. Rao, Laser-induced thrombosis in zebrafish larvae: a novel genetic screening method for thrombosis. Methods Mol Med, 2006. 129: p. 187–95.

39. Kissa, K., et al., Live imaging of emerging hematopoietic stem cells and early thymus colonization. Blood, 2008. 111(3): p. 1147–1156.

40. Day, S.M., et al., Murine thrombosis models. Thromb Haemost, 2004. 92(3): p. 486–94.

41. Sachs, U.J.H. and B. Nieswandt, In Vivo Thrombus Formation in Murine Models. Circ Res, 2007. 100(7): p. 979–991.

42. Falati, S., et al., Real-time in vivo imaging of platelets, tissue factor and fibrin during arterial thrombus formation in the mouse. Nat Med, 2002. 8(10): p. 1175–81.

43. Konopatskaya, O., et al., PKCα regulates platelet granule secretion and thrombus formation in mice. The Journal of Clinical Investigation, 2009. 119(2): p. 399–407.

44. Ni, H., et al., Persistence of platelet thrombus formation in arterioles of mice lacking both von Willebrand factor and fibrinogen. The Journal of Clinical Investigation, 2000. 106(3): p. 385–392.

45. Nagy, B., Jr, et al., Impaired activation of platelets lacking protein kinase C-{theta} isoform. Blood, 2009. 113(11): p. 2557–2567.

46. Kalia, N., et al., Critical Role of FcRγ -Chain, LAT, PLCγ2 and Thrombin in Arteriolar Thrombus Formation upon Mild, Laser-Induced Endothelial Injury In Vivo. Microcirculation, 2008. 15: p. 325–335.

47. Davidson, A.J. and L.I. Zon, The definitive (and primitive) guide to zebrafish hematopoiesis. Oncogene, 2004. 23(43): p. 7233–7246.

48. Moens, C.B., et al., Reverse genetics in zebrafish by TILLING. Briefings in Functional Genomics and Proteomics, 2008. 7(6): p. 454–459.

49. Solnica-Krezel, L., A.F. Schier, and W. Driever, Efficient Recovery of ENU-Induced Mutations From the zebrafish Germline. Genetics, 1994. 136(4): p. 1401–1420.

50. Amsterdam, A. and N. Hopkins, Mutagenesis strategies in zebrafish for identifying genes involved in development and disease. Trends in Genetics, 2006. 22(9): p. 473–478.

51. Patton, E.E. and L.I. Zon, The art and design of genetic screens: zebrafish. Nat Rev Genet, 2001. 2(12): p. 956–966.

52. Wienholds, E., et al., Efficient Target-Selected Mutagenesis in zebrafish. Genome Research, 2003. 13(12): p. 2700–2707.

53. Ekker, S.C. and J.D. Larson, Morphant technology in model developmental systems. Genesis, 2001. 30(3): p. 89–93.

54. Summerton, J., Morpholino antisense oligomers: the case for an RNase H-independent structural type. Biochim Biophys Acta, 1999. 1489: p. 141–158.

55. Corey, D. and J. Abrams, Morpholino antisense oligonucleotides: tools for investigating vertebrate development. Genome Biology, 2001. 2(5): p. reviews1015.1 - reviews1015.3.

56. Braasch, D. and D. Corey, Locked nucleic acid: fine tuning the recognition of DNA and RNA. Chem Biol, 2001. 8: p. 1–7.

57. Gene Tools LLC. http://www.gene-tools.com/

58. Nasevicius, A. and S. Ekker, Effective targeted gene 'knockdown' in zebrafish. Nat Genet, 2000. 26: p. 216–220.

59. Morcos, P.A., Achieving targeted and quantifiable alteration of mRNA splicing with Morpholino oligos. Biochem Biophys Res Commun, 2007. 358(2): p. 521–7.

60. Kim, S., et al., Vivo-Morpholino knockdown of (alpha)IIb: A novel approach to inhibit thrombocyte function in adult zebrafish. Blood Cells, Molecules, and Diseases, 2010. 44(3): p. 169–174.

61. Morcos, P.A., Y. Li, and S. Jiang, Vivo-Morpholinos: a non-peptide transporter delivers Morpholinos into a wide array of mouse tissues. Biotechniques, 2008. 45(6): p. 613–4, 616, 618 passim.

62. George, J.N., J.P. Caen, and A.T. Nurden, Glanzmann's thrombasthenia: the spectrum of clinical disease. Blood, 1990. 75(7): p. 1383–95.

Chapter 22

Platelet Receptor Shedding

Elizabeth E. Gardiner, Mohammad Al-Tamimi, Robert K. Andrews, and Michael C. Berndt

Abstract

Receptor shedding is a mechanism for irreversible removal of transmembrane cell surface receptors by proteolysis of the receptor at a position near the extracellular surface of the plasma membrane. This process generates a soluble ectodomain fragment and a membrane-associated remnant fragment, and is distinct from loss of receptor surface expression by internalization or microparticle release or secretion of alternatively spliced soluble forms of receptors lacking a transmembrane domain. There has been an increased focus on new methods for analyzing shedding of platelet glycoprotein (GP)Ib-IX-V and GPVI because these receptors are platelet specific and are critical for the initiation of platelet adhesion and activation in thrombus formation at arterial shear rates. Platelet receptor shedding provides a mechanism for downregulating surface expression resulting in loss of ligand binding, decreasing the surface density affecting receptor cross linking and signalling and generation of proteolytic fragments that may be functional and/or provide platelet-specific biomarkers.

Key words: Platelets, Receptors, Sheddases, GPIb-IX-V, GPVI, ADAM10, ADAM17

1. Introduction

Many types of receptors expressed on platelets are known to be, or are suspected of being shed (Table 1) (1–25). Shedding is a ubiquitous control mechanism for regulating cellular transmembrane proteins. Other cells may use gene regulation as part of normal receptor turnover or replacement, but this is more limited in anucleate platelets. Ectodomain shedding provides a mechanism for rapid and irreversible downregulation of receptor expression (reviewed in refs. 1, 2). Changes of receptor expression levels over time decrease surface density leading to decreased ligand binding, receptor cross linking and signalling, and possibly also control

Jonathan M. Gibbins and Martyn P. Mahaut-Smith (eds.), *Platelets and Megakaryocytes: Volume 3, Additional Protocols and Perspectives*, Methods in Molecular Biology, vol. 788, DOI 10.1007/978-1-61779-307-3_22, © Springer Science+Business Media, LLC 2012

Table 1
Examples of platelet receptors, function, and shedding

Receptor	Shedding
1. Adhesion and activation	
Leucine-rich repeat family (GPIb-IX-V)	
GPIbα	Calmodulin-regulated ADAM17-mediated shedding
GPIbβ	of GPIbα, and ADAM10/17-mediated shedding of
GPIX	GPV (1–5)
GPV	
Immunoglobulin/immunoreceptor family	
GPVI/FcRγ	Calmodulin-regulated ADAM10-mediated GPVI shedding (4, 6–8)
2. Activation and secretion	
G-protein-coupled receptors	
Protease-activated receptors	PAR-1 is shed on endothelial cells. Whereas thrombin
PAR-1, PAR-4	cleaves within the N-terminal extracellular sequence
Purinergic ADP receptors	of this seven-transmembrane receptor generating at
P2Y$_1$, P2Y$_{12}$	the new N-terminus a tethered, activating ligand,
Weak platelet agonists	ADAM17 or related sheddase cleaves at a down-
TXA2 receptor	stream site which would prevent activation by
Epinephrine receptor	thrombin (9). It is not known if this mechanism
	occurs on platelets or regulates thrombin-depen-
	dent platelet activation
Other	
CD40L	sCD40L may be generated by MMP-2 associated with
Semaphorin 4	platelet α$_{IIb}$β$_3$ or is cleaved by ADAM10 on
CLEC-2	leukocytes; ADAM17-mediated shedding releases
	soluble ectodomain of Sema4D that can regulate
	endothelial cells or monocytes (10–15)
3. Aggregation and firm adhesion	
Integrins	
α$_{IIb}$β$_3$ (GPIIb-IIIa)	Ectodomain shedding of β$_2$ integrins has been
α$_v$β$_3$	reported. MMP-9 sheds the β$_2$ subunit from
α$_1$β$_1$,	macrophages; a soluble ligand-binding form of α$_L$β$_2$
α$_2$β$_1$,	is shed from leukocytes. The β$_3$ subunit of α$_{IIb}$β$_3$
α$_5$β$_1$,	undergoes intracellular calpain-mediated proteolysis
α$_6$β$_1$	in activated platelets (16–20)
4. Inhibitory receptors	
G-protein-coupled receptors	
Prostaglandin receptors	No known shedding
PGE$_1$ receptor	
PGI$_2$ receptor	
Immunoglobulin/immunoreceptor family	
PECAM-1	PECAM-1 is susceptible to calmodulin-regulated
LAIR-1	proteolysis, and/or shear-induced shedding with a
	role for calpain (21–23)

(continued)

Table 1
(continued)

Receptor	Shedding
5. Other receptors	
Selectin	
P-selectin	Shed or alternatively spliced, secreted forms of P-selectin in plasma are a marker of platelet or endothelial cell activation (24)
Immunoglobulin/immunoreceptor family	
FcγRIIa (CD32a)	FcγRIIa undergoes calmodulin-regulated calpain-mediated intracellular proteolysis, de-ITAM-ization, and inactivation but is not shed from platelets; however, a soluble alternatively spliced form is present in plasma (25)

platelet aging or clearance. Proteolytic fragments may also be functional and/or act as potential platelet-specific biomarkers.

In this perspective, we focus on shedding of platelet glycoprotein (GP)Ib-IX-V and GPVI, which illustrate many important features of shedding, including the type of sheddases involved, whether shedding is constitutive, how it may be differentially induced, whether shedding is activation dependent, what proteolytic fragments are generated, and how the extent of shedding can be analysed at the level of the platelet or in the plasma. Shedding of GPIb-IX-V or GPVI has also been studied using experimental models, including in vivo studies in mice or monkeys. There is also recent evidence on the diagnostic potential of GPVI expression/shedding in human cardiovascular or immune-related diseases.

2. Function of Platelets

Blood platelets, and by extension their precursors, megakaryocytes, as well as platelet-derived microparticles, play a central role in normal vascular biology, in the immune-inflammatory response and in numerous human diseases. In thrombus formation, circulating platelets in the bloodstream target the damaged or diseased vessel wall, become adherent and activated, and form aggregates that promote coagulation and prevent blood loss following injury (26).

In atherothrombotic disease, thrombus formation is triggered by vascular plaques or activated endothelium, leading to blockage of blood vessels causing heart attack or stroke (27–29). Platelets also regulate inflammation by recruiting leukocytes to sites of injury by adhesion to mural platelets and by secretion of proinflammatory factors from activated platelets. By 2010, the range of functions for platelets in humans or animal models includes regulation of vascular development, control of tumour metastasis, and key roles in immune diseases, complications associated with diabetes, and inflammatory diseases, such as rheumatoid arthritis and sepsis (30). It is likely that platelet receptor shedding will influence the onset, progression, and severity of many human diseases.

3. Platelet GPIb-IX-V and GPVI

The GPIb-IX-V complex is made up of four transmembrane glycoproteins (GPIbα, GPIbβ, GPIX, and GPV), all members of the leucine-rich repeat family (31–34) (Fig. 1). From the extracellular

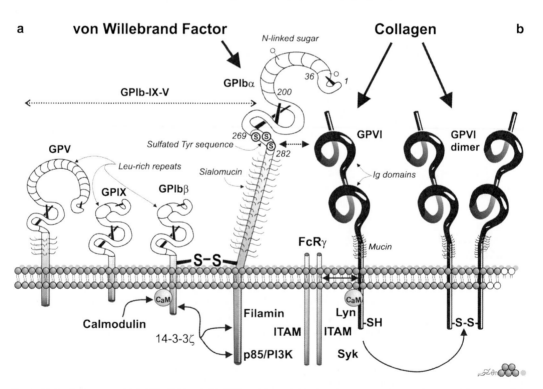

Fig. 1. Stylized diagram of (**a**) GPIb-IX-V and (**b**) GPVI residing within the platelet membrane. Structural features of the ligand-binding regions of GPIbα and GPVI are noted, with emphasis on putative regions of co-association between the ectodomains of GPIbα and GPVI that may influence platelet responses to GPVI engagement by collagen, the association of GPVI with FcRγ chain to allow intracellular signalling, as well as a Cys residue within the cytoplasmic tail of GPVI that mediates GPVI disulfide-linked homodimerization.

N-terminus, GPIbα (~135 kDa) consists of an N-terminal disulfide-looped capping sequence (residues 1–35), a series of tandem leucine-rich repeats (~24-residue amino acid sequences rich in Leu, Ile, or Val) spanning residues 36–200, a C-terminal disulfide-looped sequence (residues 201–268), an anionic sequence (residues 269–282) containing three sulphated tyrosines, an elongated sialomucin domain, a short membrane-proximal sequence containing two Cys residues linked via disulfide bonding to one or two GPIbβ subunits (35), a hydrophobic transmembrane domain, and a cytoplasmic tail of ~100 residues that contains binding sites for intracellular signalling/structural proteins: filamin A, 14-3-3ζ, the p85 subunit of phosphatidyl inositol 3-kinase (PI3-kinase) (36–38). The extracellular domain (glycocalicin; ~130 kDa) consists of the N-terminal ligand-binding globular domain (residues 1–282) and the sialomucin domain. The GPIbα ectodomain binds adhesive ligands, such as VWF and thrombospondin, counter-receptors on other cells, including P-selectin (activated endothelial cells or activated platelets) and $\alpha_M\beta_2$ (activated leukocytes), coagulation factors XII, XI, and high-molecular-weight kininogen (31, 32). The membrane-proximal sequence between the sialomucin and the first of the two Cys residues (linked to GPIbβ) contains one or more cleavage sites for platelet sheddases (Subheading 5). GPIbβ (~25 kDa) contains the N- and C-terminal capping sequences and a short leucine-rich domain, followed by a region lacking sialomucin, a transmembrane domain, and a short cytoplasmic tail (~35 residues) that binds 14-3-3ζ and calmodulin (39). GPIX (~20 kDa) is analogous to GPIbβ, but has a shorter cytoplasmic tail (~5 residues). Finally, GPV (~85 kDa) consists of the N-terminal capping sequence, a long leucine-rich repeat sequence (approximately twice the size of that in GPIbα), the C-terminal capping sequence, a short glycosylated domain, a membrane-proximal sequence, transmembrane domain, and cytoplasmic tail (~15 residues) that also binds calmodulin (39). The membrane-proximal sequence contains cleavage sites for a metalloproteinase sheddase (Subheading 5) and an upstream cleavage site for thrombin (Fig. 2a) generating different soluble GPV ectodomain fragments (~80 and ~60 kDa, respectively) (4, 40). Removal of GPV may promote thrombin's ability to activate platelets via GPIbα (41, 42).

GPVI (~62 kDa) consists of two extracellular immunoglobulin domains, a short mucin domain, a membrane-proximal sequence, a transmembrane domain, and a cytoplasmic tail (~50 residues) that binds calmodulin and constitutively activated Src family kinase, Lyn, and also contains an unpaired Cys (penultimate residue) involved in ligand-induced transient homodimerization (Fig. 1) (43–46). GPVI forms a complex with the Fc receptor γ-chain, FcRγ, which contains an immunoreceptor tyrosine-based activation motif (ITAM). On ligand-induced cross linking of GPVI/FcRγ, this ITAM enables Lyn-dependent activation of Syk kinase

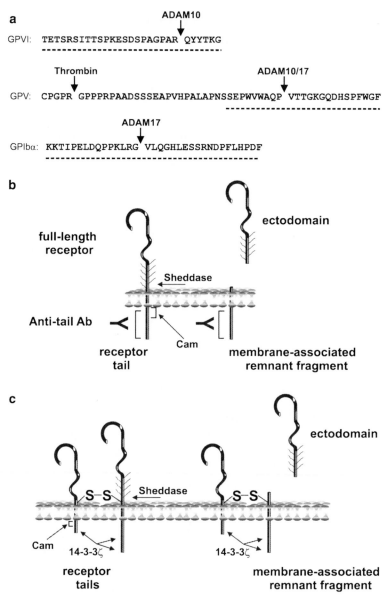

a

ADAM10

GPVI: TETSRSITTSPKESDSPAGPAR QYYTKG

Thrombin   ADAM10/17

GPV: CPGPR GPPPRPAADSSSEAPVHPALAPNSSEPWVWAQP VTTGKGQDHSPFWGF

ADAM17

GPIbα: KKTIPELDQPPKLRG VLQGHLESSRNDPFLHPDF

b

full-length
receptor

ectodomain

Sheddase

Anti-tail Ab

Cam

receptor
tail

membrane-associated
remnant fragment

c

ectodomain

S–S   Sheddase   S–S

Cam

14-3-3ζ   14-3-3ζ

receptor
tails

membrane-associated
remnant fragment

Fig. 2. (**a**) Extracellular membrane-proximal sequences of GPVI, GPV, and GPIbα showing where synthetic peptides based on these sequences (*dashed underline*) are cleaved by ADAM10 and/or ADAM17. The thrombin-cleavage site on GPV is also indicated. (**b**) Generic diagram of receptor shedding, illustrating the full-length receptor and proteolysis by a surface sheddase generating a soluble ectodomain fragment and a membrane-associated remnant fragment. Antibodies against the cytoplasmic tail recognize both full-length receptor and the membrane-associated remnant fragment. An intracellular membrane-proximal calmodulin (Cam)-binding site is indicated. (**c**) In the case of GPIbα shedding, the membrane-associated remnant remains disulfide-linked to one or two GPIbβ subunits. 14-3-3ζ-binding sites are indicated.

and promulgation of downstream signalling pathways (47, 48). The extracellular membrane-proximal sequence contains a sheddase cleavage site, generating an ~55-kDa soluble GPVI (sGPVI) fragment and ~10-kDa membrane-associated remnant fragments (4, 6–8, 21) (refer to Subheading 6).

4. Functional Roles of Platelet Receptors and Shedding

Platelet receptors act together to regulate platelet adhesion, activation, secretion, and aggregation (Table 1). In thrombus formation in flowing blood, adhesion/signalling receptors, such as GPIbα (the ligand-binding subunit of GPIb-IX-V) and GPVI, initiate adhesion of circulating platelets to VWF/thrombospondin or collagen/laminin, respectively, in the damaged vessel wall, leading to secretion of agonists, such as ADP or thromboxane A2 (TXA2) that act on specific G-protein-coupled receptors, together leading to activation of $\alpha_{IIb}\beta_3$ (GPIIb-IIIa that binds VWF or fibrinogen), $\alpha_2\beta_1$ that binds collagen, and other integrins that regulate firm adhesion and aggregation and transmit intracellular signals controlling platelet contraction (26, 29, 49–52). Shedding of adhesion receptors, GPIbα and GPVI, therefore, could significantly affect initiation and development of the thrombus. Platelet reactivity to GPVI ligands is proportional to GPVI expression levels in a graded fashion (53), suggesting that shedding could affect GPVI-dependent platelet activation. Post adhesion, shedding may be involved in breaking contacts with matrix ligands enabling platelet spreading or play a role in embolization, the breaking away of parts of the thrombus from the thrombus mass. This shedding could contribute to the passification of the activated thrombus, controlling whether the thrombus becomes occlusive. In addition to GPIb-IX-V/GPVI, thrombus development and stability may also be regulated by activating receptors, such as CD40L, semaphorin 4D, CLEC-2, and others (12, 54, 55), with CD40L and semaphorin 4D shedding from activated platelets (Table 1) (10–13).

P-selectin is present in platelet α-granules and endothelial cell Weibel–Palade bodies, and becomes expressed only on the surface of activated platelets or endothelial cells. Plasma-soluble P-selectin generated from an alternatively spliced transcript is a platelet–endothelial cell activation marker associated with cardiovascular dysfunction (24). Known binding partners include platelet GPIbα and leukocyte P-selectin glycoprotein ligand-1 (PSGL-1) (31, 56), and soluble P-selectin interacting with PSGL-1 on monocytes generates procoagulant microparticles (57). G-protein-coupled receptors on platelets respond to thrombin, protease-activated receptor-1 and -4 (PAR-1 and PAR-4) and activate platelets, control release from platelet storage granules, or respond to activating or inhibitory prostaglandins (49). Other receptors, LAIR-1 and PECAM-1, can negatively regulate the activation pathways of other receptors, including GPVI and $\alpha_{IIb}\beta_3$ (58–60).

Intrinsic coagulation Factors XII and XI are also critical for a stable occlusive thrombus, since deficiency or selective inhibition of one or both of these factors has minimal effect on bleeding but profoundly affects thrombosis in experimental models in vivo (61–65). Activated platelets express receptors that bind coagulation factors (GPIbα and gC1qR/p33) localizing coagulation at the activated platelet surface, secrete procoagulant factors that activate coagulation pathways leading to fibrin polymerization/clotting, and express procoagulant phospholipids, such as phosphatidylserine involved in efficient activation of coagulation factors and assembly of procoagulant complexes (66–68). Shedding of GPIbα that binds FXII, FXI, high-molecular-weight kininogen, and thrombin provides a mechanism for limiting coagulation at the thrombus surface. Receptor-mediated generation of procoagulant platelet-derived microparticles would also be inhibited by shedding. In this regard, GPVI-dependent microparticle generation in response to collagen exposure or inflamed synoviocytes is involved in experimental and clinical rheumatoid arthritis (69, 70).

5. Endogenous Platelet Sheddases Targeting GPIb-IX-V and GPVI

Platelet sheddases are endogenous membrane-associated proteinases that cleave receptors at an extracellular membrane-proximal site generating an ectodomain fragment and membrane-associated remnant fragment (Fig. 2b). The major sheddases for GPIb-IX-V, GPVI, and other platelet receptors are members of the *a d*isintegrin *a*nd *m*etalloproteinase (ADAM) family. ADAM10 and ADAM17 (also known as TACE: *t*umour necrosis factor-*a*lpha (TNFα) *c*onverting *e*nzyme) consist of a pro-domain, a metalloproteinase domain, a Cys-rich region, a membrane-proximal sequence, a transmembrane domain, and a cytoplasmic tail (1, 19, 71–74). An unpaired Cys in the pro-domain acts as an internal inhibitor by coordination of the active site metal ion, enabling activation of sheddase activity either by thiol-modifying agents, such as *N*-ethylmaleimide (NEM), or by proteolytic removal of the pro-domain, for example by furin or other enzymes. NEM, therefore, can directly activate sheddases without the requirement for activation of the cell (reviewed in ref. 1). ADAM10 and ADAM17 activity can also be induced by cellular activation, for example using phorbol myristate acetate (PMA), a membrane-permeable reagent that activates protein kinase C (PKC), and downstream signalling pathways (reviewed in ref. 1). Calmodulin associated with the cytoplasmic tail of the receptor plays a key role in regulating surface expression/shedding; this was originally shown for the leukocyte receptor, L-selectin, cleaved by ADAM17 (21, 75). In platelets, calmodulin binds to a membrane-proximal sequence in the cytoplasmic tail of GPVI, and the calmodulin inhibitor W7 induces shedding of GPVI (4, 6, 21). In the case of GPIb, calmodulin binds to the cytoplasmic tail of GPIbβ disulfide linked to GPIbα, and W7 also induces GPIbα shedding (4, 21, 39, 76).

Platelets also express matrix metalloproteinase (MMP) family proteins as membrane-bound or secreted forms; MMP-2 associates with $\alpha_{IIb}\beta_3$ as well as being implicated as a likely sheddase for platelet CD40L (10, 11, 77, 78). Metalloproteinase inhibitors, such as tissue inhibitors of MMP (TIMP) family members can differentially inhibit ADAM10 in vitro (TIMP-1 and TIMP-3 are inhibitory, but not TIMP-2 or TIMP-4) (15).

6. Assays for Analysing Shedding of GPIb-IX-V and GPVI

One of the methods our laboratory has used for measuring GPIb-IX-V and GPVI shedding from washed human platelets involves western blotting lysates of untreated or treated platelets using antibodies against the cytoplasmic tail that recognize both intact receptor and the membrane-associated remnant fragment. Alternatively, isolated supernatants can be blotted with antibodies against the ectodomain that can detect the shed-soluble fragment (4, 6, 25, 79). A typical experiment might involve washed platelets suspended in a divalent cation-containing buffer (Tyrode's buffer) either untreated or treated with platelet agonists, GPVI ligands, NEM (that directly activates the sheddase), PMA (that activates PKC), antibodies (see Subheading 7), or other agents for varying times. The shedding reaction may then be quenched by addition of excess EDTA or the broad-specificity metalloproteinase inhibitor GM6001, and then centrifugation to isolate platelets for preparing lysates or supernatant for analysis of shed-soluble receptor fragments. This assay enables analysis of the degree of shedding and the molecular weight of the proteolytic fragments. Loss of surface receptor from washed platelets or whole blood can also be analyzed by flow cytometry, although this does not provide information about the molecular weight or levels of fragments generated or distinguish between shedding and internalization of receptors.

The status of GPIbα or GPVI can also be estimated in platelet function assays, such as optical or impedance aggregometry using GPIbα or GPVI selective agonists (80); however, these assays are not sensitive to small changes in intact receptor and may not be useful in the setting of thrombocytopenia. To analyze shed receptors in plasma, a number of laboratories have established enzyme-linked immunosorbent assay (ELISA) methods for assessing shed-soluble GPIbα (glycocalicin), GPV, or GPVI in isolated plasma (81–84).

7. Regulation of GPVI and GPIb-IX-V Shedding by ADAM10 or ADAM17

Divalent cation-dependent proteolysis of platelet GPIbα and GPV has been known for over a decade (85). This shedding occurred on activated platelets, but was independent of platelet aggregation and calpain activity and was also distinct from loss of receptors due to release of platelet-derived microparticles. Later studies using recombinant, constitutively active forms of ADAM10 or ADAM17 and synthetic peptides based on receptor extracellular membrane-proximal sequences suggested that ADAM10 is primarily involved in cleaving of GPVI, ADAM17 cleaves GPIbα, and both ADAM10 and ADAM17 cleave GPV (4) (Fig. 2a). Mutagenesis of these sequences in GPIbα or GPVI expressed in cell lines inhibits shedding (4). These results with peptides or in cell lines are also consistent with findings in mice expressing catalytically inactive ADAM17, where there is defective GPIbα proteolysis and lower plasma glycocalicin, whereas GPVI shedding is unaffected (3, 86) (Subheading 9).

These findings, however, suggest that surface expression/shedding of GPIbα and GPVI can be differentially regulated. Western blotting with anti-cytoplasmic tail antibodies shows that a significant amount of GPIbα membrane-associated remnant fragment (ADAM17 mediated) is detectable in platelet lysates (even when isolated in the presence of metalloproteinase inhibitors), whereas there is virtually no detectable GPVI remnant fragment (ADAM10 mediated) (4). Shedding of both GPIbα and GPVI is induced by NEM, W7, PMA, and the mitochondrial-targeting compound carbonyl cyanide 3-chlorophenylhydrazone (CCCP) while GPVI shedding is not significantly induced by VWF acting at GPIbα, but is induced by GPVI-specific agonists (collagen-related peptide or snake toxins) or anti-GPVI or anti-platelet antibodies (3, 4, 6, 7, 79, 86–89). Anti-platelet antibodies that activate platelets via the platelet Fc receptor, FcγRIIa, also potently induce GPVI shedding (25). Like GPVI/FcRγ, the cytoplasmic tail of FcγRIIa contains an ITAM that activates Syk-dependent signalling pathways in platelets. Anti-GPVI monoclonal antibodies or human anti-GPVI autoantibodies can also induce metalloproteinase-dependent GPVI shedding from healthy donor platelets independent of FcγRIIa (79, 87). Other studies show that ligand-induced GPVI shedding from human platelets is also inhibited by TIMP-1 that targets ADAM10 (8) (Subheading 5). In comparison, shedding of GPIbα but not GPVI is induced by serotonin or oxidative stress (3, 4, 6, 88–91). It is also likely that other proteases generate glycocalicin, and may contribute to constitutive GPIbα shedding either by directly cleaving receptor or regulating ADAM17 activity (92) while analysis of GPVI expression on mouse platelets, where ADAM10 and/or ADAM17 activity is ablated, suggests that ADAM17 or other metalloproteinases are also capable of cleaving GPVI (93).

8. Drug-Induced Shedding

The first drug shown to directly affect platelet receptor expression was aspirin, which at high doses can induce GPIbα and GPV shedding in vitro by an acetylation-dependent mechanism and activation of platelet sheddases (94). Aspirin-induced shedding may be relevant to measuring shed platelet receptor fragments as markers in atherothrombotic disease, where there may be aspirin or other antiplatelet drugs. Other drugs could indirectly induce shedding of platelet receptors by drug-dependent anti-platelet autoantibodies, previously shown to induce shedding of GPVI via activation at the platelet Fc receptor, FcγRIIa (25) (Subheading 7).

9. Experimental Models of Shedding

Together with measurement of shed-soluble GPIbα, GPV, and GPVI in healthy human plasma (Subheading 6), investigation of platelet receptor shedding in animal models provides evidence that shedding occurs in vivo. In mice, soluble GPIbα ectodomain (glycocalicin) is found in normal plasma at ~20 μg/ml, but expression of catalytically inactive ADAM17 results in ~90% reduction in plasma glycocalicin levels and a corresponding increase of surface GPIbα (3). Using an $FeCl_3$-induced injury model, Brill and colleagues showed that oxidation-induced shedding of platelet GPIbα impaired thrombus development (88). In cynomolgus monkeys, treatment with murine anti-human GPVI antibodies or Fab fragments showed loss of platelet GPVI due to either internalization (shown by uptake of fluorescently tagged antibody, mF1232) or metalloproteinase-dependent ectodomain shedding (mF1201) (95). These antibodies were screened for the ability to cross block a human anti-GPVI autoantibody to GPVI that caused loss of platelet surface GPVI in vivo and loss of collagen responsiveness (95, 96). Injection of mF1201 into a monkey induced thrombocytopenia, loss of platelet surface GPVI, and an increase in plasma sGPVI detected by ELISA (95).

10. Shedding of GPIb-IX-V and GPVI in Human Disease

Healthy human plasma contains soluble fragments of GPIbα (glycocalicin), GPV, and GPVI (sGPVI) (81, 82, 84, 87, 97, 98). The glycocalicin index (GCI) – plasma glycocalicin level corrected for the platelet count – may act as a surrogate marker of peripheral platelet turnover; however, GCI is also significantly elevated in immunothrombocytopenic patients with ongoing megakaryocyte destruction (84). Glycocalicin is shed constitutively and circulates at relatively high levels in plasma from healthy individuals (1–3 μg/ml), and the half-life of the fragment in plasma remains undefined. The half-life of sGPVI is also unknown, although limited temporal data suggests that it is stable, similar to other immunoglobulin-family proteins in plasma (97). Compared with glycocalicin, levels of sGPVI in plasma are relatively low (~20 ng/ml), and platelet levels of membrane-associated remnant fragment are not detectable on healthy platelets (4, 25, 81, 87). Plasma-soluble GPV, generated by metalloproteinase(s) or thrombin acting at distinct cleavage sites, can also be measured by ELISA and may be a platelet-specific marker in platelet-related diseases (82, 98, 99).

Plasma levels of sGPVI were markedly elevated (~150 ng/ml compared with normal ~15 ng/ml) in a patient with immune thrombocytopenic purpura (ITP) due to an anti-GPVI autoantibody (87, 97). Surface expression of intact GPVI was decreased, and the membrane-associated proteolytic fragment was detected by the anti-GPVI cytoplasmic tail antibody (not detectable on platelets from healthy individuals), corresponding to a selective defect in GPVI-dependent platelet aggregation. Another patient with lupus nephritis and an anti-GPVI autoantibody had low levels of plasma sGPVI levels, despite a relatively normal platelet count, even though an IgG3 fraction from patient serum activated and induced shedding from normal platelets in vitro (100). In other clinical studies, plasma sGPVI levels appear to correlate with platelet surface levels, and there is a statistical elevation in acute coronary syndrome (83). Elevated surface levels of GPVI have been reported to be a risk factor in cardiovascular disease (101), and in other studies plasma sGPVI in patients with Alzheimer's disease was lower than controls (102). Since GPVI shedding is inducible by activation-dependent or activation-independent pathways, the question is how diagnostically useful plasma sGPVI will be as a platelet-specific platelet activation marker, also considering the potential effects of drugs, autoantibodies, or other causes. This requires further prospective studies and establishing temporal profiles of sGPVI levels as disease progresses.

11. Platelet Storage and Apoptosis

Platelets survive in the circulation on average for 7–10 days; however, little is known about the key determinants of platelet lifespan. Platelets that have been experimentally damaged, aged platelets, or platelets stored for transfusion can have a storage lesion characterized by distinct morphological changes, with evidence of degranulation and expression of P-selectin and profound metalloproteinase-dependent proteolysis of GPIbα (86). However, inhibitors of PI3-kinase-dependent Rap1 or p38 mitogen-activated protein kinase (MAPK) that inhibit platelet activation ameliorate the deterioration of platelet quality (89, 103). Inhibiting ADAM17-mediated shedding of GPIbα and GPV markedly improved post-transfusion recovery of both mitochondria-injured and in vitro aged mouse platelets and stored human platelets (89). Metalloproteinase inhibition also prevented proteolysis of GPIbα on damaged platelets, improving the haemostatic function in vivo (86). There is a clear role for GPIbα in the clearance of platelets that have been chilled during storage; however, this effect is linked to clustering of GPIbα at temperatures below 16°C and the importance of metalloproteinase activity is undetermined (104, 105). GPIbα is shed from platelets stored at 22°C, however, and selective reversible inhibition of metalloproteinase activity may extend the shelf life and improve quality. Finally, the disulfide-link to GPIbβ of membrane-associated remnant tail fragment of GPIbα (Fig. 2c) could maintain its membrane association following proteolyis and could therefore maintain binding of 14-3-3 proteins that have a known role in regulating cell death/survival pathways in other cells (106–108); proteolysed GPIb (lacking the capacity to bind extracellular ligands) may not only regulate platelet clearance by sequestering 14-3-3, but also serve as a unique marker of platelet patency.

12. Conclusion

In summary, platelet receptor shedding provides potential markers, readouts, and quality control parameters for platelets. Analysis of shed fragments or proteolytic status of GPIbα, GPVI, and/or other platelet receptors provides the opportunity to develop a veritable barometer of functional status and disorders.

References

1. Andrews, R. K., Karunakaran, D., Gardiner, E. E., and Berndt, M. C. (2007) Platelet receptor proteolysis: a mechanism for down-regulating platelet reactivity. *Arterioscler. Thromb. Vasc. Biol. 27*, 1511–1520.

2. Berndt, M. C., Karunakaran, D., Gardiner, E. E., and Andrews, R. K. (2007) Programmed autologous cleavage of platelet receptors. *J. Thromb. Haemost. 5 (Suppl 1)*, 212–219.

3. Bergmeier, W., Piffath, C. L., Cheng, G., Dole, V. S., Zhang, Y., von Andrian, U. H., and Wagner, D. D. (2004) Tumor necrosis factor-α-converting enzyme (ADAM17) mediates GPIbα shedding from platelets *in vitro* and *in vivo*. *Circ. Res. 95*, 677–683.

4. Gardiner, E. E., Karunakaran, D., Shen, Y., Arthur, J. F., Andrews, R. K., and Berndt, M. C. (2007) Controlled shedding of platelet glycoprotein (GP)VI and GPIb-IX-V by ADAM family metalloproteinases. *J. Thromb. Haemost. 5*, 1530–1537.

5. Rabie, T., Strehl, A., Ludwig, A., and Nieswandt, B. (2005) Evidence for a role of ADAM17 (TACE) in the regulation of platelet glycoprotein V. *J. Biol. Chem. 280*, 14462–14468.

6. Gardiner, E. E., Arthur, J. F., Kahn, M. L., Berndt, M. C., and Andrews, R. K. (2004) Regulation of platelet membrane levels of glycoprotein VI by a platelet-derived metalloproteinase. *Blood 104*, 3611–3617.

7. Bergmeier, W., Rabie, T., Strehl, A., Piffath, C. L., Prostredna, M., Wagner, D. D., and Nieswandt, B. (2004) GPVI down-regulation in murine platelets through metalloproteinase-dependent shedding. *Thromb. Haemost. 91*, 951–958.

8. Stephens, G., Yan, Y., Jandrot-Perrus, M., Villeval, J. L., Clemetson, K. J., and Phillips, D. R. (2005) Platelet activation induces metalloproteinase-dependent GPVI cleavage to down-regulate platelet reactivity to collagen. *Blood 105*, 186–191.

9. Ludeman, M. J., Zheng, Y. W., Ishii, K., and Coughlin, S. R. (2004) Regulated shedding of PAR1 N-terminal exodomain from endothelial cells. *J. Biol. Chem. 279*, 18592–18599.

10. Choi, W. S., Jeon, O. H., Kim, H. H., and Kim, D. S. (2008) MMP-2 regulates human platelet activation by interacting with integrin $\alpha_{IIb}\beta_3$. *J. Thromb. Haemost. 6*, 517–523.

11. Choi, W. S., Jeon, O. H., and Kim, D. S. (2010) CD40 ligand shedding is regulated by interaction of MMP-2 and platelet integrin $\alpha_{IIb}\beta_3$. *J. Thromb. Haemost. 8*, 1364–1371.

12. Zhu, L., Bergmeier, W., Wu, J., Jiang, H., Stalker, T. J., Cieslak, M., Fan, R., Boumsell, L., Kumanogoh, A., Kikutani, H., Tamagnone, L., Wagner, D. D., Milla, M. E., and Brass, L. F. (2007) Regulated surface expression and shedding support a dual role for semaphorin 4D in platelet responses to vascular injury. *Proc. Natl Acad. Sci. USA 104*, 1621–1626.

13. Furman, M. I., Krueger, L. A., Linden, M. D., Barnard, M. R., Frelinger, A. L., 3 rd, and Michelson, A. D. (2004) Release of soluble CD40L from platelets is regulated by glycoprotein IIb/IIIa and actin polymerization. *J. Am. Coll. Cardiol. 43*, 2319–2325.

14. Matthies, K. M., Newman, J. L., Hodzic, A., and Wingett, D. G. (2006) Differential regulation of soluble and membrane CD40L proteins in T cells. *Cell Immunol. 241*, 47–58.

15. Amour, A., Knight, C. G., Webster, A., Slocombe, P. M., Stephens, P. E., Knauper, V., Docherty, A. J., and Murphy, G. (2000) The *in vitro* activity of ADAM-10 is inhibited by TIMP-1 and TIMP-3. *FEBS Lett. 473*, 275–279.

16. Du, X., Saido, T. C., Tsubuki, S., Indig, F. E., Williams, M. J., and Ginsberg, M. H. (1995) Calpain cleavage of the cytoplasmic domain of the integrin β_3 subunit. *J. Biol. Chem. 270*, 26146–26151.

17. Vaisar, T., Kassim, S. Y., Gomez, I. G., Green, P. S., Hargarten, S., Gough, P. J., Parks, W. C., Wilson, C. L., Raines, E. W., and Heinecke, J. W. (2009) MMP-9 sheds the β_2 integrin subunit (CD18) from macrophages. *Mol. Cell Proteomics 8*, 1044–1060.

18. Owen, C. A. (2008) Leukocyte cell surface proteinases: regulation of expression, functions, and mechanisms of surface localization. *Int. J. Biochem. Cell Biol. 40*, 1246–1272.

19. Evans, B. J., McDowall, A., Taylor, P. C., Hogg, N., Haskard, D. O., and Landis, R. C. (2006) Shedding of lymphocyte function-associated antigen-1 (LFA-1) in a human inflammatory response. *Blood 107*, 3593–3599.

20. Hemler, M. E. (2006) Shedding of heterodimeric leukocyte integrin. *Blood 107*, 3417–3418.

21. Gardiner, E. E., Arthur, J. F., Berndt, M. C., and Andrews, R. K. (2005) Role of calmodulin in platelet receptor function. *Curr. Med. Chem. Cardiovasc. Hematol. Agents 3*, 283–287.

22. Wong, M. X., Harbour, S. N., Wee, J. L., Lau, L. M., Andrews, R. K., and Jackson, D. E.

(2004) Proteolytic cleavage of platelet endothelial cell adhesion molecule-1 (PECAM-1/CD31) is regulated by a calmodulin-binding motif. *FEBS Lett. 568*, 70–78.

23. Naganuma, Y., Satoh, K., Yi, Q., Asazuma, N., Yatomi, Y., and Ozaki, Y. (2004) Cleavage of platelet endothelial cell adhesion molecule-1 (PECAM-1) in platelets exposed to high shear stress. *J. Thromb. Haemost. 2*, 1998–2008.

24. Dunlop, L. C., Skinner, M. P., Bendall, L. J., Favaloro, E. J., Castaldi, P. A., Gorman, J. J., Gamble, J. R., Vadas, M. A., and Berndt, M. C. (1992) Characterization of GMP-140 (P-selectin) as a circulating plasma protein. *J. Exp. Med. 175*, 1147–1150.

25. Gardiner, E. E., Karunakaran, D., Arthur, J. F., Mu, F. T., Powell, M. S., Baker, R. I., Hogarth, P. M., Kahn, M. L., Andrews, R. K., and Berndt, M. C. (2008) Dual ITAM-mediated proteolytic pathways for irreversible inactivation of platelet receptors: de-ITAM-izing FcγRIIa. *Blood 111*, 165–174.

26. Andrews, R. K., and Berndt, M. C. (2004) Platelet physiology and thrombosis. *Thromb. Res. 114*, 447–453.

27. Libby, P. (2008) Role of inflammation in atherosclerosis associated with rheumatoid arthritis. *Am. J. Med. 121*, S21-31.

28. Reininger, A. J., Bernlochner, I., Penz, S. M., Ravanat, C., Smethurst, P., Farndale, R. W., Gachet, C., Brandl, R., and Siess, W. (2010) A 2-step mechanism of arterial thrombus formation induced by human atherosclerotic plaques. *J. Am. Coll. Cardiol. 55*, 1147–1158.

29. Schulz, C., Penz, S., Hoffmann, C., Langer, H., Gillitzer, A., Schneider, S., Brandl, R., Seidl, S., Massberg, S., Pichler, B., Kremmer, E., Stellos, K., Schonberger, T., Siess, W., and Gawaz, M. (2008) Platelet GPVI binds to collagenous structures in the core region of human atheromatous plaque and is critical for atheroprogression *in vivo*. *Basic Res. Cardiol. 103*, 356–367.

30. Leslie, M. (2010) Cell biology. Beyond clotting: the powers of platelets. *Science 328*, 562–564.

31. Andrews, R. K., Berndt, M. C., and Lopez, J. A. (2006) The Glycoprotein Ib-IX-V Complex, in *Platelets* (Michelson, A. D., Ed.) 2nd ed, pp 145–164, Academic Press, San Diego.

32. Andrews, R. K., Gardiner, E. E., Shen, Y., Whisstock, J. C., and Berndt, M. C. (2003) Glycoprotein Ib-IX-V. *Int. J. Biochem. Cell Biol. 35*, 1170–1174.

33. Andrews, R. K., Lopez, J. A., and Berndt, M. C. (1997) Molecular mechanisms of platelet adhesion and activation. *Int. J. Biochem. Cell Biol. 29*, 91–105.

34. Lopez, J. A., Andrews, R. K., Afshar-Kharghan, V., and Berndt, M. C. (1998) Bernard-Soulier syndrome. *Blood 91*, 4397–4418.

35. Luo, S. Z., Mo, X., Afshar-Kharghan, V., Srinivasan, S., Lopez, J. A., and Li, R. (2007) Glycoprotein Ibα forms disulfide bonds with 2 glycoprotein Ibβ subunits in the resting platelet. *Blood 109*, 603–609.

36. Mu, F. T., Andrews, R. K., Arthur, J. F., Munday, A. D., Cranmer, S. L., Jackson, S. P., Stomski, F. C., Lopez, A. F., and Berndt, M. C. (2008) A functional 14-3-3ζ-independent association of PI3-kinase with glycoprotein Ibα, the major ligand-binding subunit of the platelet glycoprotein Ib-IX-V complex. *Blood 111*, 4580–4587.

37. Mu, F. T., Cranmer, S. L., Andrews, R. K., and Berndt, M. C. (2010) Functional association of phosphoinositide-3-kinase with platelet glycoprotein Ibα, the major ligand-binding subunit of the glycoprotein Ib-IX-V complex. *J. Thromb. Haemost. 8*, 324–330.

38. Munday, A. D., Berndt, M. C., and Mitchell, C. A. (2000) Phosphoinositide 3-kinase forms a complex with platelet membrane glycoprotein Ib-IX-V complex and 14-3-3ζ. *Blood 96*, 577–584.

39. Andrews, R. K., Munday, A. D., Mitchell, C. A., and Berndt, M. C. (2001) Interaction of calmodulin with the cytoplasmic domain of the platelet membrane glycoprotein Ib-IX-V complex. *Blood 98*, 681–687.

40. Berndt, M. C., and Phillips, D. R. (1981) Interaction of thrombin with platelets: purification of the thrombin substrate. *Ann. N.Y. Acad. Sci. 370*, 87–95.

41. Ramakrishnan, V., DeGuzman, F., Bao, M., Hall, S. W., Leung, L. L., and Phillips, D. R. (2001) A thrombin receptor function for platelet glycoprotein Ib-IX unmasked by cleavage of glycoprotein V. *Proc. Natl Acad. Sci. USA 98*, 1823–1828.

42. Ramakrishnan, V., Reeves, P. S., DeGuzman, F., Deshpande, U., Ministri-Madrid, K., DuBridge, R. B., and Phillips, D. R. (1999) Increased thrombin responsiveness in platelets from mice lacking glycoprotein V. *Proc. Natl Acad. Sci. USA 96*, 13336–13341.

43. Andrews, R. K., Suzuki-Inoue, K., Shen, Y., Tulasne, D., Watson, S. P., and Berndt, M. C. (2002) Interaction of calmodulin with the cytoplasmic domain of platelet glycoprotein VI. *Blood 99*, 4219–4221.

44. Arthur, J. F., Shen, Y., Kahn, M. L., Berndt, M. C., Andrews, R. K., and Gardiner, E. E. (2007) Ligand binding rapidly induces disulfide-dependent dimerization of glycoprotein VI on the platelet plasma membrane. *J. Biol. Chem.* 282, 30434–30441.

45. Suzuki-Inoue, K., Tulasne, D., Shen, Y., Bori-Sanz, T., Inoue, O., Jung, S. M., Moroi, M., Andrews, R. K., Berndt, M. C., and Watson, S. P. (2002) Association of Fyn and Lyn with the proline-rich domain of glycoprotein VI regulates intracellular signaling. *J. Biol. Chem.* 277, 21561–21566.

46. Schmaier, A. A., Zou, Z., Kazlauskas, A., Emert-Sedlak, L., Fong, K. P., Neeves, K. B., Maloney, S. F., Diamond, S. L., Kunapuli, S. P., Ware, J., Brass, L. F., Smithgall, T. E., Saksela, K., and Kahn, M. L. (2009) Molecular priming of Lyn by GPVI enables an immune receptor to adopt a hemostatic role. *Proc. Natl Acad. Sci. USA* 106, 21167–21172.

47. Nieswandt, B., and Watson, S. P. (2003) Platelet-collagen interaction: is GPVI the central receptor? *Blood* 102, 449–461.

48. Jung, S. M., and Moroi, M. (2008) Platelet glycoprotein VI. *Adv. Exp. Med. Biol.* 640, 53–63.

49. Xiang, Y. Z., Kang, L. Y., Gao, X. M., Shang, H. C., Zhang, J. H., and Zhang, B. L. (2008) Strategies for antiplatelet targets and agents. *Thromb. Res.* 123, 35–49.

50. Kleinschnitz, C., Pozgajova, M., Pham, M., Bendszus, M., Nieswandt, B., and Stoll, G. (2007) Targeting platelets in acute experimental stroke: impact of glycoprotein Ib, VI, and IIb/IIIa blockade on infarct size, functional outcome, and intracranial bleeding. *Circulation* 115, 2323–2330.

51. Massberg, S., Gawaz, M., Gruner, S., Schulte, V., Konrad, I., Zohlnhofer, D., Heinzmann, U., and Nieswandt, B. (2003) A crucial role of glycoprotein VI for platelet recruitment to the injured arterial wall *in vivo*. *J. Exp. Med.* 197, 41–49.

52. Stoll, G., Kleinschnitz, C., and Nieswandt, B. (2008) Molecular mechanisms of thrombus formation in ischemic stroke: novel insights and targets for treatment. *Blood* 112, 3555–3562.

53. Chen, H., Locke, D., Liu, Y., Liu, C., and Kahn, M. L. (2002) The platelet receptor GPVI mediates both adhesion and signaling responses to collagen in a receptor density-dependent fashion. *J. Biol. Chem.* 277, 3011–3019.

54. May, F., Hagedorn, I., Pleines, I., Bender, M., Vogtle, T., Eble, J., Elvers, M., and Nieswandt, B. (2009) CLEC-2 is an essential platelet-activating receptor in hemostasis and thrombosis. *Blood* 114, 3464–3472.

55. Andre, P., Prasad, K. S., Denis, C. V., He, M., Papalia, J. M., Hynes, R. O., Phillips, D. R., and Wagner, D. D. (2002) CD40L stabilizes arterial thrombi by a β_3 integrin–-dependent mechanism. *Nat. Med.* 8, 247–252.

56. Romo, G. M., Dong, J. F., Schade, A. J., Gardiner, E. E., Kansas, G. S., Li, C. Q., McIntire, L. V., Berndt, M. C., and Lopez, J. A. (1999) The glycoprotein Ib-IX-V complex is a platelet counterreceptor for P-selectin. *J. Exp. Med.* 190, 803–814.

57. Hrachovinova, I., Cambien, B., Hafezi-Moghadam, A., Kappelmayer, J., Camphausen, R. T., Widom, A., Xia, L., Kazazian, H. H., Jr., Schaub, R. G., McEver, R. P., and Wagner, D. D. (2003) Interaction of P-selectin and PSGL-1 generates microparticles that correct hemostasis in a mouse model of hemophilia A. *Nat. Med.* 9, 1020–1025.

58. Jones, C. I., Garner, S. F., Moraes, L. A., Kaiser, W. J., Rankin, A., Ouwehand, W. H., Goodall, A. H., and Gibbins, J. M. (2009) PECAM-1 expression and activity negatively regulate multiple platelet signaling pathways. *FEBS Lett.* 583, 3618–3624.

59. Tomlinson, M. G., Calaminus, S. D., Berlanga, O., Auger, J. M., Bori-Sanz, T., Meyaard, L., and Watson, S. P. (2007) Collagen promotes sustained glycoprotein VI signaling in platelets and cell lines. *J. Thromb. Haemost.* 5, 2274–2283.

60. Crockett, J., Newman, D. K., and Newman, P. J. (2010) PECAM-1 is a negative regulator of laminin-induced platelet activation. *J. Thromb. Haemost.* 8, 1584–1593.

61. Kleinschnitz, C., Braeuninger, S., Pham, M., Austinat, M., Nolte, I., Renne, T., Nieswandt, B., Bendszus, M., and Stoll, G. (2008) Blocking of platelets or intrinsic coagulation pathway-driven thrombosis does not prevent cerebral infarctions induced by photothrombosis. *Stroke* 39, 1262–1268.

62. Kleinschnitz, C., Stoll, G., Bendszus, M., Schuh, K., Pauer, H. U., Burfeind, P., Renne, C., Gailani, D., Nieswandt, B., and Renne, T. (2006) Targeting coagulation factor XII provides protection from pathological thrombosis in cerebral ischemia without interfering with hemostasis. *J. Exp. Med.* 203, 513–518.

63. Renne, T., Pozgajova, M., Gruner, S., Schuh, K., Pauer, H. U., Burfeind, P., Gailani, D., and Nieswandt, B. (2005) Defective thrombus formation in mice lacking coagulation factor XII. *J. Exp. Med.* 202, 271–281.

64. Decrem, Y., Rath, G., Blasioli, V., Cauchie, P., Robert, S., Beaufays, J., Frere, J. M.,

Feron, O., Dogne, J. M., Dessy, C., Vanhamme, L., and Godfroid, E. (2009) Ir-CPI, a coagulation contact phase inhibitor from the tick *Ixodes ricinus*, inhibits thrombus formation without impairing hemostasis. *J. Exp. Med.* 206, 2381–2395.

65. Hagedorn, I., Schmidbauer, S., Pleines, I., Kleinschnitz, C., Kronthaler, U., Stoll, G., Dickneite, G., and Nieswandt, B. (2010) Factor XIIa inhibitor recombinant human albumin Infestin-4 abolishes occlusive arterial thrombus formation without affecting bleeding. *Circulation* 121, 1510–1517.

66. Peerschke, E. I., Murphy, T. K., and Ghebrehiwet, B. (2003) Activation-dependent surface expression of gC1qR/p33 on human blood platelets. *Thromb. Haemost.* 89, 331–339.

67. Muller, F., Mutch, N. J., Schenk, W. A., Smith, S. A., Esterl, L., Spronk, H. M., Schmidbauer, S., Gahl, W. A., Morrissey, J. H., and Renne, T. (2009) Platelet polyphosphates are proinflammatory and procoagulant mediators *in vivo*. *Cell* 139, 1143–1156.

68. Walsh, P. N. (2004) Platelet coagulation-protein interactions. *Semin. Thromb. Hemost.* 30, 461–471.

69. Boilard, E., Nigrovic, P. A., Larabee, K., Watts, G. F., Coblyn, J. S., Weinblatt, M. E., Massarotti, E. M., Remold-O'Donnell, E., Farndale, R. W., Ware, J., and Lee, D. M. (2010) Platelets amplify inflammation in arthritis *via* collagen-dependent microparticle production. *Science* 327, 580–583.

70. Zimmerman, G. A., and Weyrich, A. S. (2010) Immunology. Arsonists in rheumatoid arthritis. *Science* 327, 528–529.

71. Brocker, C. N., Vasiliou, V., and Nebert, D. W. (2009) Evolutionary divergence and functions of the ADAM and ADAMTS gene families. *Hum. Genomics* 4, 43–55.

72. Wijeyewickrema, L. C., Berndt, M. C., and Andrews, R. K. (2005) Snake venom probes of platelet adhesion receptors and their ligands. *Toxicon* 45, 1051–1061.

73. Janes, P. W., Saha, N., Barton, W. A., Kolev, M. V., Wimmer-Kleikamp, S. H., Nievergall, E., Blobel, C. P., Himanen, J. P., Lackmann, M., and Nikolov, D. B. (2005) ADAM meets Eph: an ADAM substrate recognition module acts as a molecular switch for ephrin cleavage in *trans*. *Cell* 123, 291–304.

74. Murphy, G. (2009) Regulation of the proteolytic disintegrin metalloproteinases, the 'sheddases'. *Semin. Cell Dev. Biol.* 20, 138–145.

75. Kahn, J., Walcheck, B., Migaki, G. I., Jutila, M. A., and Kishimoto, T. K. (1998) Calmodulin regulates L-selectin adhesion molecule expression and function through a protease-dependent mechanism. *Cell* 92, 809–818.

76. Mo, X., Nguyen, N. X., Mu, F.-T., Luo, S.-Z., Fan, H., Andrews, R. K., Berndt, M. C., and Li, R. (2010) Transmembrane and trans-subunit regulation of ectodomain shedding of platelet glycoprotein Ibα. *J. Biol. Chem.* 285, 32096–32104.

77. Santos-Martinez, M. J., Medina, C., Jurasz, P., and Radomski, M. W. (2008) Role of metalloproteinases in platelet function. *Thromb. Res.* 121, 535–542.

78. Reinboldt, S., Wenzel, F., Rauch, B. H., Hohlfeld, T., Grandoch, M., Fischer, J. W., and Weber, A. A. (2009) Preliminary evidence for a matrix metalloproteinase-2 (MMP-2)-dependent shedding of soluble CD40 ligand (sCD40L) from activated platelets. *Platelets* 20, 441–444.

79. Al-Tamimi, M., Mu, F. T., Arthur, J. F., Shen, Y., Moroi, M., Berndt, M. C., Andrews, R. K., and Gardiner, E. E. (2009) Antiglycoprotein VI monoclonal antibodies directly aggregate platelets independently of FcγRIIa and induce GPVI ectodomain shedding. *Platelets* 20, 75–82.

80. Gardiner, E. E., Arthur, J. F., Shen, Y., Karunakaran, D., Moore, L. A., Am Esch, J. S., Andrews, R. K., and Berndt, M. C. (2010) GPIbα-selective activation of platelets induces platelet signaling events comparable to GPVI activation events. *Platelets* 21, 244–252.

81. Al-Tamimi, M., Mu, F. T., Moroi, M., Gardiner, E. E., Berndt, M. C., and Andrews, R. K. (2009) Measuring soluble platelet glycoprotein VI in human plasma by ELISA. *Platelets* 20, 143–149.

82. Aleil, B., Meyer, N., Wolff, V., Kientz, D., Wiesel, M. L., Gachet, C., Cazenave, J. P., and Lanza, F. (2006) Plasma glycoprotein V levels in the general population: normal distribution, associated parameters and implications for clinical studies. *Thromb. Haemost.* 96, 505–511.

83. Bigalke, B., Stellos, K., Weig, H. J., Geisler, T., Seizer, P., Kremmer, E., Potz, O., Joos, T., May, A. E., Lindemann, S., and Gawaz, M. (2009) Regulation of platelet glycoprotein VI (GPVI) surface expression and of soluble GPVI in patients with atrial fibrillation (AF) and acute coronary syndrome (ACS). *Basic Res. Cardiol.* 104, 352–357.

84. Steinberg, M. H., Kelton, J. G., and Coller, B. S. (1987) Plasma glycocalicin. An aid in the classification of thrombocytopenic disorders. *N. Engl. J. Med.* 317, 1037–1042.

85. Fox, J. E. (1994) Shedding of adhesion receptors from the surface of activated platelets. *Blood Coagul. Fibrinol. 5*, 291–304.

86. Bergmeier, W., Burger, P. C., Piffath, C. L., Hoffmeister, K. M., Hartwig, J. H., Nieswandt, B., and Wagner, D. D. (2003) Metalloproteinase inhibitors improve the recovery and hemostatic function of *in vitro*-aged or -injured mouse platelets. *Blood 102*, 4229–4235.

87. Gardiner, E. E., Al-Tamimi, M., Mu, F. T., Karunakaran, D., Thom, J. Y., Moroi, M., Andrews, R. K., Berndt, M. C., and Baker, R. I. (2008) Compromised ITAM-based platelet receptor function in a patient with immune thrombocytopenic purpura. *J. Thromb. Haemost. 6*, 1175–1182.

88. Brill, A., Chauhan, A. K., Canault, M., Walsh, M. T., Bergmeier, W., and Wagner, D. D. (2009) Oxidative stress activates ADAM17/TACE and induces its target receptor shedding in platelets in a p38-dependent fashion. *Cardiovasc. Res. 84*, 137–144.

89. Canault, M., Duerschmied, D., Brill, A., Stefanini, L., Schatzberg, D., Cifuni, S. M., Bergmeier, W., and Wagner, D. D. (2010) p38 mitogen-activated protein kinase activation during platelet storage: consequences for platelet recovery and hemostatic function in vivo. *Blood 115*, 1835–1842.

90. Arthur, J. F., Gardiner, E. E., Matzaris, M., Taylor, S. G., Wijeyewickrema, L., Ozaki, Y., Kahn, M. L., Andrews, R. K., and Berndt, M. C. (2005) Glycoprotein VI is associated with GPIb-IX-V on the membrane of resting and activated platelets. *Thromb. Haemost. 93*, 716–723.

91. Duerschmied, D., Canault, M., Lievens, D., Brill, A., Cifuni, S. M., Bader, M., and Wagner, D. D. (2009) Serotonin stimulates platelet receptor shedding by tumor necrosis factor-α-converting enzyme (ADAM17). *J. Thromb. Haemost. 7*, 1163–1171.

92. Wang, Z., Shi, Q., Yan, R., Liu, G., Zhang, W., and Dai, K. (2010) The role of calpain in the regulation of ADAM17-dependent GPIbα ectodomain shedding. *Arch. Biochem. Biophys. 495*, 136–143.

93. Bender, M., Hofmann, S., Stegner, D., Chalaris, A., Bösl, M., Braun, A., Scheller, J., Rose-John, S., and Nieswandt, B. (2010) Differentially regulated GPVI ectodomain shedding by multiple platelet-expressed proteinases. *Blood 116*, 3347–3355.

94. Aktas, B., Pozgajova, M., Bergmeier, W., Sunnarborg, S., Offermanns, S., Lee, D., Wagner, D. D., and Nieswandt, B. (2005) Aspirin induces platelet receptor shedding via ADAM17 (TACE). *J. Biol. Chem. 280*, 39716–39722.

95. Takayama, H., Hosaka, Y., Nakayama, K., Shirakawa, K., Naitoh, K., Matsusue, T., Shinozaki, M., Honda, M., Yatagai, Y., Kawahara, T., Hirose, J., Yokoyama, T., Kurihara, M., and Furusako, S. (2008) A novel antiplatelet antibody therapy that induces cAMP-dependent endocytosis of the GPVI/Fc receptor γ-chain complex. *J. Clin. Invest. 118*, 1785–1795.

96. Sugiyama, T., Okuma, M., Ushikubi, F., Sensaki, S., Kanaji, K., and Uchino, H. (1987) A novel platelet aggregating factor found in a patient with defective collagen-induced platelet aggregation and autoimmune thrombocytopenia. *Blood 69*, 1712–1720.

97. Gardiner, E. E., Thom, J. Y., Al-Tamimi, M., Hughes, A., Berndt, M. C., Andrews, R. K., and Baker, R. I. (2010) Restored platelet function after romiplostim treatment in a patient with immune thrombocytopenic purpura. *Br. J. Haematol. 149*, 625–628.

98. Blann, A. D., Lanza, F., Galajda, P., Gurney, D., Moog, S., Cazenave, J. P., and Lip, G. Y. (2001) Increased platelet glycoprotein V levels in patients with coronary and peripheral atherosclerosis - the influence of aspirin and cigarette smoking. *Thromb. Haemost. 86*, 777–783.

99. Gurney, D., Lip, G. Y., and Blann, A. D. (2002) A reliable plasma marker of platelet activation: does it exist? *Am. J. Hematol. 70*, 139–144.

100. Nurden, P., Tandon, N., Takizawa, H., Couzi, L., Morel, D., Fiore, M., Pillois, X., Loyau, S., Jandrot-Perrus, M., and Nurden, A. T. (2009) An acquired inhibitor to the GPVI platelet collagen receptor in a patient with lupus nephritis. *J. Thromb. Haemost. 7*, 1541–1549.

101. Bigalke, B., Langer, H., Geisler, T., Lindemann, S., and Gawaz, M. (2007) Platelet glycoprotein VI: a novel marker for acute coronary syndrome. *Semin. Thromb. Hemost. 33*, 179–184.

102. Laske, C., Leyhe, T., Stransky, E., Eschweiler, G. W., Bueltmann, A., Langer, H., Stellos, K., and Gawaz, M. (2008) Association of platelet-derived soluble glycoprotein VI in plasma with Alzheimer's disease. *J. Psychiatr. Res. 42*, 746–751.

103. Schubert, P., Thon, J. N., Walsh, G. M., Chen, C. H., Moore, E. D., Devine, D. V., and Kast, J. (2009) A signaling pathway contributing to platelet storage lesion development: targeting PI3-kinase-dependent Rap1

activation slows storage-induced platelet deterioration. *Transfusion 49,* 1944–1955.

104. Hoffmeister, K. M., Felbinger, T. W., Falet, H., Denis, C. V., Bergmeier, W., Mayadas, T. N., von Andrian, U. H., Wagner, D. D., Stossel, T. P., and Hartwig, J. H. (2003) The clearance mechanism of chilled blood platelets. *Cell 112,* 87–97.

105. Hoffmeister, K. M., Josefsson, E. C., Isaac, N. A., Clausen, H., Hartwig, J. H., and Stossel, T. P. (2003) Glycosylation restores survival of chilled blood platelets. *Science 301,* 1531–1534.

106. Fu, H., Subramanian, R. R., and Masters, S. C. (2000) 14-3-3 proteins: structure, function, and regulation. *Annu. Rev. Pharmacol. Toxicol. 40,* 617–647.

107. Feng, S., Christodoulides, N., and Kroll, M. H. (1999) The glycoprotein Ib/IX complex regulates cell proliferation. *Blood 93,* 4256–4263.

108. Bialkowska, K., Zaffran, Y., Meyer, S. C., and Fox, J. E. (2003) 14-3-3ζ mediates integrin-induced activation of Cdc42 and Rac. Platelet glycoprotein Ib-IX regulates integrin-induced signaling by sequestering 14-3-3ζ. *J. Biol. Chem. 278,* 33342–33350.

Chapter 23

Endogenous Inhibitory Mechanisms and the Regulation of Platelet Function

Chris I. Jones, Natasha E. Barrett, Leonardo A. Moraes, Jonathan M. Gibbins, and Denise E. Jackson

Abstract

The response of platelets to changes in the immediate environment is always a balance between activatory and inhibitory signals, the cumulative effect of which is either activation or quiescence. This is true of platelets in free flowing blood and of their regulation of haemostasis and thrombosis. In this review, we consider the endogenous inhibitory mechanisms that combine to regulate platelet activation. These include those derived from the endothelium (nitric oxide, prostacyclin, CD39), inhibitory receptors on the surface of platelets (platelet endothelial cell adhesion molecule-1, carcinoembryonic antigen cell adhesion molecule 1, G6b-B – including evidence for the role of Ig-ITIM superfamily members in the negative regulation of ITAM-associated GPVI platelet–collagen interactions and GPCR-mediated signalling and in positive regulation of "outside-in" integrin $\alpha_{IIb}\beta_3$-mediated signalling), intracellular inhibitory receptors (retinoic X receptor, glucocorticoid receptor, peroxisome proliferator-activated receptors, liver X receptor), and emerging inhibitory pathways (canonical Wnt signalling, Semaphorin 3A, endothelial cell specific adhesion molecule, and junctional adhesion molecule-A).

Key words: Platelet immunoreceptors, Platelet inhibition, Thrombus formation, Platelet–collagen interactions, Immunoreceptor tyrosine-based activatory motif, Immunoreceptor tyrosine-based inhibitory motif

1. Introduction

A plethora of activatory receptors and cell signalling mechanisms control the activation of platelets at sites of tissue injury, and in a pathological setting these may lead to the development of thrombosis. The platelet activation process involves multiple waves of regulation. This is principally through the involvement

Jonathan M. Gibbins and Martyn P. Mahaut-Smith (eds.), *Platelets and Megakaryocytes: Volume 3, Additional Protocols and Perspectives*, Methods in Molecular Biology, vol. 788, DOI 10.1007/978-1-61779-307-3_23, © Springer Science+Business Media, LLC 2012

of positive feedback loops that are mediated through a range of pro-thrombotic factors which include ADP, thromboxane (TX) A_2, serotonin, epinephrine, and tachykinins that are released or generated by activated platelets. Redundancy in activatory mechanisms and crosstalk between these are essential to enable a rapid and complete response following tissue injury. The counterpoint to this being precise regulation that is required to prevent inappropriate activation and uncontrolled platelet thrombus formation.

While in the context of haemostasis, the function of a platelet may be perceived as a "one-way process" leading ultimately to thrombus formation and protection from blood loss, the control of the function of these cells is much like the regulation of cellular processes in many other cell types, with functional response representing the product of positive and negative signals. In the absence of vessel injury or disease, inhibitory signals prevail ensuring that platelets remain quiescent in the circulation, but when platelets encounter injury or disease sites, activatory signals tip the balance in the favour of activation. The control of the platelet contribution to haemostasis and thrombosis may, therefore, be considered like a rheostat, rather than an "on/off switch".

Factors that activate platelets may be considered to be strong, such as collagens and thrombin, or weak such as most of the secondary (or positive feedback) mediators, including ADP, TXA_2, and serotonin. Modulation of the availability or receptor activation by these weaker platelet agonists has formed the basis of successful anti-thrombotic strategies through adjustment of the platelet rheostat enabling controlled suppression of platelet reactivity.

In recent years, it has emerged that the endogenous inhibition of platelet function is also regulated by powerful and less-powerful mechanisms. Powerful inhibitory signals to platelets are provided by the healthy arterial endothelium through the production and release of bioactive nitric oxide (NO) and prostacyclin (PGI_2) into flowing blood. When the integrity of the vascular endothelium is breached due to injury or blood vessel disease, such signals in the vicinity become diminished. An increasing family of receptors, such as platelet–endothelial cell adhesion molecule-1 (PECAM-1) and carcinoembryonic antigen cell adhesion molecule 1 (CEACAM1), largely adhesive in nature, are responsible for weaker inhibition of platelet function, although in combination their role may be substantial. The precise role of such receptors that require cell–cell contact in order to signal in health and disease is currently unclear, but this may be important in limiting the development or spread of a thrombus at the site of injury, thrombus stabilisation, and resolution of the injury. It is perhaps pertinent to note that such receptors are likely to be functional in localities, where the endothelium is damaged or diseased, and therefore where local concentrations of NO and PGI_2 are low.

A third and surprising arm to the inhibitory mechanisms available within platelets has emerged recently, involving selected intracellular receptors whose roles would normally be associated with the regulation of gene transcription. In the absence of a nucleus, non-genomic regulation of platelet signalling is becoming apparent. The implications of these mechanisms in normal physiology are currently uncertain.

The nature of a specific injury or blood vessel damage is likely to impact on the balance of activatory and inhibitory platelet signalling. Determining the role of specific mechanisms in different physiological circumstances remains a challenging area of research that requires resolution towards establishing whether targeting inhibitory as well as activatory fine-tuning mechanisms may provide more efficacious anti-thrombotic therapies.

In this chapter, we introduce the currently characterised or emerging endogenous mechanisms for the inhibition of platelet function.

2. Endothelium-Derived Inhibitory Mechanisms

2.1. Soluble Mediators

2.1.1. Nitric Oxide

NO, a free radical produced by healthy endothelial cells, induces vasodilation (smooth muscle relaxation) and inhibition of circulating platelets (1). NO is synthesized from the amino acid l-arginine by nitric oxide synthases (NOS), a family of enzymes which can be either constitutively active or inducible. Once formed, NO diffuses through the endothelial cell membrane and is released into the blood stream from where it crosses the platelet plasma membrane to bind to and activate the cytosolic enzyme soluble guanylyl cyclase (sGC) resulting in increased production of cyclic guanosine $3',5'$-monophosphate (cGMP) from GTP (2, 3). The immediate consequences of increased cGMP are the direct activation of protein kinase G (PKG) (4–6), to which it binds, the indirect activation of PKA via inhibition of phosphodiesterase 3 (PDE-3), and the subsequent increase in cAMP (7–9). The activation of PKG is considered the primary mode by which NO inhibits platelets and leads to the phosphorylation of numerous targets, including inositol-1,4,5-trisphosphate receptors (IP_3R) (10), PDE-5 (11), vasodilator-stimulated phosphoprotein (VASP) (12, 13), RAP1b (14, 15), and the thromboxane A_2 receptor (16). The corollary of which is the diminution of activatory signalling and the inhibition of platelet activation.

In recent years, two controversies have developed around the inhibition of platelets by NO. The first concerns the ability of platelets to synthesize NO. Numerous reports have indicated that platelets express both eNOS (17) and to a lesser extent iNOS (18), and generate NO in response to stimulation with collagen, ADP, thrombin, and vWF (19–24). Platelet-derived NO has been demonstrated to inhibit several platelet functions, including granule secretion (25), platelet aggregation (26, 27), thrombus growth (28), and platelet adhesion (29). Against this, however, mice lacking eNOS do not show any thrombotic phenotype (30), nor have enhanced aggregation to ADP (31), elevated surface levels of P-selectin, or increased production of thromboxane B_2 (31, 32). Furthermore, Gambaryan et al. have cast doubt on earlier findings indicting the presence of NOS in platelets (33). Using a highly purified platelet preparation, Gambaryan et al. suggest that neither eNOS nor iNOS are present in human or mouse platelets and that previous reports of their presence may be due to erythrocyte or leukocyte contamination and the pitfalls of commercial antibodies (33). This study does not address the generation of NO by platelets, which has been reported by numerous studies (34) and cannot be explained purely by contamination of the platelet sample. The possibility, therefore, exists of alternative NOS-like activity within platelets that has yet to be identified (35).

The second controversy centres on the ability of NO to inhibit platelets in a cGMP-independent manner via nitrosylation (addition of an NO group) of substrates other than the heme ion in sGC, or nitration, particularly tyrosine nitration. NO-treated platelets show the modification of a range of molecules, including $\alpha_{IIb}\beta_3$, protein disulphide isomerase, and α-actinin (36–38). Furthermore, some inhibition of platelet function by NO survives in the presence of 1H-(1, 2, 4)oxadiazolo(4,3-a)quinoxalin-1-one (ODQ), a selective sGC inhibitor (25, 36, 37, 39–41). The evidence against the functional relevance of these cGMP-independent effects comes from knockout mice lacking sGC (42). The addition of NO to the platelets of these mice, even at millimolar concentrations, failed to cause an increase in cGMP or any inhibition of platelet function (42). Clearly, this strongly suggests that NO inhibits platelets predominantly through cGMP-dependent mechanisms, although it does not rule out the possibility of more subtle effects resulting from measurable nitrosylation and nitration.

2.1.2. Prostacyclin

Like NO, endothelium-derived PGI_2 is also a potent inhibitor of platelets. Indeed NO and PGI_2 are able to act synergistically to inhibit platelet aggregation, though not adhesion (43–45). PGI_2, produced by the vascular endothelium (46, 47), is an inhibitory (48, 49) member of the family of prostanoids derived from the fatty acid arachidonic acid (50), whereas another family member TXA_2, produced by platelets, has a local stimulatory effect (51, 52). Synthesis of both prostanoids is closely mirrored, sharing the key enzymes and pathways. Arachidonic acid is converted first to PGG_2 and then onto PGH_2 by prostaglandin G/H synthase, more commonly known as cyclo-oxygenase (COX) (53, 54). Of the COX isoforms, COX-1 is a constitutive enzyme found in most mammalian cells, including mature platelets, whereas COX-2 is an inducible form that is found primarily in macrophages and inflammatory cells (55). PGH_2 is further converted to the various prostanoids by cell-specific enzymes. For example, PGI_2 is generated by the action of prostacyclin synthase (PGIS) in vascular endothelium (56) and smooth muscle cells (57), whereas TXA_2 is generated by the action of thromboxane synthase in platelets (58, 59). Aspirin (acetylsalicylic acid), a widely used anti-platelet drug, acetylates and inhibits both forms of COX, thus dosage is the key to targeting the therapeutic window such that production of TXA_2 is reduced without the detrimental loss of PGI_2.

Once synthesized, PGI_2 is released into the blood stream. Here, PGI_2 is able to bind to the prostacyclin receptor (IP) on the surface of platelets and arterial smooth muscle cells, inhibiting aggregation (60, 61) and causing vasodilation (62), respectively. Activation of the IP receptor on the surface of platelets induces production of cAMP (63, 64) (via $G\alpha_s$ (65)), resulting in activation of PKA (66), the subsequent inhibition of several pathways, including PKC activation, calcium release, and platelet inhibition (63, 64, 67).

2.2. Non-soluble Mediators

2.2.1. CD39

CD39, or nucleoside triphosphate diphosphohydrolase 1 (NTPDase 1), is an ecto-nucleosidase (68), i.e. an extracellular membrane-anchored glycoprotein, found on the surface of a variety of cells including endothelial cells (69, 70) and platelets (71), where it is able to inhibit platelet activation by hydrolysis of both ADP and ATP (released by activated platelets) to AMP (72, 73). The AMP produced is degraded to adenosine by another platelet/endothelial surface protein, CD73 (74, 75). Adenosine, itself a strong inhibitor of platelets (76), binds to the P1 receptor on platelets causing an increase in cAMP levels (77). The importance of CD39 in platelet inhibition has been demonstrated in studies, whereby platelet inhibition was maintained even when NO and prostacyclin effects were blocked (70, 73, 76). Unlike NO and PGI_2, which have short half lives, CD39, as a transmembrane protein, is relatively long lived.

3. Inhibitory Platelet Cell Surface Receptors

The potential inhibitory role of adhesion receptors initially came to the fore following the discovery that the platelet receptor PECAM-1 possesses an immunoreceptor tyrosine-based inhibitory motif (ITIM) (78–80). The central role of immune-like signalling in activation pathways in the platelet in response to collagen raised the possibility that platelets may also share immune receptor like inhibitory mechanisms. The biological functions of ITIM-containing receptors are complex and in some contexts they stimulate activatory signalling. This dichotomy also appears in the roles of PECAM-1 in different aspects of platelet function, inhibiting early signals that cause platelet activation and aggregation yet enhancing some integrin outside-in regulating functions. The family of inhibitory ITIM-containing platelet receptors is slowly increasing, with the recent additions and characterisation of CEACAM1 and G6b-B (81–85).

3.1. Platelet Endothelial Cell Adhesion Molecule-1

PECAM-1 is a 130-kDa transmembrane glycoprotein possessing a 574 amino acid residue extracellular domain, a 19 amino acid transmembrane domain, and a cytoplasmic domain that is subject to alternative splicing (86, 87), and has become recognised for its ability to stimulate cell signalling. Its expression is restricted to haematopoietic (platelets, monocytes, neutrophils, T lymphocyte subsets) and endothelial cells (88–91), in which it has become associated with a range of biological processes. On the platelet surface, the expression level of PECAM-1 has been estimated in various studies to be between 5,000 and 8,800 copies per cell (92–95). Following platelet activation, surface exposure of PECAM-1, which is also localised in the intracellular, alpha-granule compartment of platelets, is increased (92, 93, 96).

The principal ligand for PECAM-1 is believed to be PECAM-1 itself through homophilic interactions on adjacent cells, although potential additional heterophilic counter-receptors, such as CD38 and integrin $\alpha_v\beta_3$, have been reported (97). While classified as an adhesion receptor, PECAM-1 on its own is not sufficient to support adhesive interactions. Indeed platelets are capable of aggregation in the absence of PECAM-1 expression (98), and platelet aggregation defects, such as Glanzmann thrombasthenia, persist in the presence of PECAM-1 (99).

PECAM-1 was expression cloned from an endothelial cDNA library and, due to sequence homology to receptors such as ICAM1 and VCAM1, was assigned to the Ig superfamily of receptors (86), although later observations that PECAM-1 becomes tyrosine phosphorylated upon stimulation of platelets with thrombin, thrombin receptor activatory peptide (TRAP), collagen, and collagen-related peptide (CRP) led to suspicions that this receptor engages in cell signalling. A closer examination of the sequence of the cytoplasmic tail of this receptor led to the identification of the presence of an ITIM ((L/V/I/S/T) XYXX(L/V)) and a related Immunoreceptor Tyrosine-based Switch Motif (ITSM, TxYxx(V/I)) (100, 101). These motifs incorporates two tyrosine residues (residues 663 and 686) that become tyrosine phosphorylated following stimulation of platelets with thrombin, TRAP, GPVI, and GPIb (78, 96, 102). This is largely dependent on integrin $\alpha_{IIb}\beta_3$-dependent platelet aggregation, although low-level phosphorylation may be detected in the absence of aggregation or following exposure of platelets to shear stress (96, 103).

Transient over expression studies in COS cells revealed that PECAM-1 could be tyrosine phosphorylated by either Src-related kinases or Csk-related kinases, but not Syk, Itk, or Pyk2 (104). Consistent with this finding, in primary platelets, Src-related kinases Fyn, Lyn, Src, Yes, and Hck were shown to co-immunoprecipitate with PECAM-1 (96). In addition, Fyn is an essential component involved in stretch and shear flow-driven PECAM-1 tyrosine phosphorylation in endothelial cells (105). In contrast, inhibition of antigen receptor signalling by PECAM-1 was dependent on the presence of functional ITIMs, SHP-2, and p56lck (106). Taken together, these studies highlight the importance of Src-related kinases in PECAM-1 tyrosine phosphorylation and downstream signalling events.

The phosphorylation of the tyrosine residues within the PECAM-1 ITIM and ITSM results in the recruitment of specific Src-homology 2 (SH2) domain containing signalling proteins, such as the protein-tyrosine phosphatases, SHP-2 and SHP-1, which are associated with activation of phosphatase activity (78, 100). SHP-2 is more likely to be responsible for instigating PECAM-1 signalling due to fivefold higher affinity for the tyrosine-phosphorylated ITIM (107). A number of other proteins have been shown to be capable of interaction with the PECAM-1 cytoplasmic tail, such as the Ser/Thr phosphatase PP2A, the lipid phosphatase SHIP, and phospholipase Cγ1, although the relevance of these interactions, several of which have not been shown within platelets or with full-length endogenous protein, is uncertain (108, 109). Progress has been recently made in understanding how the recruitment of SHP-2 to PECAM-1 down regulates GPVI-stimulated activatory signals. In contrast to experiments where PECAM-1 is co-clustered with an activatory receptor (102, 110), when PECAM-1 is activated by antibody cross linking prior to stimulation with collagen, GPVI-proximal signalling is unaffected (111), but phosphatidylinositol 3-kinase (PI3-K) signalling is reduced, as noted by inhibition of the phosphorylation of the downstream kinase PKB/Akt (111). Central to the activation of PI3-K downstream of GPVI is its recruitment to a signalling complex with the transmembrane adapter protein linker for activation of T cells (LAT) which is located within lipid rafts, ordered lipid domains within the plasma membrane that are enriched in phosphoinositide substrates of PI3-K. Upon recruitment of SHP-2 to PECAM-1, the p85 regulatory subunit of PI3-K also becomes associated. Active SHP-2 destabilises the interaction of PI3-K with GPVI-driven LAT signalling complexes resulting in the redistribution of PI3-K away from LAT (111). Since approximately 80% of PECAM-1 is located outside of lipid rafts (112), this results in redistribution of PI3-K away from its substrate and therefore diminished PI3-K signalling.

The direct stimulation of PECAM-1 tyrosine phosphorylation and signalling in vitro is possible through the binding of selected antibodies that is enhanced through cross linking using a secondary antibody (111, 113, 114) or through incubation with a high concentration of bivalent recombinant PECAM-1 ectodomain chimera (103), which has enabled the effects of PECAM-1 signalling on platelet function to be assessed.

The stimulation of PECAM-1 signalling prior to the incubation of platelets with collagen, CRP (a GPVI-specific ligand), and convulxin (Cvx, a snake venom protein that also binds and activates GPVI) was found to result in diminished platelet aggregation that is overcome at higher concentrations of the platelet agonists (103, 113). This was accompanied by reduced levels of platelet secretion, inositol phosphates production, calcium mobilisation, and total protein tyrosine phosphorylation. Consistent with an inhibitory role for PECAM-1 in platelets, platelets from PECAM-1-deficient mice exhibit hyper-reactivity in response to collagen or CRP; aggregation responses are enhanced at low agonist concentrations, and they display a lower threshold for the release of serotonin or ATP from dense granules and increased thrombus formation (thrombus volume) in vitro on immobilised Type I collagen under arterial flow conditions (98, 103). Using intravital microscopy and a laser injury model of thrombosis on cremaster muscle arterioles, Falati et al. observed exaggerated thrombotic responses in PECAM-1-deficient mice, resulting in larger and more stable thrombi (115). Thrombi, which were approximately 35% larger also formed more rapidly. Consistent with this, modestly increased thrombus formation was observed using a carotid artery model of thrombosis induced by application of $FeCl_3$ (115).

The levels of platelet PECAM-1 between individuals varies widely (three- to fourfold) within the human population with levels up to 20,000 in around 20% of the population (116). Recent analysis of the relationship between platelet PECAM-1 surface expression and platelet reactivity provided compelling evidence for the involvement of this receptor in the modulation of platelet reactivity. Using flow cytometry, the surface levels of platelet glycoprotein receptors and platelet responses to CRP, ADP, and TRAP were measured (P-selectin exposure, fibrinogen binding) using platelets from 89 subjects selected from 500 donors stratified as either high, medium, or low responders to platelet stimulation (114). An inverse relationship was observed between levels of PECAM-1 expression and the level of platelet responses to CRP. PECAM-1 expression accounted for between 8 and 10% of total variability in responses. To place this in context, levels of GPVI accounted for approximately 20% of variability in response to this agonist. Interestingly, levels of surface PECAM-1 were also associated with the responsiveness of platelets to ADP, accounting for between 6 and 9% of total variability in response. PECAM-1 levels showed no relationship to responses to TRAP, although subsequent experiments established that unlike thrombin, TRAP-mediated platelet activation is not inhibited by PECAM-1, suggesting the potential involvement of GPIb in PECAM-1-mediated inhibition of thrombin responses (thrombin binds to GPIb, but TRAP does not). Taken together, these data lend substance to the notion that PECAM-1 plays an important role in the regulation of platelet reactivity and that PECAM-1 may inhibit platelet activation stimulated through different receptors or pathways.

Upon binding to collagen and clustering, the GPVI-FcR γ-chain collagen receptor initiates activatory signalling through the tyrosine phosphorylation of an immunoreceptor tyrosine-based activatory motif (ITAM) present within the cytoplasmic tail of the FcR γ-chain (117–119). The GPIb-IX-V platelet adhesion receptor has also been reported to couple to the FcR γ-chain (120), and FcγRIIa possesses an intrinsic ITAM within its cytoplasmic tail (121, 122). In common with GPVI, these receptors can also stimulate platelet activation through an immunoreceptor-like "ITAM" signalling mechanism, and this may be stimulated through antibody-mediated receptor clustering. Co-clustering of these receptors with PECAM-1 using specific antibodies also results in diminished platelet activation (102, 110). While co-clustering may not represent endogenous mechanisms driven through PECAM-1 ligation, this does suggest that ITIM-mediated signalling through PECAM-1 may cause the general inhibition of ITAM signalling.

Antibody-mediated activation of PECAM-1 has been shown to reduce platelet activation stimulated by non-ITAM receptor ligands, such as thrombin, indicating that PECAM-1 may serve as a general inhibitor of platelet activation. Inhibition of thrombin-mediated aggregation, secretion, and intracellular calcium mobilisation was found to be less potent than for the inhibition of GPVI-mediated platelet activation, although hyper-reactivity to thrombin, ADP, or PAR-4 is not evident in PECAM-1-deficient platelets (123, 124). The reason for this discrepancy is unclear. Whether there is an alternative mechanism for the inhibition of GPCR-mediated platelet activation is still to be confirmed.

PECAM-1 signalling has long been associated with the modulation of integrin function. Indeed antibody-mediated cross linking of PECAM-1 has been shown to cause an increase in adhesion mediated by β_1, β_2, or β_3 integrins in a range of cell types, including endothelial cells, T cells, neutrophils, eosinophils, and natural killer cells (125–131). PECAM-1 antibodies have also been shown to increase integrin-mediated adhesion in CD34+ haematopoietic progenitor cells (132). In addition, engagement of PECAM-1 with bivalent anti-domain 6 PECAM-1 antibodies leads to alteration in conformational changes of integrin $\alpha_{IIb}\beta_3$ to modulate integrin-dependent adhesion and aggregation of human platelets (133). Given the important role of integrin function in platelet aggregation, it may be considered surprising that PECAM-1 inhibits and does not activate platelets. The answer to this paradox is beginning to become apparent in that PECAM-1 may have seemingly opposing roles in different aspects of platelet function. As described above, PECAM-1 cross linking results in the inhibition of signalling that lies upstream of integrin affinity modulation and ligand binding. Following integrin engagement, PECAM-1 is, however, able to enhance outside-in integrin signalling. This is exemplified in platelets that lack PECAM-1, which display impaired functions stimulated through outside-in signalling through integrin $\alpha_{IIb}\beta_3$, including clot retraction, spreading on fibrinogen, and FAK phosphorylation (124). PECAM-1 may, therefore, have a double life – suppressing platelet function, which is potentially more important at sites of low NO and PGI_2 bioavailability, but once the signal threshold for platelet activation has been reached it is able to help to drive integrin-mediated events. Whether this represents the actions of different subpopulations of PECAM-1 and the means by which PECAM-1 regulates integrin signalling has yet to be established, although Rap1, a molecule that is highly implicated in integrin activation, has been reported to be activated following PECAM-1 cross linking (134). Further work is required to unravel these seemingly opposing actions.

Apart from its role in regulation of thrombosis, PECAM-1 is considered to be a multifunctional receptor involved in regulation of a variety of biological processes, including autoimmunity, apoptosis, IgE-mediated anaphylaxis, integrin-mediated cell adhesion, cell migration, transmigration, immunological responsiveness, endothelial cell permeability, angiogenesis, and macrophage phagocytosis (135–145). Many of these biological processes are modulated by PECAM-1 requiring an intact ITIM or ITSM, recruitment of SH2 domain-containing protein-tyrosine phosphatases or adaptors, and Src-related kinases (106, 136, 146–148).

3.2. Carcinoembryonic Antigen Cell Adhesion Molecule 1

Carcinoembryonic antigen (CEA)-related cell adhesion molecule 1 contains 4 Ig-domains, a transmembrane domain followed by a long (ITIM containing) or short cytoplasmic domain. CEACAM1 displays a broad expression that is conserved in human and rodents. It is expressed on the surface of granulocytes, leukocytes, monocytes, dendritic cells, and epithelial and endothelial cells. CEACAMs preferentially form homodimers (i.e. CEACAM1 binds itself via a homophilic interaction involving Ig-Domain 1) or heterodimers with CEA and CEACAM6. Until recently, the role of CEACAM1 in platelet function was not defined. Previous studies examined CEACAM1 expression in rat platelets and concluded that it was predominantly in intracellular pools and may represent an activation antigen (149). However, based upon recent studies, CEACAM1 was found to be expressed on the surface and in intracellular pools of murine and human platelets.

CEACAM1 expression was also shown to be differentially regulated by platelet activation (81).

From a functional perspective, CEACAM1 was shown to negatively regulate platelet collagen GPVI-FcRγ chain signalling. Specifically, *Ceacam1*-/- platelets displayed enhanced Type I collagen and CRP-mediated platelet aggregation, increased platelet adhesion on Type I collagen, and elevated CRP-mediated alpha and dense granule secretion. *Ceacam1*-/- platelets formed larger thrombi when perfused over an immobilised Type I collagen matrix under arterial flow compared to wild-type platelets. Furthermore, using intravital microscopy of ferric chloride-injured mesenteric arterioles, thrombi formed in *Ceacam1*-/- arterioles were larger, more stable, and more dependent on GPVI receptor than wild-type arterioles in vivo. In addition, *Ceacam1*-/- mice were shown to be more susceptible to Type I collagen-induced pulmonary thromboembolism than wild-type mice. Taken together, these studies highlighted that CEACAM1, like PECAM-1, is a negative regulator of platelet GPVI–collagen interactions that limits the surface coverage of platelets at the initial contact phase of platelet thrombus formation and growth (81).

3.3. G6b-B

Through a proteomics study of platelet membrane proteins and a gene expression study, the platelet-restricted expression of the transmembrane orphan receptor G6b was simultaneously discovered by two groups (83, 85). Seven G6b splice variants have been identified (G6b-A-G), although only two of these are predicted to possess transmembrane domains (150). The generation of a variant-specific antibody enabled the G6b-B variant to be identified as the form found in platelets (83). As a 224 amino acid residue protein, G6b-B has a predicted molecular mass of 23.2 kDa, although it migrates as a larger protein (approximately 32 kDa) on SDS-PAGE due to glycosylation (83, 85, 150). It possesses a 125 amino acid extracellular domain and a 76 amino acid cytoplasmic tail that contains 2 ITIMs. While a ligand to G6b has yet to be identified, in cells transfected to express the receptor, treatment with pervanadate to inhibit phosphatases resulted in tyrosine phosphorylation and the binding of SHP-1 and SHP-2 to G6b (150). While both ITIM tyrosine residues were phosphorylated and mediated interaction with the tyrosine phosphatases, only Tyr211 was found to be essential for these interactions. In non-stimulated platelets, G6b is tyrosine phosphorylated and associates weakly with SHP-1 (83). Tyrosine phosphorylation is increased upon platelet activation stimulated with CRP or thrombin, which coincides with increased association with the phosphatase. Pretreatment of platelets with an antibody raised against the extracellular domain of G6b has been shown to cause inhibition of platelet aggregation stimulated with CRP or thrombin. The authors (82) conclude that this is due to receptor clustering, phosphorylation, and SHP-1 binding, although this remains to be confirmed and an adhesive role for G6b on platelets ruled out. Interestingly, there appear to be differences in the mode of action for G6b in these experiments in comparison to studies with PECAM-1 and CEACAM1. Activation of PECAM-1 signalling by antibody cross linking results in inhibition of CRP or thrombin-induced calcium mobilisation (113), whereas G6b antibody-mediated platelet inhibition is maintained in ionomycin-treated platelets (82), suggestive of a calcium-independent mechanism. Given the current absence of a G6b-B-deficient mouse, Mori et al. have used a reconstituted system in DT40 cells using a luciferase transcription reporter assay to establish that G6b-B is able to inhibit constitutive and agonist-induced signalling stimulated through GPVI and the ITAM-containing C-type lectin-like receptor 2 (CLEC-2, through which the snake venom protein rhodocytin activates platelets) (84). Mutation of the ITIM tyrosine residues confirmed their essential role in the inhibitory effects observed. Curiously, inhibition was maintained in the absence of SHP-1, SHP-2, and the lipid phosphatase SHIP, suggesting the role of alternative effectors. The characterised activatory and inhibitory roles for e.g. SHP-2 may, however, cloud interpretation of such experiments, particularly since the experimental model used does not assay the acute signalling encountered in platelets.

4. Inhibitory Intracellular Receptor Mechanisms

A number of recent reports have implicated selected intracellular receptors in the inhibitory regulation of platelet function (151, 152). This is surprising, since these receptors are conventionally associated with the regulation of gene transcription, which is of little relevance to anucleate platelets. In the context of the platelet, and likely other cells too, these receptors appear to possess non-genomic functions. Platelets contain a number of intracellular receptors, including the retinoic X receptor (RXR) (153), the glucocorticoid receptor (GR) (152), the liver X receptor (LXR) (154), and the peroxisome proliferator-activated receptor (PPAR) isoforms PPARγ (155) and PPARβ/δ (156).

4.1. RXR and Peroxisome Proliferator-activated Receptor Gamma (PPARγ)

The treatment of platelets with RXR ligands, such as 9-*cis*-retinoic acid and methoprene acid, results in inhibition of platelet aggregation in response to ADP or TXA$_2$ (153). Platelet function is unaffected by all-*trans*-retinoic acid, an isomer of 9-*cis*-retinoic acid, that does not bind to RXR. The ability of RXR to co-immunoprecipitate with Gα$_q$ from platelets suggests a potential role for this receptor in the inhibition of G protein signalling. The functions of RXR are, however, intertwined with those of the PPARs with which they form heterodimers. The treatment of platelets with PPARγ ligands, such as the synthetic anti-diabetes drug rosiglitazone or the endogenous ligand 15-deoxy-$\Delta^{12,14}$-prostaglandin J_2 (15d-PGJ_2), results in the inhibition of platelet aggregation in response to collagen, accompanied by reduced intracellular mobilisation of calcium and α-granule secretion (157). This is consistent with clinical studies that have shown glitazone-class drugs to diminish the levels of markers of platelet reactivity, including P-selectin exposure on the platelet surface and sCD40L release (158–160). The actions of these ligands, including the reduction of thrombus formation under arterial flow conditions in vitro, were shown to be prevented by the PPARγ antagonist GW9662 (157). While non-genomic mechanisms of action for PPARγ are not well-characterised, PPARγ was shown to associate with, and inhibit signalling through, early components within the GPVI signalling pathway, leading to reduced phosphorylation and activation of LAT, PLCγ2, and PKB/Akt (157). The interaction of PPARγ with Syk and LAT is inhibited in the presence of rosiglitazone, suggesting a potential positive signalling role for PPARγ in the GPVI signalling pathway that is inhibited upon ligand-induced disruption of PPARγ signalling complexes.

4.2. PPARβ/δ

PPARβ/δ is less well-studied, but is known to be ubiquitously expressed and important in the regulation of lipid metabolism, cellular proliferation, and the inflammatory response (151, 161). PPARβ/δ has also been demonstrated to decrease plaque formation and attenuate the progression of atherosclerosis (162). Little is known about the role of this receptor in platelets, although PGI_2, an important antithrombotic and endogenous platelet hormone, is a ligand for PPARβ/δ (163). The treatment of platelets with PPARβ/δ ligands GW0742 and L-165041 results in the inhibition of platelet aggregation in response to a variety of platelet agonists, and GW0742 is able to synergise with NO to inhibit platelet aggregation (151). PPARβ/δ ligands are now in the early stages of clinical trials, and at this point in their development there is no clinical data showing the effects of these ligands on platelet function or thrombosis in humans.

4.3. Liver X Receptor α and β

LXRα and β are transcription factors that have been implicated in regulating inflammatory conditions and atherogenesis through the regulation of cholesterol homeostasis. Indeed LXR ligands have anti-inflammatory effects and are atheroprotective (164–166). LXRβ is present in platelets and localised to the cytosol (154). Treatment of platelets with the LXR agonist GW3965 results in inhibition of platelet activation stimulated by collagen, CRP, or thrombin causing reduced levels of aggregation, secretion, and binding of fibrinogen to the platelet surface (154). The administration of GW3965 to mice prior to laser-induced injury to cremaster arterioles results in the formation of smaller and less-stable thrombi. In common with PPAR, LXR was found to interact with components of the GPVI signalling pathway, and diminished signalling. Indeed such interactions were enhanced upon treatment with GW3965 (154). It is possible that the mechanisms and functions of PPARγ and LXR in platelets may overlap or interact, since 15d-PGJ$_2$ and GW3965 modulate physical interactions between the receptors.

4.4. Glucocorticoid Receptor

The glucocorticoid receptor (GR) is also a member of the nuclear hormone receptor superfamily along with oestrogen receptors, PPARγ and RXR. Glucocorticoids represent a major class of endogenous anti-inflammatory hormones that regulate many aspects of immunity and metabolism (167). Natural and synthetic glucocorticoids bind to the GR, producing major conformational changes that result in nuclear translocation or transrepression of transcription factors, including nuclear factor κB (168). Although clinically not known to increase bleeding time, human platelets do express the glucocorticoid receptor in addition to sex hormone receptors (152). The treatment of platelets with the glucocorticoid ligand prednisolone results in the inhibition of platelet aggregation in response to ADP or TXA$_2$, which may be selectively reversed by the glucocorticoid antagonist mifepristone (RU468). Although the consequences of platelet glucocorticoid receptors are still not well-established, prednisolone does appear to act in a rapid and acute manner through the GR in platelets (152).

It is possible that more of the 48 known human nuclear receptors may be present in human platelets and participate in the regulation of platelet activation and aggregation. Due to its accessibility and abundance, the human platelet represents an excellent model to study non-genomic mechanisms regulated by these receptors.

5. Emerging Platelet Inhibitory Mechanisms

As our understanding of platelet function increases, it is likely that further inhibitory mechanisms will come to light. Recent work suggests inhibitory roles for both β-catenin downstream of canonical Wnt signalling, Semaphorin 3A (Sema3A), and adhesion receptors, including endothelial cell-specific adhesion molecule (ESAM) and junctional adhesion molecule-A (JAM-A).

5.1. Canonical Wnt Signalling

In nucleated cells, β-catenin fulfils a dual role as part of cell adhesion, connecting cadherins to the cytoskeleton via α-catenin, and in regulating transcription (169, 170). Cytoplasmic concentration of β-catenin is regulated by a complex of glycogen synthase kinase 3β (GSK3β), casein kinase 1α (CK1α), axin-1, FRAT-1, and adenomatous polyposis coli (APC) (171, 172). Upon binding to this complex β-catenin is sequentially phosphorylated first by CK1α and then GSK3β (173, 174). The phosphorylated β-catenin is then targeted for ubiquitination and proteosomal degradation (175–177). In unstimulated cells, therefore, β-catenin is mainly found in complex with adheren junctions with little existing free within the cytoplasm. Upon binding of Wnt ligands to the frizzled (Fzd)–lipoprotein receptor-related protein 5/6 (LRP5/6) receptor complex on the surface of cells, the cytoplasmic protein dishevelled becomes activated and negatively regulates the GSK3β, CK1α, axin-1, FRAT-1, and APC complex, preventing the phosphorylation of β-catenin (170, 178). This results in an increase in cytoplasmic β-catenin and, in nucleated cells, an increase in β-catenin in the nucleus.

Steele et al. have recently demonstrated the presence in platelets of all of the components of the Wnt signalling pathway outlined above and have shown that the canonical agonist Wnt3 can be released by TRAP-stimulated platelets (179). Furthermore, they demonstrate that the Wnt3a ligand inhibits a range of platelet functions, including adhesion, dense granule secretion, aggregation, activation of RhoA, and altered shape change (179). Yet again, this is an example of a pathway traditionally thought of as a mechanism for regulating transcription having a non-genomic acute effect in the anucleate platelet. How β-catenin exerts this inhibitory effect is as yet unknown.

5.2. Semaphorin 3A

Semaphorins are a family of soluble and membrane-bound proteins containing a conserved 500 amino acid Sema domain near their amino terminus. Sema3A (Collapsin-1) binds to neurpilin 1 on the cell surface. Plexin A1, which forms a complex with neurpilin 1, then transduces the signal into the cell (180, 181). The effect of Sema3A binding is primarily actin reorganisation (182) which in neuronal cells results in growth cone collapse (182, 183). Following on from the demonstration that Sema3A regulates integrin function and, therefore, adhesion and migration of endothelial cells on extracellular matrix (184), Kashiwagi et al. reported that platelets contain both neuropilin-1 and plexin-A1 and demonstrated the binding of Sema3A to platelets (185). The binding of Sema3A to platelets inhibited the activation of $\alpha_{IIb}\beta_3$ by ADP, thrombin, and Cvx, leading to an inhibition of aggregation, and platelet adhesion and spreading on immobilised fibrinogen. Platelet spreading independent of $\alpha_{IIb}\beta_3$ was also inhibited. While agonist-induced intracellular calcium flux and the levels of cyclic nucleotides were not affected by Sema3A, the elevation of F-actin, phosphorylation of cofilin, and activation of RAC1 that normally accompany platelet activation were inhibited (185). The exact mechanisms by which Sema3A inhibits platelet function have not been fully elucidated; however, the results outlined above indicate that its effects are mediated though RAC1-dependent actin rearrangement, although it is unlikely that this is the sole mechanism (185). As Sema3A is produced by endothelial cells, we can probably add it to NO, PGI$_2$, and CD39 on the list of endothelium-derived regulators of platelet function.

5.3. Potential Inhibitory Adhesion Receptors

Platelets express members of the CTX family of adhesion receptors, including ESAM (186) and JAM-A (187). Like PECAM-1 and CEACAM1, they mediate adhesive interactions in trans, although neither receptors possess an ITIM.

5.3.1. ESAM

ESAM, a 55 kDa type I transmembrane glycoprotein, possesses two extracellular Ig domains and a long cytoplasmic tail in which a PDZ domain is located (188). The gene that codes for this protein, which localises to sites of contact between endothelial cells, has been deleted in ESAM knockout mice, which paradoxically resulted in increased stability of cell–cell junctions indicating that this protein may serve to destabilise cell contacts in a regulated manner (189). ESAM localises to points of platelet–platelet contact following activation, suggestive of an adhesive role. Platelets from ESAM-deficient mice, however, were more sensitive to ADP or thrombin receptor stimulation, forming more stable aggregates at lower concentrations. Interestingly, calcium mobilisation, α-granule secretion, integrin $\alpha_{IIb}\beta_3$ activation, and spreading were unaffected, although clot retraction, a function that requires outside-in signalling following integrin $\alpha_{IIb}\beta_3$ ligation, was delayed (190). ESAM-deficient mice were found to achieve more stable haemostasis following transection of the tail, and larger thrombi were formed in vivo using a laser injury model of thrombosis (190). While the signalling that ESAM controls in platelets has yet to be characterised, interaction of the PDZ domain with the adapter protein NHERF-1 has been suggested to mediate signalling protein complex formation. These data point towards a role for ESAM in modulating thrombus growth, limiting thrombus size and stability.

5.3.2. JAM-A

JAM-A is also recruited to cell–cell junctions, and its expression in platelets led to the hypothesis that it may perform an adhesive role during thrombus formation. Ablation of JAM-A in mice reveals a platelet phenotype that is not dissimilar to that observed in ESAM-deficient mice: reduced tail bleeding times, increased thrombus formation in cremaster muscle arterioles using a laser injury model of thrombosis, and faster time to carotid artery occlusion using a FeCl$_3$-induced vessel injury (191). Collagen-epinephrine-induced thromboembolism was increased in these mice, suggesting that this defect is due to platelet and not endothelial JAM-A (192). In vitro, platelets from these mice exhibited greater levels of aggregation, although secretion was unaffected. Clot retraction was enhanced in JAM-A-deficient platelets, indicative of elevated levels of outside-in signalling through integrin $\alpha_{IIb}\beta_3$, which is accompanied by elevated levels of ERK1 and p38 MAP kinase signalling (192).

Strong similarities, therefore, exist between the functions of ESAM and JAM-A in platelets, and potentially their mechanisms of action, although further work is required to confirm this and to determine the physiological scenarios in which the function of these receptors is important.

6. Summary

Platelets are an important clinical target for cardiovascular diseases with anti-platelet therapy providing an effective strategy for reducing the risk of disease. Current anti-platelet drugs, however, focus on the inhibition of activatory mechanisms. As outlined in this chapter, there are numerous endogenous mechanisms for attenuating platelet activation, with novel mechanisms continuing to be identified. Interventions targeting inhibitory signalling pathways may offer a potent approach for therapeutically regulating platelet activation to prevent thrombosis.

References

1. Furchgott, R. F., and Zawadzki, J. V. (1980) The obligatory role of endothelial cells in the relaxation of arterial smooth muscle by acetylcholine, *Nature 288*, 373–376.

2. Bellamy, T. C., and Garthwaite, J. (2002) The receptor-like properties of nitric oxide-activated soluble guanylyl cyclase in intact cells, *Mol Cell Biochem 230*, 165–176.

3. Munzel, T., Feil, R., Mulsch, A., Lohmann, S. M., Hofmann, F., and Walter, U. (2003) Physiology and pathophysiology of vascular signaling controlled by guanosine $3',5'$-cyclic monophosphate-dependent protein kinase, *Circulation 108*, 2172–2183.

4. Eigenthaler, M., Ullrich, H., Geiger, J., Horstrup, K., Honig-Liedl, P., Wiebecke, D., and Walter, U. (1993) Defective nitrovasodilator-stimulated protein phosphorylation and calcium regulation in cGMP-dependent protein kinase-deficient human platelets of chronic myelocytic leukemia, *J Biol Chem 268*, 13526–13531.

5. Tsudo, M., Kozak, R. W., Goldman, C. K., and Waldmann, T. A. (1987) Contribution of a p75 interleukin 2 binding peptide to a high-affinity interleukin 2 receptor complex, *Proc Natl Acad Sci USA 84*, 4215–4218.

6. Eigenthaler, M., Nolte, C., Halbrugge, M., and Walter, U. (1992) Concentration and regulation of cyclic nucleotides, cyclic-nucleotide-dependent protein kinases and one of their major substrates in human platelets. Estimating the rate of cAMP-regulated and cGMP-regulated protein phosphorylation in intact cells, *Eur J Biochem 205*, 471–481.

7. Maurice, D. H., and Haslam, R. J. (1990) Molecular basis of the synergistic inhibition of platelet function by nitrovasodilators and activators of adenylate cyclase: inhibition of cyclic AMP breakdown by cyclic GMP, *Mol Pharmacol 37*, 671–681.

8. Nolte, C., Eigenthaler, M., Horstrup, K., Honig-Liedl, P., and Walter, U. (1994) Synergistic phosphorylation of the focal adhesion-associated vasodilator-stimulated phosphoprotein in intact human platelets in response to cGMP- and cAMP-elevating platelet inhibitors, *Biochem Pharmacol 48*, 1569–1575.

9. Jensen, B. O., Selheim, F., Doskeland, S. O., Gear, A. R., and Holmsen, H. (2004) Protein kinase A mediates inhibition of the thrombin-induced platelet shape change by nitric oxide, *Blood 104*, 2775–2782.

10. Cavallini, L., Coassin, M., Borean, A., and Alexandre, A. (1996) Prostacyclin and sodium nitroprusside inhibit the activity of the platelet inositol 1,4,5-trisphosphate receptor and promote its phosphorylation, *J Biol Chem 271*, 5545–5551.

11. Mullershausen, F., Friebe, A., Feil, R., Thompson, W. J., Hofmann, F., and Koesling, D.

(2003) Direct activation of PDE5 by cGMP: long-term effects within NO/cGMP signaling, *J Cell Biol 160*, 719–727.

12. Butt, E., Abel, K., Krieger, M., Palm, D., Hoppe, V., Hoppe, J., and Walter, U. (1994) cAMP- and cGMP-dependent protein kinase phosphorylation sites of the focal adhesion vasodilator-stimulated phosphoprotein (VASP) in vitro and in intact human platelets, *J Biol Chem 269*, 14509–14517.

13. Waldmann, R., Nieberding, M., and Walter, U. (1987) Vasodilator-stimulated protein phosphorylation in platelets is mediated by cAMP- and cGMP-dependent protein kinases, *Eur J Biochem 167*, 441–448.

14. Miura, Y., Kaibuchi, K., Itoh, T., Corbin, J. D., Francis, S. H., and Takai, Y. (1992) Phosphorylation of smg p21B/rap1B p21 by cyclic GMP-dependent protein kinase, *FEBS Lett 297*, 171–174.

15. Danielewski, O., Schultess, J., and Smolenski, A. (2005) The NO/cGMP pathway inhibits Rap 1 activation in human platelets via cGMP-dependent protein kinase I, *Thromb Haemost 93*, 319–325.

16. Reid, H. M., and Kinsella, B. T. (2003) The alpha, but not the beta, isoform of the human thromboxane A2 receptor is a target for nitric oxide-mediated desensitization. Independent modulation of Tp alpha signaling by nitric oxide and prostacyclin, *J Biol Chem 278*, 51190–51202.

17. Sase, K., and Michel, T. (1995) Expression of constitutive endothelial nitric oxide synthase in human blood platelets., *Life Sciences*, 2049–2055.

18. Mehta, J. L., Chen, L. Y., Kone, B. C., Mehta, P., and Turner, P. (1995) Identification of constitutive and inducible forms of nitric oxide synthase in human platelets., *J Lab Clin Med 125*, 370–377.

19. Freedman, J. E., Loscalzo, J., Barnard, M. R., Alpert, C., Keaney, J. F., and Michelson, A. D. (1997) Nitric oxide released from activated platelets inhibits platelet recruitment, *J Clin Invest 100*, 350–356.

20. Malinski, T., Radomski, M. W., Taha, Z., and Moncada, S. (1993) Direct electrochemical measurement of nitric oxide released from human platelets, *Biochem Biophys Res Commun 194*, 960–965.

21. Lantoine, F., Brunet, A., Bedioui, F., Devynck, J., and Devynck, M. A. (1995) Direct measurement of nitric oxide production in platelets: relationship with cytosolic Ca2+ concentration, *Biochem Biophys Res Commun 215*, 842–848.

22. Riba, R., Sharifi, M., Farndale, R. W., and Naseem, K. M. (2005) Regulation of platelet guanylyl cyclase by collagen: evidence that Glycoprotein VI mediates platelet nitric oxide synthesis in response to collagen, *Thromb Haemost 94*, 395–403.

23. Stojanovic, A., Marjanovic, J. A., Brovkovych, V. M., Peng, X., Hay, N., Skidgel, R. A., and Du, X. (2006) A phosphoinositide 3-kinase-AKT-nitric oxide-cGMP signaling pathway in stimulating platelet secretion and aggregation, *J Biol Chem 281*, 16333–16339.

24. Riba, R., Oberprieler, N. G., Roberts, W., and Naseem, K. M. (2006) Von Willebrand factor activates endothelial nitric oxide synthase in blood platelets by a glycoprotein Ib-dependent mechanism, *J Thromb Haemost 4*, 2636–2644.

25. Morrell, C. N., Matsushita, K., Chiles, K., Scharpf, R. B., Yamakuchi, M., Mason, R. J., Bergmeier, W., Mankowski, J. L., Baldwin, W. M., 3 rd, Faraday, N., and Lowenstein, C. J. (2005) Regulation of platelet granule exocytosis by S-nitrosylation, *Proc Natl Acad Sci USA 102*, 3782–3787.

26. Radomski, M. W., Palmer, R. M., and Moncada, S. (1990) An L-arginine/nitric oxide pathway present in human platelets regulates aggregation, *Proc Natl Acad Sci USA 87*, 5193–5197.

27. Radomski, M. W., Palmer, R. M., and Moncada, S. (1990) Characterization of the L-arginine:nitric oxide pathway in human platelets, *Br J Pharmacol 101*, 325–328.

28. Storey, R. F., and Heptinstall, S. (1999) Laboratory investigation of platelet function, *Clin Lab Haematol 21*, 317–329.

29. Williams, R. H., and Nollert, M. U. (2004) Platelet-derived NO slows thrombus growth on a collagen type III surface, *Thromb J 2*, 11.

30. Li, W., Mital, S., Ojaimi, C., Csiszar, A., Kaley, G., and Hintze, T. H. (2004) Premature death and age-related cardiac dysfunction in male eNOS-knockout mice, *J Mol Cell Cardiol 37*, 671–680.

31. Iafrati, M. D., Vitseva, O., Tanriverdi, K., Blair, P., Rex, S., Chakrabarti, S., Varghese, S., and Freedman, J. E. (2005) Compensatory mechanisms influence hemostasis in setting of eNOS deficiency, *Am J Physiol Heart Circ Physiol 288*, H1627-1632.

32. Freedman, J. E., Sauter, R., Battinelli, E. M., Ault, K., Knowles, C., Huang, P. L., and Loscalzo, J. (1999) Deficient platelet-derived nitric oxide and enhanced hemostasis in mice lacking the NOSIII gene, *Circ Res 84*, 1416–1421.

33. Gambaryan, S., Kobsar, A., Hartmann, S., Birschmann, I., Kuhlencordt, P. J., Muller-Esterl, W., Lohmann, S. M., and Walter, U. (2008) NO-synthase-/NO-independent regulation of human and murine platelet soluble guanylyl cyclase activity, *J Thromb Haemost 6*, 1376–1384.

34. Naseem, K. M., and Riba, R. (2008) Unresolved roles of platelet nitric oxide synthase, *J Thromb Haemost 6*, 10–19.

35. Naseem, K. M. (2008) eNOS, iNOS or no NOS, that is the question!, *J Thromb Haemost 6*, 1373–1375.

36. Marcondes, S., Cardoso, M. H., Morganti, R. P., Thomazzi, S. M., Lilla, S., Murad, F., De Nucci, G., and Antunes, E. (2006) Cyclic GMP-independent mechanisms contribute to the inhibition of platelet adhesion by nitric oxide donor: a role for alpha-actinin nitration, *Proc Natl Acad Sci USA 103*, 3434–3439.

37. Oberprieler, N. G., Roberts, W., Riba, R., Graham, A. M., Homer-Vanniasinkam, S., and Naseem, K. M. (2007) cGMP-independent inhibition of integrin alphaIIbbeta3-mediated platelet adhesion and outside-in signalling by nitric oxide, *FEBS Lett 581*, 1529–1534.

38. Shah, C. M., Bell, S. E., Locke, I. C., Chowdrey, H. S., and Gordge, M. P. (2007) Interactions between cell surface protein disulphide isomerase and S-nitrosoglutathione during nitric oxide delivery, *Nitric Oxide 16*, 135–142.

39. Gordge, M. P., Hothersall, J. S., and Noronha-Dutra, A. A. (1998) Evidence for a cyclic GMP-independent mechanism in the antiplatelet action of S-nitrosoglutathione, *Br J Pharmacol 124*, 141–148.

40. Crane, M. S., Rossi, A. G., and Megson, I. L. (2005) A potential role for extracellular nitric oxide generation in cGMP-independent inhibition of human platelet aggregation: biochemical and pharmacological considerations, *Br J Pharmacol 144*, 849–859.

41. Tsikas, D., Ikic, M., Tewes, K. S., Raida, M., and Frolich, J. C. (1999) Inhibition of platelet aggregation by S-nitroso-cysteine via cGMP-independent mechanisms: evidence of inhibition of thromboxane A2 synthesis in human blood platelets, *FEBS Lett 442*, 162–166.

42. Dangel, O., Mergia, E., Karlisch, K., Groneberg, D., Koesling, D., and Friebe, A. (2010) Nitric oxide-sensitive guanylyl cyclase is the only nitric oxide receptor mediating platelet inhibition, *J Thromb Haemost 8*, 1343–1352.

43. Radomski, M. W., Palmer, R. M., and Moncada, S. (1987) The role of nitric oxide and cGMP in platelet adhesion to vascular endothelium, *Biochem Biophys Res Commun 148*, 1482–1489.

44. Radomski, M. W., Palmer, R. M., and Moncada, S. (1987) The anti-aggregating properties of vascular endothelium: interactions between prostacyclin and nitric oxide, *Br J Pharmacol 92*, 639–646.

45. Lidbury, P. S., Antunes, E., de Nucci, G., and Vane, J. R. (1989) Interactions of iloprost and sodium nitroprusside on vascular smooth muscle and platelet aggregation, *Br J Pharmacol 98*, 1275–1280.

46. Weksler, B. B., Marcus, A. J., and Jaffe, E. A. (1977) Synthesis of prostaglandin I2 (prostacyclin) by cultured human and bovine endothelial cells, *Proc Natl Acad Sci USA 74*, 3922–3926.

47. Ingerman-Wojenski, C., Silver, M. J., Smith, J. B., and Macarak, E. (1981) Bovine endothelial cells in culture produce thromboxane as well as prostacyclin, *J Clin Invest 67*, 1292–1296.

48. Moncada, S., Gryglewski, R., Bunting, S., and Vane, J. R. (1976) An enzyme isolated from arteries transforms prostaglandin endoperoxides to an unstable substance that inhibits platelet aggregation, *Nature 263*, 663–665.

49. Gryglewski, R. J., Bunting, S., Moncada, S., Flower, R. J., and Vane, J. R. (1976) Arterial walls are protected against deposition of platelet thrombi by a substance (prostaglandin X) which they make from prostaglandin endoperoxides, *Prostaglandins 12*, 685–713.

50. Needleman, P., Turk, J., Jakschik, B. A., Morrison, A. R., and Lefkowith, J. B. (1986) Arachidonic acid metabolism, *Annu Rev Biochem 55*, 69–102.

51. Hamberg, M., Svensson, J., and Samuelsson, B. (1975) Thromboxanes: a new group of biologically active compounds derived from prostaglandin endoperoxides, *Proc Natl Acad Sci USA 72*, 2994–2998.

52. FitzGerald, G. A. (1991) Mechanisms of platelet activation: thromboxane A2 as an amplifying signal for other agonists, *Am J Cardiol 68*, 11B–15B.

53. Marnett, L. J., Rowlinson, S. W., Goodwin, D. C., Kalgutkar, A. S., and Lanzo, C. A. (1999) Arachidonic acid oxygenation by COX-1 and COX-2. Mechanisms of catalysis and inhibition, *J Biol Chem 274*, 22903–22906.

54. Smith, W. L., DeWitt, D. L., and Garavito, R. M. (2000) Cyclooxygenases: structural,

cellular, and molecular biology, *Annu Rev Biochem 69*, 145–182.

55. Patrignani, P., Sciulli, M. G., Manarini, S., Santini, G., Cerletti, C., and Evangelista, V. (1999) COX-2 is not involved in thromboxane biosynthesis by activated human platelets, *J Physiol Pharmacol 50*, 661–667.

56. Spisni, E., Bartolini, G., Orlandi, M., Belletti, B., Santi, S., and Tomasi, V. (1995) Prostacyclin (PGI2) synthase is a constitutively expressed enzyme in human endothelial cells, *Exp Cell Res 219*, 507–513.

57. Smith, W. L., DeWitt, D. L., and Allen, M. L. (1983) Bimodal distribution of the prostaglandin I2 synthase antigen in smooth muscle cells, *J Biol Chem 258*, 5922–5926.

58. Needleman, P., Moncada, S., Bunting, S., Vane, J. R., Hamberg, M., and Samuelsson, B. (1976) Identification of an enzyme in platelet microsomes which generates thromboxane A2 from prostaglandin endoperoxides, *Nature 261*, 558–560.

59. Hsu, P. Y., Tsai, A. L., Kulmacz, R. J., and Wang, L. H. (1999) Expression, purification, and spectroscopic characterization of human thromboxane synthase, *J Biol Chem 274*, 762–769.

60. Dutta-Roy, A. K., and Sinha, A. K. (1987) Purification and properties of prostaglandin E1/prostacyclin receptor of human blood platelets, *J Biol Chem 262*, 12685–12691.

61. Tsai, A. L., Hsu, M. J., Vijjeswarapu, H., and Wu, K. K. (1989) Solubilization of prostacyclin membrane receptors from human platelets, *J Biol Chem 264*, 61–67.

62. Jones, R. L., Qian, Y., Wong, H. N., Chan, H., and Yim, A. P. (1997) Prostanoid action on the human pulmonary vascular system, *Clin Exp Pharmacol Physiol 24*, 969–972.

63. Tateson, J. E., Moncada, S., and Vane, J. R. (1977) Effects of prostacyclin (PGX) on cyclic AMP concentrations in human platelets, *Prostaglandins 13*, 389–397.

64. Gorman, R. R., Bunting, S., and Miller, O. V. (1977) Modulation of human platelet adenylate cyclase by prostacyclin (PGX), *Prostaglandins 13*, 377–388.

65. Kobayashi, T., Ushikubi, F., and Narumiya, S. (2000) Amino acid residues conferring ligand binding properties of prostaglandin I and prostaglandin D receptors. Identification by site-directed mutagenesis, *J Biol Chem 275*, 24294–24303.

66. Siess, W. (1989) Molecular mechanisms of platelet activation, *Physiol Rev 69*, 58–178.

67. Armstrong, R. A. (1996) Platelet prostanoid receptors, *Pharmacol Ther 72*, 171–191.

68. Robson, S. C., Sevigny, J., and Zimmermann, H. (2006) The E-NTPDase family of ectonucleotidases: Structure function relationships and pathophysiological significance, *Purinergic Signal 2*, 409–430.

69. Kansas, G. S., Wood, G. S., and Tedder, T. F. (1991) Expression, distribution, and biochemistry of human CD39. Role in activation-associated homotypic adhesion of lymphocytes, *J Immunol 146*, 2235–2244.

70. Marcus, A. J., Broekman, M. J., Drosopoulos, J. H., Islam, N., Alyonycheva, T. N., Safier, L. B., Hajjar, K. A., Posnett, D. N., Schoenborn, M. A., Schooley, K. A., Gayle, R. B., and Maliszewski, C. R. (1997) The endothelial cell ecto-ADPase responsible for inhibition of platelet function is CD39, *J Clin Invest 99*, 1351–1360.

71. Koziak, K., Sevigny, J., Robson, S. C., Siegel, J. B., and Kaczmarek, E. (1999) Analysis of CD39/ATP diphosphohydrolase (ATPDase) expression in endothelial cells, platelets and leukocytes, *Thromb Haemost 82*, 1538–1544.

72. Plesner, L. (1995) Ecto-ATPases: identities and functions, *Int Rev Cytol 158*, 141–214.

73. Marcus, A. J., Safier, L. B., Hajjar, K. A., Ullman, H. L., Islam, N., Broekman, M. J., and Eiroa, A. M. (1991) Inhibition of platelet function by an aspirin-insensitive endothelial cell ADPase. Thromboregulation by endothelial cells, *J Clin Invest 88*, 1690–1696.

74. Zimmermann, H. (1992) 5′-Nucleotidase: molecular structure and functional aspects, *Biochem J 285 (Pt 2)*, 345–365.

75. Dwyer, K. M., Robson, S. C., Nandurkar, H. H., Campbell, D. J., Gock, H., Murray-Segal, L. J., Fisicaro, N., Mysore, T. B., Kaczmarek, E., Cowan, P. J., and d'Apice, A. J. (2004) Thromboregulatory manifestations in human CD39 transgenic mice and the implications for thrombotic disease and transplantation, *J Clin Invest 113*, 1440–1446.

76. Marcus, A. J., Broekman, M. J., Drosopoulos, J. H., Islam, N., Pinsky, D. J., Sesti, C., and Levi, R. (2003) Heterologous cell-cell interactions: thromboregulation, cerebroprotection and cardioprotection by CD39 (NTPDase-1), *J Thromb Haemost 1*, 2497–2509.

77. Burnstock, G. (2002) Purinergic signaling and vascular cell proliferation and death, *Arterioscler Thromb Vasc Biol 22*, 364–373.

78. Jackson, D. E., Ward, C. M., Wang, R., and Newman, P. J. (1997) The protein-tyrosine phosphatase SHP-2 binds platelet/endothelial cell adhesion molecule-1 (PECAM-1) and forms a distinct signaling complex during platelet aggregation, *J. Biol. Chem. 272*, 6986–6993.

79. Newman, P. J. (1999) Switched at birth: a new family for PECAM-1, *J Clin Invest 103*, 5–9.

80. Gibbins, J. M. (2002) The negative regulation of platelet function: extending the role of the ITIM, *Trends Cardiovasc Med 12*, 213–219.

81. Wong, C., Liu, Y., Yip, J., Chand, R., Wee, J. L., Oates, L., Nieswandt, B., Reheman, A., Ni, H., Beauchemin, N., and Jackson, D. E. (2009) CEACAM1 negatively regulates platelet-collagen interactions and thrombus growth in vitro and in vivo, *Blood 113*, 1818–1828.

82. Newland, S. A., Macaulay, I. C., Floto, R. A., de Vet, E. C., Ouwehand, W. H., Watkins, N. A., Lyons, P. A., and Campbell, R. D. (2007) The novel inhibitory receptor G6B is expressed on the surface of platelets and attenuates platelet function in vitro, *Blood 109*, 4806–4809.

83. Senis, Y. A., Tomlinson, M. G., Garcia, A., Dumon, S., Heath, V. L., Herbert, J., Cobbold, S. P., Spalton, J. C., Ayman, S., Antrobus, R., Zitzmann, N., Bicknell, R., Frampton, J., Authi, K. S., Martin, A., and Wakelam, M. J. O. (2007) A comprehensive proteomics and genomics analysis reveals novel transmembrane proteins in human platelets and mouse megakaryocytes including G6b-B, a novel immunoreceptor tyrosine-based inhibitory motif protein, *Molecular & Cellular Proteomics 6*, 548–564.

84. Mori, J., Pearce, A. C., Spalton, J. C., Grygielska, B., Eble, J. A., Tomlinson, M. G., Senis, Y. A., and Watson, S. P. (2008) G6b-B Inhibits Constitutive and Agonist-induced Signaling by Glycoprotein VI and CLEC-2, *Journal of Biological Chemistry 283*, 35419–35427.

85. Macaulay, I. C., Tijssen, M. R., Thijssen-Timmer, D. C., Gusnanto, A., Steward, M., Burns, P., Langford, C. F., Ellis, P. D., Dudbridge, F., Zwaginga, J. J., Watkins, N. A., van der Schoot, C. E., and Ouwehand, W. H. (2007) Comparative gene expression profiling of in vitro differentiated megakaryocytes and erythroblasts identifies novel activatory and inhibitory platelet membrane proteins, *Blood 109*, 3260–3269.

86. Newman, P. J., Berndt, M. J., Gorski, J., White, G. C., Lyman, S., Paddock, C., and Muller, W. A. (1990) PECAM-1 (CD31) cloning and relation to adhesion molecules of the immunoglobulin gene superfamily., *Science 247*, 1219–1222.

87. Kirschbaum, N. E., Gumina, R. J., and Newman, P. J. (1994) Organization of the gene for human platelet/endothelial cell adhesion molecule-1 shows alternatively spliced isoforms and a functionally complex cytoplasmic domain, *Blood 84*, 4028–4037.

88. Ohto, H., Maeda, H., Shibata, Y., Chen, R. F., Ozaki, Y., Higashihara, M., Takeuchi, A., and Tohyama, H. (1985) A novel leukocyte differentiation antigen: two monoclonal antibodies TM2 and TM3 define a 120-kd molecule present on neutrophils, monocytes, platelets, and activated lymphoblasts, *Blood 66*, 873–881.

89. Goyert, S. M., Ferrero, E. M., Seremetis, S. V., Winchester, R. J., Silver, J., and Mattison, A. C. (1986) Biochemistry and expression of myelomonocytic antigens, *J Immunol 137*, 3909–3914.

90. Lyons, A. B., Cooper, S. J., Cole, S. R., and Ashman, L. K. (1988) Human myeloid differentiation antigens identified by monoclonal antibodies to the myelomonocytic leukemia cell line RC-2A, *Pathology 20*, 137–146.

91. Cabanas, C., Sanchez-Madrid, F., Bellon, T., Figdor, C. G., Te Velde, A. A., Fernandez, J. M., Acevedo, A., and Bernabeu, C. (1989) Characterization of a novel myeloid antigen regulated during differentiation of monocytic cells, *Eur J Immunol 19*, 1373–1378.

92. Newman, P. J. (1994) The role of PECAM-1 in vascular cell biology, *Ann N Y Acad Sci 714*, 165–174.

93. Wu, X. W., and Lian, E. C. (1997) Binding properties and inhibition of platelet aggregation by a monoclonal antibody to CD31 (PECAM-1), *Arterioscler Thromb Vasc Biol 17*, 3154–3158.

94. Metzelaar, M. J., Korteweg, J., Sixma, J. J., and Nieuwenhuis, H. K. (1991) Biochemical characterization of PECAM-1 (CD31 antigen) on human platelets, *Thromb Haemost 66*, 700–707.

95. Mazurov, A. V., Vinogradov, D. V., Kabaeva, N. V., Antonova, G. N., Romanov, Y. A., Vlasik, T. N., Antonov, A. S., and Smirnov, V. N. (1991) A monoclonal antibody, VM64, reacts with a 130 kDa glycoprotein common to platelets and endothelial cells: heterogeneity in antibody binding to human aortic endothelial cells, *Thromb Haemost 66*, 494–499.

96. Cicmil, M., Thomas, J. M., Sage, T., Barry, F. A., Leduc, M., Bon, C., and Gibbins, J. M. (2000) Collagen, Convulxin, and Thrombin Stimulate Aggregation-independent Tyrosine Phosphorylation of CD31 in Platelets. Evidence for the involvement of Src family kinases., *J. Biol. Chem. 275*, 27339–27347.

97. Buckley, C. D., Doyonnas, R., Newton, J. P., Blystone, S. D., Brown, E. J., Watt, S. M., and Simmons, D. L. (1996) Identification of avb3 as a heterotypic ligand for CD31/PECAM-1., *J.Cell Sci. 109*, 437–445.

98. Patil, S., Newman, D. K., and Newman, P. J. (2001) Platelet endothelial cell adhesion molecule-1 serves as an inhibitory receptor that modulates platelet responses to collagen, *Blood 97*, 1727–1732.

99. Cramer, E. M., Berger, G., and Berndt, M. C. (1994) Platelet alpha-granule and plasma membrane share two new components: CD9 and PECAM-1, *Blood 84*, 1722–1730.

100. Jackson, D. E., Kupcho, K. R., and Newman, P. J. (1997) Characterization of Phosphotyrosine Binding Motifs in the Cytoplasmic Domain of Platelet/Endothelial Cell Adhesion Molecule-1 (PECAM-1) that are required for the Cellular Association and Activation of the Protein-tyrosine Phosphatase, SHP-2, *J. Biol. Chem. 272*, 24868–24875.

101. Vivier, E., and Daeron, M. (1997) Immunoreceptor tyrosine-based inhibitory motifs., *Immunol. Today 18*, 286–291.

102. Rathore, V., Stapleton, M. A., Hillery, C. A., Montgomery, R. R., Nichols, T. C., Merricks, E. P., Newman, D. K., and Newman, P. J. (2003) PECAM-1 negatively regulates GPIb/V/IX signaling in murine platelets, *Blood 102*, 3658–3664.

103. Jones, K. L., Hughan, S. C., Dopheide, S. M., Farndale, R. W., Jackson, S. P., and Jackson, D. E. (2001) Platelet endothelial cell adhesion molecule-1 is a negative regulator of platelet-collagen interactions, *Blood 98*, 1456–1463.

104. Cao, M. Y., Huber, M., Beauchemin, N., Famiglietti, J., Albelda, S. M., and Veillette, A. (1998) Regulation of mouse PECAM-1 tyrosine phosphorylation by the Src and Csk families of protein-tyrosine kinases, *J Biol Chem 273*, 15765–15772.

105. Chiu, Y. J., McBeath, E., and Fujiwara, K. (2008) Mechanotransduction in an extracted cell model: Fyn drives stretch- and flow-elicited PECAM-1 phosphorylation, *J Cell Biol 182*, 753–763.

106. Newman, D. K., Hamilton, C., and Newman, P. J. (2001) Inhibition of antigen-receptor signaling by Platelet Endothelial Cell Adhesion Molecule-1 (CD31) requires functional ITIMs, SHP-2, and p56(lck), *Blood 97*, 2351–2357.

107. Hua, C. T., Gamble, J. R., Vadas, M. A., and Jackson, D. E. (1998) Recruitment and Activation of SHP-1 Protein-tyrosine Phosphatase by Human Platelet Endothelial Cell Adhesion Molecule-1 (PECAM-1). Identification of Immunoreceptor Tyrosine-Based Inhibitory Motif-Like Binding Motifs and Substrates, *J. Biol. Chem. 273*, 28332–28340.

108. Pumphrey, N. J., Taylor, V. T., Freeman, S., Douglas, M. R., Bradfield, P. F., Young, S. P., Lord, J. M., Wakelam, M. J. O., Bird, I. N., Salmon, M., and Buckley, C. D. (1999) Differetial association of cytoplasmic signalling molecules SHP-1, SHP-2, SHIP and phosphlipase C-γl with PECAM-1/CD31, *FEBS Lett. 450*, 77–83.

109. Relou, I. A. M., Gorter, G., Ferreira, I. A., van Rijn, H. J. M., and Akkerman, J. W. N. (2003) Platelet endothelial cell adhesion molecule-1 (PECAM-1) inhibits low density lipoprotein-induced signaling in platelets, *Journal of Biological Chemistry 278*, 32638–32644.

110. Thai le, M., Ashman, L. K., Harbour, S. N., Hogarth, P. M., and Jackson, D. E. (2003) Physical proximity and functional interplay of PECAM-1 with the Fc receptor Fc gamma RIIa on the platelet plasma membrane, *Blood 102*, 3637–3645.

111. Moraes, L. A., Barrett, N. E., Jones, C. I., Holbrook, L. M., Spyridon, M., Sage, T., Newman, D. K., and Gibbins, J. M. (2010) PECAM-1 regulates collagen-stimulated platelet function by modulating the association of PI3 Kinase with Gab1 and LAT, *J Thromb Haemost*.

112. Sardjono, C. T., Harbour, S. N., Yip, J. C., Paddock, C., Tridandapani, S., Newman, P. J., and Jackson, D. E. (2006) Palmitoylation at Cys(595) is essential for PECAM-I localisation into membrane microdomains and for efficient PECAM-I-mediated cytoprotection, *Thrombosis and Haemostasis 96*, 756–766.

113. Cicmil, M., Thomas, J. M., Leduc, M., Bon, C., and Gibbins, J. M. (2002) PECAM-1 signalling inhibits the activation of human platelets, *Blood 99*, 137–144.

114. Jones, C. I., Garner, S. F., Moraes, L. A., Kaiser, W. J., Rankin, A., Ouwehand, W. H., Goodall, A. H., and Gibbins, J. M. (2009) PECAM-1 expression and activity negatively regulate multiple platelet signaling pathways, *FEBS Lett 583*, 3618–3624.

115. Falati, S., Patil S., Gross, P. L., Stapleton, M., Merrill-Skoloff G., Barrett, N. E., Pixton, K. L., Weiler H., Cooley B., Newman D.K., Newman, P. J., Furie, B. C., Furie B., and Gibbins, J. M. (2006) Platelet PECAM-1 inhibits thrombus formation in vivo, *Blood 107*, 535–541.

116. Novinska, M. S., Rathoare, V., Newman, D. K., and Newman, P. J. (2007) PECAM-1, *Platelets*, Chapter 11, pages 221–230.

117. Gibbins, J., Asselin, J., Farndale, R., Barnes, M., Law, C. L., and Watson, S. P. (1996) Tyrosine phosphorylation of the Fc receptor gamma-chain in collagen-stimulated platelets, *J Biol Chem 271*, 18095–18099.

118. Gibbins, J. M., Okuma, M., Farndale, R., Barnes, M., and Watson, S. P. (1997) Glycoprotein VI is the collagen receptor in platelets which underlies tyrosine phosphorylation of the Fc receptor γ-chain, *FEBS Letters 413*, 255–259.

119. Poole, A., Gibbins, J. M., Turner, M., vanVugt, M. J., vandeWinkel, J. G. J., Saito, T., Tybulewicz, V. L. J., and Watson, S. P. (1997) The Fc receptor gamma-chain and the tyrosine kinase Syk are essential for activation of mouse platelets by collagen, *16*, 2333–2341.

120. Falati, S., Edmead, C. E., and Poole, A. W. (1999) Glycoprotein Ib-V-IX, a receptor for von Willebrand factor, couples physically and functionally to the Fc receptor gamma- chain, Fyn, and Lyn to activate human platelets, *Blood 94*, 1648–1656.

121. Reth, M. (1989) Antigen receptor tail clue, *Nature 338*, 383–384.

122. Odin, J. A., Edberg, J. C., Painter, C. J., Kimberly, R. P., and Unkeless, J. C. (1991) Regulation of phagocytosis and [Ca2+]i flux by distinct regions of an Fc receptor, *Science 254*, 1785–1788.

123. Dhanjal, T. S., Ross, E. A., Auger, J. M., McCarty, O. J., Hughes, C. E., Senis, Y. A., Buckley, C. D., and Watson, S. P. (2007) Minimal regulation of platelet activity by PECAM-1, *Platelets 18*, 56–67.

124. Wee, J. L., and Jackson, D. E. (2005) The Ig-ITIM superfamily member PECAM-1 regulates the "outside-in" signaling properties of integrin alpha(IIb)beta(3) in platelets, *Blood 106*, 3816–3823.

125. Berman, M. E., and Muller, W. A. (1995) Ligation of Platelet Endothelial-Cell Adhesion Molecule-1 (Pecam-1/Cd31) on Monocytes and Neutrophils Increases Binding- Capacity of Leukocyte Cr3 (Cd11b/Cd18), *J. Immunol. 154*, 299–307.

126. Berman, M. E., Xie, Y., and Muller, W. A. (1996) Roles of platelet endothelial cell adhesion molecule-1 (PECAM- 1,CD31) in natural killer cell transendothelial migration and beta(2) integrin activation, *J. Immunol. 156*, 1515–1524.

127. Chiba, R., Nakagawa, N., Kuroasawa, K., Tanaka, Y., Saito, Y., and Iwamoto, I. (1999) Ligation of CD31(PECAM-1) on endothelial cells increases adhesive function of αvβ3 integrin and enhances b1 integrin mediated adhesion of eosinophilis to endothelial cells, *Blood 94*, 1319–1329.

128. Pellegatta, F., Chierchia, S. L., and Zocchi, M. R. (1998) Functional association of platelet endothelial cell adhesion molecule-1 and phosphoinositide 3-kinase in human neutrophils, *Journal of Biological Chemistry 273*, 27768–27771.

129. Piali, L., Albelda, S. M., Baldwin, H. S., Hammel, P., Gisler, R. H., and Imhof, B. A. (1993) Murine platelet endothelial cell adhesion molecule (PECAM-1/CD31) modulates b2 integrins on lymphokine-activated killer cells., *Eur.J.Immunol. 23*, 2464–2471.

130. Tanaka, Y., Albelda, S. M., Horgan, K. J., van Seventer, G. A., Shimizu, Y., Newman, W., Hallam, J., Newman, P. J., Buck, C. A., and Shaw, S. (1992) CD31 expressed on distinctive T cell subsets is a preferential amplifier of b1 integrin-mediated adhesion, *J Exp Med 176*, 245–253.

131. Zhao, T. M., and Newman, P. J. (2001) Integrin activation by regulated dimerization and oligomerization of platelet endothelial cell adhesion molecule (PECAM)-1 from within the cell, *Journal of Cell Biology 152*, 65–73.

132. Leavesley, D. I., Oliver, J. M., Swart, B. W., Berndt, M. C., Haylock, D. N., and Simmons, P. J. (1994) Signals from Platelet Endothelial-Cell Adhesion Molecule Enhance the Adhesive Activity of the Very Late Antigen-4 Integrin of Human Cd34(+) Hematopoietic Progenitor Cells, *J. Immunol. 153*, 4673–4683.

133. Varon, D., Jackson, D. E., Shenkman, B., Dardik, R., Tamarin, I., Savion, N., and Newman, P. J. (1998) Platelet/endothelial cell adhesion molecule-1 serves as a costimulatory agonist receptor that modulates integrin-dependent adhesion and aggregation of human platelets, *Blood 91*, 500–507.

134. Reedquist, K. A., Ross, E., Koop, E. A., Wolthuis, R. M. F., Zwartkruis, F. J. T., van Kooyk, Y., Salmon, M., Buckley, C. D., and Bos, J. L. (2000) The small GTPase, Rap1, mediates CD31-induced integrin adhesion, *Journal of Cell Biology 148*, 1151–1158.

135. Wilkinson, R., Lyons, A. B., Roberts, D., Wong, M. X., Bartley, P. A., and Jackson, D. E. (2002) Platelet endothelial cell adhesion molecule-1 (PECAM-1/CD31) acts as a regulator of B-cell development, B-cell antigen receptor (BCR)-mediated activation, and autoimmune disease, *Blood 100*, 184–193.

136. Gao, C., Sun, W., Christofidou-Solomidou, M., Sawada, M., Newman, D. K., Bergom, C., Albelda, S. M., Matsuyama, S., and Newman, P. J. (2003) PECAM-1 functions as a specific

and potent inhibitor of mitochondrial-dependent apoptosis, *Blood 102*, 169–179.

137. Wong, M. X., Roberts, D., Bartley, P. A., and Jackson, D. E. (2002) Absence of platelet endothelial cell adhesion molecule-1 (CD31) leads to increased severity of local and systemic IgE-mediated anaphylaxis and modulation of mast cell activation, *J Immunol 168*, 6455–6462.

138. Graesser, D., Solowiej, A., Bruckner, M., Osterweil, E., Juedes, A., Davis, S., Ruddle, N. H., Engelhardt, B., and Madri, J. A. (2002) Altered vascular permeability and early onset of experimental autoimmune encephalomyelitis in PECAM-1-deficient mice, *J Clin Invest 109*, 383–392.

139. Albelda, S. M., Muller, W. A., Buck, C. A., and Newman, P. J. (1991) Molecular and cellular properties of PECAM-1 (endoCAM/CD31): a novel vascular cell-cell adhesion molecule, *J Cell Biol 114*, 1059–1068.

140. Albelda, S. M., Oliver, P. D., Romer, L. H., and Buck, C. A. (1990) EndoCAM: a novel endothelial cell-cell adhesion molecule, *J Cell Biol 110*, 1227–1237.

141. Schimmenti, L. A., Yan, H. C., Madri, J. A., and Albelda, S. M. (1992) Platelet endothelial cell adhesion molecule, PECAM-1, modulates cell migration, *J Cell Physiol 153*, 417–428.

142. Muller, W. A., Weigl, S. A., Deng, X., and Phillips, D. M. (1993) PECAM-1 is required for transendothelial migration of leukocytes, *J Exp Med 178*, 449–460.

143. DeLisser, H. M., Christofidou-Solomidou, M., Strieter, R. M., Burdick, M. D., Robinson, C. S., Wexler, R. S., Kerr, J. S., Garlanda, C., Merwin, J. R., Madri, J. A., and Albelda, S. M. (1997) Involvement of endothelial PECAM-1/CD31 in angiogenesis, *Am J Pathol 151*, 671–677.

144. Chimini, G. (2002) Apoptosis: repulsive encounters, *Nature 418*, 139–141.

145. Ferrero, E., Ferrero, M. E., Pardi, R., and Zocchi, M. R. (1995) The platelet endothelial cell adhesion molecule-1 (PECAM1) contributes to endothelial barrier function, *FEBS Lett 374*, 323–326.

146. Gratzinger, D., Barreuther, M., and Madri, J. A. (2003) Platelet-endothelial cell adhesion molecule-1 modulates endothelial migration through its immunoreceptor tyrosine-based inhibitory motif, *Biochem Biophys Res Commun 301*, 243–249.

147. O'Brien, C. D., Cao, G., Makrigiannakis, A., and DeLisser, H. M. (2004) Role of immunoreceptor tyrosine-based inhibitory motifs

of PECAM-1 in PECAM-1-dependent cell migration, *Am J Physiol Cell Physiol 287*, C1103-1113.

148. Zhu, J. X., Cao, G., Williams, J. T., and Delisser, H. M. (2010) SHP-2 phosphatase activity is required for PECAM-1-dependent cell motility, *Am J Physiol Cell Physiol 299*, C854-865.

149. Hansson, M., Odin, P., Johansson, S., and Obrink, B. (1990) Comparison and functional characterization of C-CAM, glycoprotein IIb/IIIa and integrin beta 1 in rat platelets, *Thromb Res 58*, 61–73.

150. de Vet, E., Aguado, B., and Campbell, R. D. (2001) G6b, a novel immunoglobulin superfamily member encoded in the human major histocompatibility complex, interacts with SHP-1 and SHP-2, *Journal of Biological Chemistry 276*, 42070–42076.

151. Ali, F. Y., Davidson, S. J., Moraes, L. A., Traves, S. L., Paul-Clark, M., Bishop-Bailey, D., Warner, T. D., and Mitchell, J. A. (2006) Role of nuclear receptor signaling in platelets: antithrombotic effects of PPARbeta, *Faseb J 20*, 326–328.

152. Moraes, L. A., Paul-Clark, M. J., Rickman, A., Flower, R. J., Goulding, N. J., and Perretti, M. (2005) Ligand-specific glucocorticoid receptor activation in human platelets, *Blood 106*, 4167–4175.

153. Moraes, L. A., Swales, K. E., Wray, J. A., Damazo, A., Gibbins, J. M., Warner, T. D., and Bishop-Bailey, D. (2007) Nongenomic signaling of the retinoid X receptor through binding and inhibiting Gq in human platelets, *Blood 109*, 3741–3744.

154. Spyridon M, Moraes, L. A., Jones, C. I., Sage, T., Sasikumar, P., Bucci, G., and Gibbins, J. M. (2010) LXR as a novel anti-thrombotic target, *Under revision*.

155. Akbiyik, F., Ray, D. M., Gettings, K. F., Blumberg, N., Francis, C. W., and Phipps, R. P. (2004) Human bone marrow megakaryocytes and platelets express PPARgamma, and PPARgamma agonists blunt platelet release of CD40 ligand and thromboxanes, *Blood 104*, 1361–1368.

156. Ali, F. Y., Davidson, S. J., Moraes, L. A., Traves, S. L., Paul-Clark, M., Bishop-Bailey, D., Warner, T. D., and Mitchell, J. A. (2005) Role of nuclear receptor signaling in platelets: antithrombotic effects of PPAR beta, *Faseb Journal 19*, 326-+.

157. Moraes, L. A., Spyridon, M., Kaiser, W. J., Jones, C. I., Sage, T., Atherton, R. E. L., and Gibbins, J. M. (2010) Non-genomic effects of PPAR gamma ligands: inhibition of

GPVI-stimulated platelet activation, *J Thromb Haemost 8*, 577–587.

158. Berger, J. P., Akiyama, T. E., and Meinke, P. T. (2005) PPARs: therapeutic targets for metabolic disease, *Trends in Pharmacological Sciences 26*, 244–251.

159. Irons, B. K., Greene, R. S., Mazzolini, T. A., Edwards, K. L., and Sleeper, R. B. (2006) Implications of rosiglitazone and pioglitazone on cardiovascular risk in patients with type 2 diabetes mellitus, *Pharmacotherapy 26*, 168–181.

160. Marx, N., Duez, H., Fruchart, J. C., and Staels, B. (2004) Peroxisome proliferator-activated receptors and atherogenesis - Regulators of gene expression in vascular cells, *Circulation Research 94*, 1168–1178.

161. Moraes, L. A., Piqueras, L., and Bishop-Bailey, D. (2006) Peroxisome proliferator-activated receptors and inflammation, *Pharmacol Ther 110*, 371–385.

162. Lee, C. H., Chawla, A., Urbiztondo, N., Liao, D., Boisvert, W. A., Evans, R. M., and Curtiss, L. K. (2003) Transcriptional repression of atherogenic inflammation: modulation by PPARdelta, *Science 302*, 453–457.

163. Forman, B. M., Chen, J., and Evans, R. M. (1997) Hypolipidemic drugs, polyunsaturated fatty acids, and eicosanoids are ligands for peroxisome proliferator-activated receptors alpha and delta, *Proc Natl Acad Sci USA 94*, 4312–4317.

164. Joseph, S. B., Castrillo, A., Laffitte, B. A., Mangelsdorf, D. J., and Tontonoz, P. (2003) Reciprocal regulation of inflammation and lipid metabolism by liver X receptors, *Nature Medicine 9*, 213–219.

165. Joseph, S. B., McKilligin, E., Pei, L. M., Watson, M. A., Collins, A. R., Laffitte, B. A., Chen, M. Y., Noh, G., Goodman, J., Hagger, G. N., Tran, J., Tippin, T. K., Wang, X. P., Lusis, A. J., Hsueh, W. A., Law, R. E., Collins, J. L., Willson, T. M., and Tontonoz, P. (2002) Synthetic LXR ligand inhibits the development of atherosclerosis in mice, *Proceedings of the National Academy of Sciences of the United States of America 99*, 7604–7609.

166. Tangirala, R. K., Bischoff, E. D., Joseph, S. B., Wagner, B. L., Walczak, R., Laffitte, B. A., Daige, C. L., Thomas, D., Heyman, R. A., Mangelsdorf, D. J., Wang, X. P., Lusis, A. J., Tontonoz, P., and Schulman, I. G. (2002) Identification of macrophage liver X receptors as inhibitors of atherosclerosis, *Proceedings of the National Academy of Sciences of the United States of America 99*, 11896–11901.

167. Goulding, N. J. (2004) The molecular complexity of glucocorticoid actions in inflammation - a four-ring circus, *Curr Opin Pharmacol 4*, 629–636.

168. Schaaf, M. J., and Cidlowski, J. A. (2002) Molecular mechanisms of glucocorticoid action and resistance, *J Steroid Biochem Mol Biol 83*, 37–48.

169. Huang, H., and He, X. (2008) Wnt/beta-catenin signaling: new (and old) players and new insights, *Curr Opin Cell Biol 20*, 119–125.

170. Logan, C. Y., and Nusse, R. (2004) The Wnt signaling pathway in development and disease, *Annu Rev Cell Dev Biol 20*, 781–810.

171. Hart, M. J., de los Santos, R., Albert, I. N., Rubinfeld, B., and Polakis, P. (1998) Downregulation of beta-catenin by human Axin and its association with the APC tumor suppressor, beta-catenin and GSK3 beta, *Curr Biol 8*, 573–581.

172. Kishida, S., Yamamoto, H., Ikeda, S., Kishida, M., Sakamoto, I., Koyama, S., and Kikuchi, A. (1998) Axin, a negative regulator of the wnt signaling pathway, directly interacts with adenomatous polyposis coli and regulates the stabilization of beta-catenin, *J Biol Chem 273*, 10823–10826.

173. Liu, C., Li, Y., Semenov, M., Han, C., Baeg, G. H., Tan, Y., Zhang, Z., Lin, X., and He, X. (2002) Control of beta-catenin phosphorylation/degradation by a dual-kinase mechanism, *Cell 108*, 837–847.

174. Yost, C., Torres, M., Miller, J. R., Huang, E., Kimelman, D., and Moon, R. T. (1996) The axis-inducing activity, stability, and subcellular distribution of beta-catenin is regulated in Xenopus embryos by glycogen synthase kinase 3, *Genes Dev 10*, 1443–1454.

175. Aberle, H., Bauer, A., Stappert, J., Kispert, A., and Kemler, R. (1997) beta-catenin is a target for the ubiquitin-proteasome pathway, *EMBO J 16*, 3797–3804.

176. Latres, E., Chiaur, D. S., and Pagano, M. (1999) The human F box protein beta-Trcp associates with the Cul1/Skp1 complex and regulates the stability of beta-catenin, *Oncogene 18*, 849–854.

177. Liu, C., Kato, Y., Zhang, Z., Do, V. M., Yankner, B. A., and He, X. (1999) beta-Trcp couples beta-catenin phosphorylation-degradation and regulates Xenopus axis formation, *Proc Natl Acad Sci USA 96*, 6273–6278.

178. Macdonald, B. T., Semenov, M. V., and He, X. (2007) SnapShot: Wnt/beta-catenin signaling, *Cell 131*, 1204.

179. Steele, B. M., Harper, M. T., Macaulay, I. C., Morrell, C. N., Perez-Tamayo, A., Foy, M., Habas, R., Poole, A. W., Fitzgerald, D. J., and

Maguire, P. B. (2009) Canonical Wnt signaling negatively regulates platelet function, *Proc Natl Acad Sci USA 106*, 19836–19841.

180. Takahashi, T., Fournier, A., Nakamura, F., Wang, L. H., Murakami, Y., Kalb, R. G., Fujisawa, H., and Strittmatter, S. M. (1999) Plexin-neuropilin-1 complexes form functional semaphorin-3A receptors, *Cell 99*, 59–69.

181. Tamagnone, L., Artigiani, S., Chen, H., He, Z., Ming, G. I., Song, H., Chedotal, A., Winberg, M. L., Goodman, C. S., Poo, M., Tessier-Lavigne, M., and Comoglio, P. M. (1999) Plexins are a large family of receptors for transmembrane, secreted, and GPI-anchored semaphorins in vertebrates, *Cell 99*, 71–80.

182. Fournier, A. E., Nakamura, F., Kawamoto, S., Goshima, Y., Kalb, R. G., and Strittmatter, S. M. (2000) Semaphorin3A enhances endocytosis at sites of receptor-F-actin colocalization during growth cone collapse, *J Cell Biol 149*, 411–422.

183. Fan, J., Mansfield, S. G., Redmond, T., Gordon-Weeks, P. R., and Raper, J. A. (1993) The organization of F-actin and microtubules in growth cones exposed to a brain-derived collapsing factor, *J Cell Biol 121*, 867–878.

184. Serini, G., Valdembri, D., Zanivan, S., Morterra, G., Burkhardt, C., Caccavari, F., Zammataro, L., Primo, L., Tamagnone, L., Logan, M., Tessier-Lavigne, M., Taniguchi, M., Puschel, A. W., and Bussolino, F. (2003) Class 3 semaphorins control vascular morphogenesis by inhibiting integrin function, *Nature 424*, 391–397.

185. Kashiwagi, H., Shiraga, M., Kato, H., Kamae, T., Yamamoto, N., Tadokoro, S., Kurata, Y., Tomiyama, Y., and Kanakura, Y. (2005) Negative regulation of platelet function by a secreted cell repulsive protein, semaphorin 3A, *Blood 106*, 913–921.

186. Nasdala, I., Wolburg-Buchholz, K., Wolburg, H., Kuhn, A., Ebnet, K., Brachtendorf, G., Samulowitz, U., Kuster, B., Engelhardt, B., Vestweber, D., and Butz, S. (2002) A transmembrane tight junction protein selectively expressed on endothelial cells and platelets, *Journal of Biological Chemistry 277*, 16294–16303.

187. Sobocka, M. B., Sobocki, T., Banerjee, P., Weiss, C., Rushbrook, J. I., Norin, A. J., Hartwig, J., Salifu, M. O., Markell, M. S., Babinska, A., Ehrlich, Y. H., and Kornecki, E. (2000) Cloning of the human platelet F11 receptor: a cell adhesion molecule member of the immunoglobulin superfamily involved in platelet aggregation, *Blood 95*, 2600–2609.

188. Hirata, K., Ishida, T., Penta, K., Rezaee, M., Yang, E., Wohlgemuth, J., and Quertermous, T. (2001) Cloning of an immunoglobulin family adhesion molecule selectively expressed by endothelial cells, *Journal of Biological Chemistry 276*, 16223–16231.

189. Wegmann, F., Petri, B., Khandoga, A. G., Moser, C., Khandoga, A., Volkery, S., Li, H., Nasdala, I., Brandau, O., Fassler, R., Butz, S., Krombach, F., and Vestweber, D. (2006) ESAM supports neutrophil extravasation, activation of Rho, and VEGF-induced vascular permeability, *Journal of Experimental Medicine 203*, 1671–1677.

190. Stalker, T. J., Wu, J., Morgans, A., Traxler, E. A., Wang, L., Chatterjee, M. S., Lee, D., Quertermous, T., Hall, R. A., Hammer, D. A., Diamond, S. L., and Brass, L. F. (2009) Endothelial cell specific adhesion molecule (ESAM) localizes to platelet-platelet contacts and regulates thrombus formation in vivo, *Journal of Thrombosis and Haemostasis 7*, 1886–1896.

191. Naik, M. U., Stalker, T. J., Brass, L. F., and Naik, U. P. (2008) Junctional Adhesion Molecule a Helps Maintain Integrin alpha IIb beta 3 in Resting State, *Blood 112*, 48–48.

192. Naik, M. U., Stalker, T. J., Brass, L. F., and Naik, U. P. (2009) Platelet Junctional Adhesion Molecule-A Regulates Thrombosis by Negatively Regulating Outside-in Signaling through Integrin alpha IIb beta 3, *Blood 114*, 69–70.

Chapter 24

Platelet Proteomics: State of the Art and Future Perspective

Yotis Senis and Ángel García

Abstract

Platelets pose unique challenges to cell biologists due to their lack of nucleus and low levels of messenger RNA. Platelets cannot be cultured in great abundance or manipulated using common recombinant DNA technologies. As a result, platelet research has lagged behind that of nucleated cells. The advent of mass spectrometry and its application to protein biochemistry brought with it great hopes for the platelet community that are now being realized. This technology is ideally suited for identifying low-abundance proteins, protein–protein interactions, and post-translational modifications in complex protein mixtures. Over the past 10 years, proteomics has delivered in many ways, providing platelet biologists with a comprehensive list of proteins expressed in platelets, information on post-translational modifications, protein interactions and sub-cellular localization. Several novel and important platelet membrane proteins, including CLEC-2, CD148, G6b-B, G6f, and Hsp47, have been identified using proteomics-based approaches. New, more sensitive instrumentation and novel approaches are making it increasingly possible to identify ever lower amounts of proteins. In this chapter we highlight some of the major achievements of platelet proteomics to date, discussing challenges and how they were overcome. We also discuss new frontiers and applications of proteomics to platelets and microparticles in health and disease, as we strive to better understand the molecular mechanisms underlying the platelet response to vascular injury.

Key words: Platelet proteomics, 2-DE, Mass spectrometry, Secretome, Membrane proteome, Signalling pathways, Interactome, Phosphorylation, Glycosylation, Platelet-related diseases

1. Introduction: Pioneering Platelet Proteomics

Platelets are small fragments of megakaryocytes that are generated with a complete repertoire of proteins for carrying out their physiological functions. Platelets do not contain a nucleus and have very low levels of mRNA; thus, de novo protein synthesis is minimal. As a result, identification and analysis of novel platelet proteins has been hampered and dependent on the development of antibodies. Platelet biologists have been at a clear disadvantage and lag behind

Jonathan M. Gibbins and Martyn P. Mahaut-Smith (eds.), *Platelets and Megakaryocytes: Volume 3, Additional Protocols and Perspectives*, Methods in Molecular Biology, vol. 788, DOI 10.1007/978-1-61779-307-3_24, © Springer Science+Business Media, LLC 2012

research on nucleated cells as commonly used molecular biology-based techniques such as generation of cDNA expression and SAGE libraries have limited application to platelets. Over the past 10 years, there has been a large advancement in the identification of novel platelet proteins through the application of mass spectrometry. Mass spectrometry is ideally suited for studying the protein composition of platelets, as it can identify vanishingly low levels of proteins (*attomoles!*) in complex protein mixtures. The disadvantage of such a sensitive technique is the identification of background or contaminating proteins. Validation of specific proteins by immunological-based approaches such as western blotting, flow cytometry, and confocal microscopy is, therefore, essential. Mass spectrometry has lead to an explosion in our knowledge of the protein makeup of platelets, which in turn has led to a better understanding of the molecular mechanisms regulating platelet function.

The primary aim of initial platelet proteomics studies was to map the entire platelet proteome. No small task, but at the same time a good starting point to test the strengths and limitations of the technology and refine strategies if necessary. Realizing the complexities of such a massive undertaking, the aim quickly evolved to investigating specific compartments of the platelet proteome, such as the secretome and membrane proteome. Masking of low abundance proteins by highly abundant cytoskeletal and secreted proteins (actin, tubulin, filamin, fibrinogen, VWF, and thrombospondin) was clearly a problem, as was the conspicuous absence of transmembrane proteins. Partitioning the platelet proteome into more manageable compartments was a way of circumventing some of these issues. Compartments of particular interest are the secretome, the membrane proteome, and the phosphoproteome (Fig. 1). The secretome is of physiological importance as platelets contain a variety of biologically active proteins and other small molecules that enhance platelet activation, thrombus generation and wound repair. The membrane proteome was also of great interest as surface receptors, channels and transporters are essential for regulating platelet function. Similarly, the platelet phosphoproteome is also of great importance, as many of the main platelet surface receptors signal through a series of tyrosine phosphorylation events. It is, therefore, important to identify changes in phosphorylation of signalling and cytoskeletal proteins.

Another pitfall that was quickly realized was that non-platelet proteins were being identified, such as B and T cell-specific proteins, raising the question of the purity of platelet preparations. The mass spectrometry technique is so sensitive that trace amounts of contaminating cells can potentially lead to the detection of non-platelet proteins. This raised the importance of validating potentially interesting hits for expression in platelets. The general work flow of platelet proteomics studies that has come out of these pioneering studies is shown in Fig. 2.

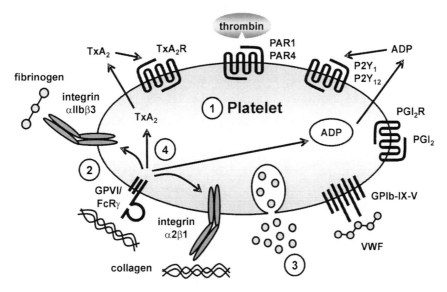

Fig. 1. Platelet proteomes. (1) Global platelet proteome, (2) Membrane proteome, (3) Secretome, (4) Phosphoproteome.

Below are summarized some of the key platelet proteomics studies that have greatly enhanced our knowledge of the platelet proteome.

1.1. The Complete Platelet Proteome

The analysis of the complete platelet proteome is an analytical challenge that is made more difficult by the broad dynamic range of protein quantities expressed in the platelet. It is difficult to estimate the total number of proteins present in platelets. A typical eukaryotic cell contains 20,000–25,000 different genes, and it has been speculated that a third of them are expressed in most cell types. However, many of these genes are expressed at very low levels. Because platelets are anucleate, it is even more difficult to estimate the total number of proteins that they contain.

Over the last decade several research groups worldwide have tried to characterize the complete platelet proteome. Studies focused on the analysis of whole cell lysates and also sub-cellular compartments (e.g. plasma membranes, organelles), signalling pathways and the secretome. In this section, we focus on the attempts to profile the proteome of platelets in a basal state. Researchers approached this goal through two major strategies related to the protein separation method that was used, either gel-based or non-gel-based. Protein identification was always by mass spectrometry.

Most gel-based platelet proteomics studies to date take advantage of the high resolution provided by two-dimensional gel electrophoresis (2-DE), where proteins are separated according to their isoelectric point (pI) and size (molecular weight). Following the electrophoresis step, gels are stained and information on the presence of proteins is provided by extensive image analysis.

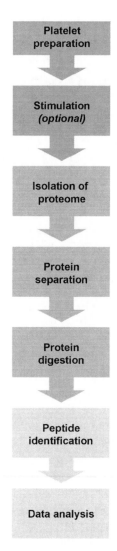

Fig. 2. General work flow of platelet proteomics studies.

When large narrow pI range 2D gels (*zoom* gels) are combined with staining with highly sensitive fluorescent dyes (e.g. SYPRO Ruby), more than 2,000 spots can be visualized if running 500 μg of platelet whole cell lysates. Protein spots of interest can be excised from the gel, and the proteins in-gel digested with trypsin, and identified by mass spectrometry, which can provide data on peptide presence and protein amino sequence. Powerful databases, such as Swiss-Prot and TrEMBL, can be used for protein identification (1).

In 2000, Marcus and collaborators reported the first analysis of the human platelet proteome by taking advantage of the power offered by the combination of 2-DE and MS. They identified 186 protein features, mostly cytoskeletal, that constituted the first pI 3–10 platelet proteome map (2). A few years later, two reports from the same research group established in the Glycobiology Institute at Oxford University (UK) constituted the broadest investigation to date on the 2-DE-based human proteome (3, 4). Both reports provided a high-resolution 2-DE proteome map comprising more than 2,300 different protein features. Proteins were separated by 2-DE using narrow pH gradients during the isoelectric focussing step – first dimension – and 9–16% PAGE gradient gels for the second dimension (18×18 cm). Gels were stained with a highly sensitive fluorescent dye (similar to Sypro Ruby), and following image analysis, the corresponding protein spots were excised, in-gel trypsin digested, and analyzed by liquid chromatography tandem mass spectrometry (LC-MS/MS). Overall, more than 1,000 proteins were identified, corresponding to 411 open reading frames (ORFs). The list was rich in proteins involved in signalling (24%) and protein synthesis and degradation (22%). The large number of signalling proteins, which included kinases, G proteins, and adapters, was somehow to be expected in view of the ability of platelets to undergo powerful and rapid activation following damage to the vasculature. The presence of a large number of proteins involved in translation, transcription, and regulation of the cell cycle seemed unusual at the time in light of the limited degree of protein synthesis that takes place in platelets. It is unclear whether these proteins are functionally relevant or were just incorporated into the platelet during the budding process from the megakaryocyte. Nevertheless, recent studies suggest protein synthesis is a relevant process in platelets (5). Strikingly, approximately 45% of the proteins identified by these two reports had never been reported in platelets previously, including 15 hypothetical proteins that had not been described in any other cell type, which emphasized the power of the approach.

As shown below, 2-DE in combination with MS has been the method of choice for signalling studies and also in platelet clinical proteomics. However, it is well known that 2-DE has certain limitations. The use of sub-cellular pre-fractionation techniques in combination with 1D-SDS-PAGE and LC-MS/MS or shotgun approaches based on multidimensional nanoscale capillary LC-MS/MS overcomes problems arising from the fact that 2-DE cannot properly resolve many high-molecular-weight, highly basic, and hydrophobic proteins (1). A recent article by Yu and colleagues illustrates how a combination of SDS-PAGE and/or LC-MS/MS can be used to elucidate the global platelet proteome (6). The authors set up a high-throughput platform for maximum exploration of the rat and human platelet proteome, which led to the identification of the largest number of proteins expressed in both rat and human platelets. Indeed, they consistently identified 837 unique proteins, making it the first comprehensive protein database so far for rat platelets.

A particular shotgun approach that has been widely applied to study the whole platelet proteome, complementing the 2-DE studies mentioned above is combined fractional diagonal chromatography (COFRADIC™) approach. Central to this method is a modification reaction that alters the retention behaviour of specific peptides on reverse-phase columns. In the initial study, Gevaert and collaborators applied this method to isolate N-terminal peptides, which were analyzed by LC-MS/MS (7). Using this technique, 264 proteins were identified in a cytosolic and membrane skeleton fraction of human platelets. In a more recent study, Gevaert and collaborators presented a COFRADIC approach in which cysteine-containing peptides were isolated from a complete platelet proteome digest and used to identify their precursor proteins after LC-MS/MS analysis. This approach led to the identification of 163 different proteins with a broad range of functions and abundance (8). In a third study from the same group, Martens and colleagues presented an integrated analysis of the whole platelet proteome by combining three different COFRADIC sorting techniques. Methionyl, cysteinyl, and amino terminal peptides were isolated and analyzed by MS/MS. Merging the peptide identifications obtained after database searching resulted in a core set of 641 platelet proteins (9). This work led to the identification of a high number of hydrophobic membrane proteins and hypothetical proteins not previously identified by 2-DE, demonstrating that gel-free methods can overcome some of the limitations of the 2-DE approach. Other applications of the above approaches are shown in the sections below that focus on the study of particular subsets of proteins.

1.2. The Platelet Membrane Proteome

It is estimated that more than half of all proteins interact either directly or indirectly with cellular membranes (10, 11). These include receptors, channels and transporters, cytoskeletal and signalling proteins, all of which play important roles in sampling and/ or responding to the extracellular surroundings. Identifying the complete repertoire of platelet membrane associated and transmembrane proteins, thus, has important implications for understanding how platelets function under pathophysiological conditions. From a clinical perspective, approximately 70% of all known drug targets are transmembrane plasma membrane (PM) proteins (12). Therefore, mapping the platelet PM proteome will undoubtedly lead to a better understanding of the molecular mechanisms regulating platelet activation and thrombosis, and identification of novel anti-thrombotic drug targets.

Membrane proteins are classified as either "integral" or "peripheral" membrane proteins (IMPs or PMPs, respectively) (Fig. 3). IMPs are permanently attached to the membrane and are divided into transmembrane proteins that span the entire membrane, and *monotopic* proteins that are permanently attached to

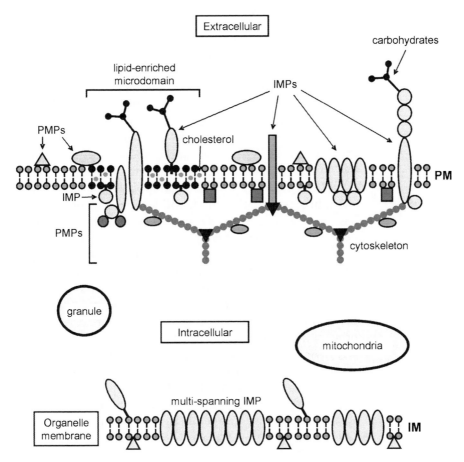

Fig. 3. Membrane proteomes. *IMPs* integral membrane proteins, *PMPs* peripheral membrane proteins, *PM* plasma membrane, *IM* intracellular membrane.

only one side of the membrane, such as glycosylphosphatidylinositol (GPI)-anchored receptors (Fig. 3). Transmembrane IMPs are further divided into *bitopic* and *polytopic* proteins that contain either a single or multiple transmembrane domains (TMDs), respectively (13). All transmembrane proteins contain a hydrophobic membrane spanning domain that has either a α-helical or β-barrel structure (13). By contrast, PMPs are temporarily associated with either lipid molecules or IMPs embedded in the membrane through a combination of hydrophobic, electrostatic, and other transient non-covalent interactions (10).

Two major difficulties in identifying α-helical IMPs are as follows: (1) their hydrophobic properties, making them difficult to solubilize and analyze by MS and (2) their low abundance (13–15). Large proteins containing multiple hydrophobic TMDs, do not solubilize well in isoelectric focusing (IEF) sample buffer and those that do are prone to precipitate at their pI (13). The limitation in the dynamic range of detection of proteomics is also an issue as many membrane proteins are typically present at low levels. Another complication is masking by highly abundant, contaminating PMPs attached to the inner surface of the PM. These factors render it essential to enrich surface proteins prior to resolving by a chromatographic technique other than gel-based IEF.

The most successful platelet membrane proteomics studies to date employed the following general strategy: (1) membrane enrichment, (2) removal of PMPs, (3) protein/peptide separation by 1D-E or non-classical 2-DE (not involving IEF), (4) band excision and in-gel trypsinization, and (5) peptide identification by LC-MS/MS. More recently, membrane enrichment followed by either COFRADIC or the commonly used shot gun approach Multidimensional Protein Identification Technology (MudPIT) has also been successfully applied to identifying platelet IMPs (16). It is now well established that membrane enrichment and protein separation by non-classical 2-DE are critical steps for successful IMP identification. Also important is removal of associated PMPs and peptide separation by either 1D or 2D liquid chromatography. Optimizing the chromatographic step used to introduce the sample into the mass spectrometer improves IMP identification and must be carefully considered when devising a strategy for identifying IMPs (14). Membrane enrichment reduces sample complexity by removing abundant PMPs, which enhances the dynamic range of the sample and increases the chances of identifying low abundance proteins. However, in the case of the PM, the associated proteins may be of interest, as important signalling and cytoskeletal proteins interact with the PM in resting and activated conditions.

Until recently, there was a dramatic under-representation of membrane proteins identified in platelets using the classical 2-DE-LC-MS/MS platform. Only the most abundant platelet PM proteins, namely, the integrin $\alpha_{IIb}\beta_3$ and GPIb-IX-V were being identified. The main reasons for these shortcomings were the use of classical 2-DE and lack of membrane enrichment.

The first major step forward in platelet membrane proteomics came in 2005 when Moebius and co-workers employed membrane-enrichment and 1D-E rather than 2-DE to improve IMP identification. These changes resulted in a sixfold increase in the number of IMPs identified, compared with previous studies (4, 8, 17). Washed human platelets were prepared and treated with neuraminidase to remove sialyic acid side chains and reduce the net negative charge on the platelets (17). Platelets were subsequently mechanically lysed by sonication and membranes enriched by discontinuous sorbital gradient centrifugation. Samples were resolved by 16-BAC/SDS-PAGE gels, silver stained, bands excised, and proteins trypsinized. Peptides were separated by nano-HPLC and identified by ESI-ion trap, ESI-Q-TOF, or ESI linear ion trap mass spectrometers. The presence of TMDs was predicted using SOUSI and TMHMM algorithms. This study was ground breaking in that

83 PM proteins were identified, several of which had never before been identified in platelets, include G6b-A and several hypothetical and unknown proteins (17). In addition, 48 TMD-containing proteins were identified that are localized in other cellular compartments, including mitochondria, ER, and vesicles. This was the first time low/intermediate abundance membrane proteins were identified, including GPVI (3,000–5,000 copies/platelet) and the integrin subunits α2, α5, and α6, which are expressed at 1,000–3,000 copies each. A shortcoming of this study was the lack of G protein-coupled receptors. This is likely due to low abundance, low solubility, and long tryptic fragments resulting from a lack of basic amino acids in TMDs, which cannot be analyzed classically by reverse phase separation.

The next major advance in mapping the platelet membrane proteome came from the study by Senis and co-workers (18). Senis et al. followed a similar strategy to that of Moebius et al., in that membranes were first enriched then resolved by a gel-based technique (18). However, different membrane enrichment strategies were used and proteins were resolved by 1D-E rather than 16-BAC/SDS-PAGE. Three membrane enrichment strategies used were (1) WGA affinity chromatography, (2) biotin-avidin affinity chromatography, and (3) free-flow electrophoresis (FFE). Tryptic fragments were identified by nano-HPLC ion trap and Q-TOF mass spectrometers (18). A total of 21 PM and 2 intracellular membrane (IM) proteins were identified by two or more peptide hits following WGA affinity chromatography. This was comparable to the number of TMD-containing proteins identified by Lewandrowski et al. using concanavalin A to affinity enrich PM proteins (19). Although lectin affinity enrichment greatly enhanced the number of IMPs identified compared with classical 2-DE proteomics approaches, there was still a large deficit in the number of known platelet membrane proteins being identified. Possible reasons for this include the following: (1) competition between highly abundant surface glycoproteins and secreted proteins, (2) the specificity of lectins for carbohydrate side chains, and (3) low-affinity interactions in some cases.

Biotin-avidin affinity chromatography was also employed, which would in theory pull down all biotinylated surface proteins (18). The membrane-insoluble biotinylating reagent sulfo-NHS-SS-biotin was used to biotinylate surface proteins and thereby limit labelling of intracellular proteins. However, a substantial number (>100 proteins) of intracellular proteins were also identified, most of which are known PMPs. This was most likely due to the mild washing conditions that would not have removed all PMPs. NeutrAvidin-beads were used rather than avidin- or streptavidin-beads to facilitate removal of bound proteins through the reducing agent DTT. Despite outperforming WGA affinity chromatography, biotin-avidin affinity chromatography still did not identify all known platelet surface IMPs. In total, 35 PM, 14 IM, and 5 TMD-containing proteins of unknown localization were identified by two or more peptide hits.

The third membrane enrichment approach employed by Senis et al. was FFE (18). Advantages of this approach include the separation of plasma and intracellular membranes and little or no loss of IMPs from the PM fraction. Disadvantages include the requirement of specialist equipment and possible masking of low abundance IMPs and signalling proteins. PMP contamination can be reduced by washing the PM fraction with high-salt, high-pH conditions. A total of 35 PM, 30 IM, and 10 TMD-containing proteins of unknown locations were identified in the FFE-generated PM fraction by two or more peptide hits, and 31 PM, 66 IM, and 20 TMD-containing proteins of unknown location in the IM fraction. Significantly, only 2 of the 44 proteins identified in the FFE-IM fraction were known PM proteins. The presence of IM proteins in the PM fraction and vice versa is, therefore, most likely due to the presence of proteins in both membrane compartments as well as a degree of cross-contamination. The majority of the IM proteins identified are expressed in the ER.

In total, 46 PM, 68 IM, and 22 TMD-containing proteins of unknown compartmentalization were identified using these three membrane enrichment approaches by two or more unique peptides. Sixty percent of proteins were identified by more than one enrichment method. Over a third of the proteins identified by all of the enrichment methods are well known, highly expressed platelet surface proteins. As with the study by Moebius et al., only a small proportion (17%) of the identified PM proteins contained more than one predicted TMD. No seven-transmembrane G protein-coupled receptors were identified; however, three tetraspanins, CD9, Tspan9, and Tspan33, which contain four TMDs, and several multi-spanning IM proteins, including calcium-transporting ATPase type 2C, IP_3 receptors, and SERCA2A, which are predicted to contain 8, 6, and 7 TMDs, respectively, suggesting that the lack of identification of G protein-coupled receptors may be due, in part, to their low abundance, were identified. At the time, it was estimated that approximately 100 of the TMD-containing proteins reported had not previously been described in platelets on the basis of biochemical and functional data, demonstrating the advantage of using the three separate enrichment techniques.

Two receptor-like PM proteins identified in this study of particular interest are the PTP CD148 (also referred to as DEP-1, PTPRJ, and rPTPη) and the ITIM-containing receptor G6b-B. Although CD148 had previously been reported in platelets, its functional role had not been elucidated (20, 21). G6b-A and -B isoforms had previously been identified in other platelet proteomics studies, but neither had been confirmed to be expressed in platelets by other means (9, 17). G6b-B expression was confirmed by western blotting using a specific polyclonal antibody (18). CD148 was hypothesized to fulfil the role of a CD45-like PTP in platelets and regulate ITAM and integrin receptor signalling; G6b-B was hypothesized to inhibit ITAM receptor signalling. Follow-up studies of both CD148 and G6b-B have demonstrated that CD148 is a critical positive regulator of Src family kinases in platelets and that G6b-B negatively regulates GPVI and CLEC-2 signalling (22–24). These two examples highlight the usefulness and importance of proteomics data in initiating novel areas of platelet research.

The most comprehensive platelet membrane proteome study to date by Lewandrowski et al. combined aqueous two-phase partitioning of platelet membranes with protein identification by shotgun and COFRADIC approaches (16). The shotgun approaches included (1) 1D-E-LC-MS/MS and (2) MudPIT. Two-phase membrane partitioning followed by this three peptide identification strategy proved to be a highly successful approach. A total of 1,282 proteins were identified in the PM fraction, of which 498 proteins were identified using the peptide-centric COFRADIC approach on a single peptide basis, and 1,202 proteins were identified by the combined results of the two shotgun-based approaches. Four hundred and eighteen proteins were identified by the two shotgun-based approaches and COFRADIC, demonstrating the complementarity of the approaches. All proteins were identified by a minimum of two valid peptide identifications using the two shotgun approaches, which together with the search parameters used, gave a false-positive rate of <1%, making it a highly reliable and accurate data set.

Lewandrowski et al. identified 626 TMD-containing proteins, 30 of which contained 7 TMDs, including G protein-coupled receptors such as PAR1, PAR4, P2Y1, P2Y12, and CXCR4. Nearly 100 proteins were predicted to have ≥8 TMDs, including solute carriers and calcium channels. This was a vast improvement from any of the previous platelet membrane proteome studies. This is at least partially due to the carbonate extraction step used to reduce contamination with PM-associated cytoskeleton (16).

The sub-cellular distribution of proteins was estimated with GoMiner and Ontologizer algorithms. Based on these bioinformatics analyses, of the 1,282 proteins identified in the PM fraction, 371 were predicted to be PM components, of which 142 are IMPs. In addition, a range of proteins from other membrane compartments were also identified, including 199 from the ER, 148 from the Golgi apparatus, and 140 from vesicles. A comparison of the summed proteins identified in the ER, Golgi apparatus, and vesicles, with those in the PM, revealed a major overlap of 116 proteins. Presumably, these proteins are shuttled between these compartments. Many of the proteins identified were uncharacterized and were not assigned a sub-cellular location.

The majority of the proteins identified in the PM compartment were classified as possessing signal transduction activity (156 proteins) or as receptors (104 proteins) (16). Thirteen of the receptors identified were G protein-coupled receptors, 6 of which (AVPR1A, GPR92, P2Y1, P2Y12, PAR-1, and PTGIR) have only recently been identified in platelets by quantitative PCR (25). This study was also highly successful in identifying proteins involved in cell adhesion and aggregate formation, two critical functions of platelets. In total, 86 proteins implicated in cell adhesion and 47 proteins present or involved in cell junctions were identified, including all of the platelet integrins and key components of their signalling apparatus.

Besides G protein-coupled receptors, numerous other classes of multi-spanning PM proteins identified in this study, included the following: tetraspanins, which have membrane ordering functions in various cell types, and ion transporters and channels, which allow the movement of ions and small molecules across various membranes. The identification of 91 ion transporters, most of which have ≥ 8 TMDs, demonstrates the superiority of the strategy used in this study to that of previous studies for identifying platelet multi-spanning IMPs. The reason for this vast improvement is likely multi-factorial and reflects the applicability of the entire strategy rather than a specific step for identifying multi-spanning IMPs.

Another novel and informative aspect of this study was the relative quantification of platelet PM receptors. Lewandrowski et al. used a modified version of the exponentially modified Protein Abundance Index (emPAI) to generate approximate estimates of platelet protein abundance, which correlated well with published levels (16). These data can, therefore, be used to differentiate between high-, medium-, and low-abundance proteins. Interestingly, the novel platelet ITIM receptor G6b-B had a high emPAI score of 122 and is ranked among the most abundant receptors on the platelet surface (estimated ~20,000 copies/platelet). However, it should be noted that although a general correlation was found between emPAI and copy numbers of surface receptors, absolute copy numbers of individual proteins may not fit with its position in the emPAI ranking. Another potential problem with this type of semi-quantitative analysis is that different copy numbers of receptors are reported in the literature depending on the assay used.

Finally, Lewandrowski et al. used the STRING algorithm for network analysis for the complete list of 1,282 proteins. Proteins were clustered into groups, the central ones being adhesion-mediated receptors and kinases or effectors. Other groupings included the following: (1) actin- and cytoskeleton-associated proteins, including Arp2/3 complex, vinculin, and VASP, (2) vesicle-associated SNAPs and SNAREs, (3) proteins related to metabolism, (4) glycosylation, and (5) proteins of mitochondrial origin.

A total of 858 protein–protein interactions were revealed by STRING. However, a large number of proteins were not associated with other membrane or soluble proteins.

1.3. The Secretome

A critical platelet response to activation is granule secretion. The general term given to all proteins and small molecules released from activated platelets is "secretome" or "releasate" (Fig. 1). Secretome proteins may be derived from α-granules, dense granules, microparticles/microvesicles, and exosomes. Identifying the protein composition of these various compartments is essential for defining the complete platelet secretome.

The two main classes of platelet secretory granules are α- and dense granules, both of which are derived from megakaryocyte multivesicular bodies. α-granules contain a variety of proteins synthesized in megakaryocytes and packaged into α-granules during platelet production. Other α-granule proteins are taken up from the plasma by endocytosis or receptor-mediated uptake and transport. The α-granule membrane also contains transmembrane glycoproteins that are incorporated into the plasma membrane upon granule secretion. α-Granule proteins contribute to a variety of physiological processes, including the following: coagulation and fibrinolysis; platelet activation, adhesion, and aggregation; neovascularisation and wound repair; inflammation and host defence. Interestingly, recent evidence demonstrates that α-granules are in fact heterogeneous in terms of protein content; however, the reason and mechanism underlying this heterogeneity is not known (26).

Dense granules derive their name from their opaque appearance by electron microscopy. They do not contain secreted proteins, but rather small molecules that enhance platelet activation and the prothrombotic response, including ADP, ATP, calcium, serotonin, and polyphosphate. Similar to α-granules, dense granule membranes also contain transmembrane proteins that get incorporated into the plasma membrane upon granule secretion and contribute to thrombosis, inflammation, and wound repair.

The importance of α- and dense granules to platelet function is illustrated by storage pool deficiencies (SPDs), encompassing a range of rare genetic disorders characterized by variable degrees of reduction in the numbers and content of α-granules (α-SPD, Grey platelet syndrome [GPS] and Quebec platelet disorder), dense granules (δ-SPD, Hermansky–Pudlak Syndrome and Chediak–Higashi Syndrome), or both (α, δ-SPD) (27). Platelets from these patients exhibit reduced functional responses to various agonists, the physiological consequences being mild to moderate bleeding in response to injury. Thus, elucidating the protein constituents of α- and dense granules and mechanisms of formation, packaging, and release has important implications for understanding the platelet response to injury and the basis of these and other pathological conditions.

The first study characterizing the platelet secretome was by Coppinger and colleagues in 2004 (28). The secretome of thrombin-stimulated platelets was prepared by centrifugation and microparticles removed by ultracentrifugation. Proteins were identified by MudPIT and tandem mass spectrometry. Of the 81 proteins identified in this study, 37% were known to be secreted or shed from activated platelets and 35% were known to be released from various other cell types. No transmembrane or signalling proteins were identified; however, several cytoskeletal proteins were identified. Interestingly, cytoskeletal proteins have also been identified in several subsequent platelet secretome studies, raising the possibility of alternative functions of these proteins (29, 30). Three novel platelet released proteins identified in this study, namely, secretagranin III, cyclophilin A, and calumenin, were show to be present in human atherosclerotic plaques, providing a potentially novel link between platelets and atherosclerosis.

In a follow-up study by the same group, aspirin pre-treatment reduced the amount of proteins released by ADP-, collagen-, and TRAP-activated platelets (31). Normalized, log-transformed spectral counts revealed that different distinct and overlapping secretome profiles are generated with different agonists. More abundant proteins were released by all three agonists, whereas less abundant proteins were only released by TRAP. It was suggested that aspirin inhibits a second stage of secretion that is dependent upon thromboxane A2 and possibly outside-in integrin $\alpha_{IIb}\beta_3$ signalling.

Della Corte and co-workers employed 2D difference gel electrophoresis (2D-DIGE) to obtain a more quantitative assessment of the secretome of thrombin-activated platelets (32). Spots with at least a fivefold change in mean intensity between thrombin-activated platelets and control unstimulated platelets were excised from gels and subjected to MALDI-TOF MS analysis. Low overall identification of proteins in this study was due to experimental design, as the focus was differentially secreted proteins rather than total proteins secreted. Of the 36 differentially expressed proteins, 21 increased following platelet activation, including the cytoskeletal proteins α-tubulin and vinculin, transferrin, protein disulfide isomerase, thrombospondin, and the nuclear protein lamin A. Paradoxically, certain plasma proteins were found to decrease in the secretome of activated platelets, including apolipoprotein A, fibrinogen, immunoglobulin, and transferrin, which may be due to sample preparation.

Recently, Piersma and co-workers reproducibly identified 225 proteins in the secretome of TRAP-activated platelets (30). Their approach involved a standard GeLC approach to separate and digest proteins, LTQ-FT instrument for peptide identification and Sequest-driven search of an IPI database. Although considerable overlap was observed with datasets from earlier platelet secretome studies, many novel proteins were also identified. A diverse repertoire of cytoskeletal proteins was identified, supporting previous findings from Coppinger (28, 30). The most likely explanation for cytoskeletal contamination in both studies is as constituents of microparticles.

Recently, studies have focused on characterizing the proteomes of α-granules and microparticles in their entireties. These studies are of particular interest as these are two of the main contributors to the platelet secretome and not all α-granules are released from activated platelets. Two recently published studies by Maynard and co-workers focused on the α-granule proteomes of normal and GPS platelets (33, 34). α-granules were enriched by sucrose-gradient centrifugation from platelet lysates prepared by ultrasonication. A concern with applying the same centrifugation conditions for isolating "ghost granules" is altered migration in the sucrose density gradient due to altered densities. Thus, ghost granules with similar properties to normal α-granules will be selected for. This approach resulted in a significant enrichment of platelet α-granules; however, there remained a substantial amount of contamination from other platelet compartments, making it essential to validate protein localization to α-granules by immuno-electron microscopy. Mitochondrial-associated proteins were manually removed from the analysis.

In the first study by Maynard and co-workers of normal platelet α-granules, 284 unique proteins were identified, 44 of which appear to be new α-granule proteins (33). Twenty-six were also found in the membrane fraction, which may play roles in regulating granule release or platelet function. Several of the proteins were confirmed to be present in α-granules by immuno-EM. Sixty-five percent overlap was reported with the initial study of the platelet releasate by Coppinger and colleagues, demonstrating the complementarity of the approaches.

In their subsequent study, Maynard and co-workers used a similar approach to characterize the protein constituents of GPS ghost granules (34). Normalized peptide hits (NPHs) for soluble α-granule proteins synthesized in megakaryocytes were markedly decreased or not detected in GPS platelets, whereas NPHs for soluble, endocytosed proteins were only moderately affected. NPHs for membrane-bound proteins were similar between normal and GPS platelets, although P-selectin and Glut3 were marginally reduced, which coincided with immuno-EM of resting platelets. Proteins not previously known to be reduced in GPS were also identified in this study, including latent transforming growth factor-β-binding (LTBP1), a component of transforming growth factor-β complex. One of the main conclusions of this study is that the basis of GPS is a reduction in incorporation of endogenously synthesized megakaryocyte proteins into α-granules. However, this must be tempered by the fact that this was a single patient and

GPS comes in many different forms. Nonetheless, this study contributed significantly to our understanding of the condition and raises the possibility of applying platelet granule proteomics to the clinical diagnosis of SPDs.

Despite dense granules containing very low levels of proteins, a proteomics study recently conducted by Hernandez-Ruiz and co-workers investigated proteins in this compartment (29). Washed platelets were prepared from large amounts of expired pooled platelet-rich plasma to overcome the low protein content of dense granules. Platelets were subsequently sonicated and dense granules enriched by ultracentrifugation through a discontinuous Histodenz gradient. Proteins were identified using two methods: 2-DE and MALDI-MS, and LC-MS/MS. All 40 of the proteins identified had previously been reported in platelets and most had been shown to be released from activated platelets. Many actin-associated proteins, glycolytic enzymes, and regulatory proteins had not previously been associated with dense granules. Several are shared with α-granules, such as PF-4, β-thromboglobulin, and fibrinogen, suggesting some overlap in protein content. Interestingly, the adapter protein 14-3-3ζ was identified in dense granules. This protein interacts with a variety of proteins, including GPIbα and GPIbβ, which it links to the cytoskeleton and regulates its function (35). It is released by several other cell types and is present in atherosclerotic plaques, suggesting that it has as yet undefined functions outside of cells (29). This study demonstrates the power of proteomics to identify novel proteins in compartments not rich in proteins.

1.4. Microparticles

Microparticles (MPs) are small (~0.1 μm diameter) secretory vesicles that originate from the plasma membrane and have attracted considerable attention in vascular research. They are shed from the surface of a variety of cells following apoptosis or activation. In the vascular context, MPs are released by endothelial cells, smooth muscle cells, lymphocytes, monocytes, erythrocytes, megakaryocytes, and platelets. Their importance is emphasized by the alteration of their plasma concentration in several vascular pathologies (36). Indeed, plasma levels of MPs are markedly elevated in patients with acute coronary syndrome (ACS) (37). MPs contain a unique subset of proteins derived from the parent cell, and in recent years it has become clear that MPs have important biological functions. Numerous studies have shown that MPs, isolated either from plasma or from supernatants of stimulated cells, increase expression of adhesion molecules on endothelial cells and monocytes, stimulate the release of cytokines, alter vascular reactivity, induce angiogenesis, and may even be involved in cancer metastasis (38, 39). Because of their critical pathophysiological role, characterization of the protein composition of MPs is pivotal to the understanding of their function. Proteomics is the instrument of choice for this type of analysis. The proteome of MPs from human plasma has been mapped by 2-DE and subsequent MS (40). Overall, plasma MPs displayed distinct protein features and a greater number of protein spots than that detected in whole plasma.

Platelet-derived MPs account for over 90% of the plasma MPs from healthy individuals (41). It is now well established that activated platelets play critical roles in the evolution of atherosclerosis, from the initiation of the fatty streak through the progression of atheromatous plaque to the final atherothrombotic events that lead to ACS (42). More recently, increasing evidence has suggested that the roles of platelets in atherosclerosis may at least be partially mediated by the production of platelet-derived MPs (43, 44). Other types of MPs, leukocyte- or endothelial cell-derived, have also been investigated in patients with atherothrombosis (43); however, platelet-derived MPs must be the most important participant in atherothrombosis because they are quantitatively abundant and are generated by activated platelets (43).

The proteome of platelet-derived MPs was analyzed using 1D SDS-PAGE and LC-MS/MS (38). The authors of this study identified 578 proteins associated with the MPs released by washed platelets activated with ADP in vitro. Among the hits were surface proteins typical of platelets, such as integrin $\alpha_{IIb}\beta_3$ and P-selectin. The presence of several chemokines suggested that platelet-derived MPs may play an important role in the regulation of leukocyte migration and differentiation. Moreover, the finding of 380 proteins not previously identified in platelets suggested that the formation of MPs is accompanied by selective enrichment of specific subsets of the proteome. Interestingly, the same research group carried out a comparative analysis of MPs isolated from plasma versus platelet-derived MPs using proteomic methods based on 1D-SDS-PAGE and LC-MS/MS (41). In that way, they identified 21 proteins detected in plasma MPs that were essentially absent in platelet-derived MPs. Those included proteins associated with apoptosis, iron transport, immune response and the coagulation process. Conversely, 11 proteins – mostly cytoskeletal – were enriched in platelet-derived MPs. This was the first attempt to differentiate the proteome of plasma and platelet MPs; although there are significant differences, the number of differences is not high, probably due to the fact that platelet-derived MPs constitute the majority of the circulating MPs present in plasma. Interestingly, a recent report analyzed and characterized platelet-derived MPs by flow cytometry using antibodies against the major surface glycoproteins, the platelet activation antigen P-selectin, and a marker of procoagulant activity (phosphatidylserine exposure) (45). MPs were generated by exposure of washed platelets to thrombin receptor activating peptide (TRAP) or ionophore. The main conclusion from the study is that platelet-derived MPs are heterogeneous and highly dependent on the activation mechanism. This is also emphasized by a recent study that analyzed the TRAP-induced platelet releasate (46). This study included the soluble proteins released from granules as well as the proteins present in MPs, which are formed upon TRAP stimulation. The study, based on 1D-SDS-PAGE and LC-MS/MS, led to the identification of 225 proteins consistently present in the platelet releasate proteome. Interestingly, 40% of the data consisted of novel releasate proteins, compared to previous studies that utilized other agonists for platelet stimulation.

As more studies are performed with increasingly advanced instrumentation, methodologies and databases, increasing numbers of platelet secretome and organelle proteins will undoubtedly be identified.

1.5. The Phosphoproteome

Phosphorylation of proteins and peptides is one of the most common post-translational modifications (PTMs). The majority of proteins are at least transiently phosphorylated during their lifetime. A diverse concert of protein phosphatases and protein kinases defines the transient character of this modification, which contributes to the regulatory mechanism of numerous biological/biochemical processes, including protein folding, function, activity, interaction, location, and degradation (47).

In general, there are four types of protein phosphorylations (40): (1) O-phosphates, (2) N-phosphates, (3) acylphosphates, and (4) S-phosphates. O-phosphates are the most abundant, occurring on *serine- (pSer), threonine- (pThr), or tyrosine residues (pTyr). The transient character and low stoichiometry of phosphorylation renders the analysis of phosphorylation sites a demanding and challenging task.

As mentioned above, due to their anucleate character, platelets exhibit a limited capability of de novo protein synthesis (5). Furthermore, the sensitive regulation between inhibition and activation must respond to sudden stimuli within seconds. Accordingly, platelets are regulated by PTMs which control both protein activity and interaction. In this context, phosphorylation plays a critical role not only in the initial steps of activation and aggregation following exposure to adhesive and activating ligands, such as collagen, fibrinogen, vWF, ADP, thrombin, and thromboxane A2, but also in inhibitory pathways that are stimulated by endothelium-derived nitric oxide (NO), prostacyclin (PGI$_2$), and immunoreceptor tyrosine-based inhibitory motif (ITIM)-containing receptors. The initial steps of platelet activation are regulated by a concert of tyrosine and serine/threonine kinases and phosphatases whereby inhibitory events are strongly determined by the cyclic nucleotide dependent protein kinases A and G, which are highly expressed in human platelets.

Several studies have focused on the platelet phosphoproteome following distinct stimulation (48–50), as well as on the global phosphoproteome of resting platelets (51, 52). Different activation states of platelets were investigated using immunoprecipitation (49, 50), strong cationic exchange (SCX), and immobilized ion metal affinity chromatography (IMAC) enrichment (51). Most of the studies focusing on the platelet phosphoproteome following platelet activation are related to the analysis of signalling cascades, and will be mentioned more in detail in the corresponding section below. Regarding the analysis of the phosphoproteome of resting human platelets, the most comprehensive study to date was developed by Zahedi and colleagues in 2008 (51). In this study, platelets were isolated from fresh blood. After lysis and protein tryptic digestion, phosphopeptides were enriched by means of IMAC and SCX. Phosphopeptide-enriched samples were analyzed by nano LC-MS/MS or precursor ion scanning. In this way, 564 phosphorylation sites were identified from more than 270 proteins, of which many had not been described in platelets before. Among those were several unknown potential protein kinase A (PKA) and protein kinase G (PKG) substrates. In addition, they also found that GPIbα is phosphorylated at Ser603 in resting platelets, which may represent a novel mechanism for the regulation of signalling via one of the most important platelet receptors (GPIb-IX-V) by PKA/PKG (51).

Another relevant study was carried out by Qureshi and colleagues, who employed a combination of proteomic profiling by SDS-PAGE and LC-MS/MS and computational analyses to study the proteome and phosphoproteome of human platelets in their basal, resting state (52). The main target was to gain insights into integrin signalling. Ten independent human samples were analyzed leading to the identification of 1,507 unique platelet proteins. This is the most comprehensive platelet proteome assembled to date and includes 262 phosphorylated proteins. This proteomic dataset was used to create a platelet–protein interaction (PPI) network, with special focus on the integrin $\alpha_{IIb}\beta_3$ signalling pathway. This study provides insights into the mechanism of integrin $\alpha_{IIb}\beta_3$ activation in resting platelets and also provides an improved model for analysis and discovery of PPI dynamics and signalling pathways in the future.

1.6. The Glycoproteome

Glycosylation is one of the most common post-translational modifications of surface and secreted proteins. The enzymatic addition of saccharides to proteins serves a variety of functions, including regulating protein folding, stability, and binding affinities. Glycoproteins function as receptors, enzymes, ligands, and structural proteins. Despite the important functional roles of glycoproteins in haemostasis and thrombosis, very little work has been done to elucidate the platelet glycoproteome. Even less work has been done to identify specific glycosylation sites on platelet glycoproteins and how alteration of these sites influences protein function. This information is essential for understanding how platelet activation and thrombosis is regulated.

Glycosylation takes place co- or post-translationally in the endoplasmic reticulum and Golgi apparatus. The heterogeneous nature of glycoproteins poses unique challenges for their analyses. In addition to the diverse compositions of oligosaccharide side chains or glycans is the variety of discrete glycosylation sites on proteins. Most glycans are attached to asparagine or serine/threonine residues, which are classified as N- and O-linked glycosylations, respectively. O-linked glycosylation is generally more heterogeneous than N-linked glycosylation. The investigation of O-linked glycosylation is complicated by the fact that no specific enzyme has been identified that specifically removes O-glycans. By contrast, N-glycosidase F specifically cleaves N-linked glycans. Glycosylphosphatidylinositol (GPI) linked to the C-termini of proteins is another well-characterized form of glycosylation that mediates anchoring of proteins to the plasma membrane.

Lewandrowski et al. used two techniques to enrich platelet N-glycosylation sites: (1) lectin (concanavalin A) affinity chromatography and (2) and chemical derivatization of glycan residues and trapping of glycosylated proteins on hydrazide-functionalized resins. Forty-one different glycoproteins were identified using these techniques that were almost exclusively surface transmembrane or secreted proteins. It is not surprising that so few PM proteins were identified using these techniques, as not all surface glycoproteins are N-glycosylated and more abundant proteins will outcompete less abundant proteins for binding to either the concanavalin A or hydrazide resins. This study demonstrated that lectin affinity enrichment of membrane proteins enhances the identification of IMPs by mass spectrometry. However, a novel strategy devised by the same group termed "enhanced N-glycosylation site analysis using strong cation exchange enrichment" (ENSAS) involving aqueous 2-phase membrane partitioning followed by strong cation exchange chromatography outperformed the above approaches for identification of N-glycosylation sites (53).

2. Recent Advances in Platelet Proteomics

2.1. Trafficking in the Membrane Proteome

The platelet PM is a dynamic compartment that undergoes major changes following activation. These changes include the following: reorganization into functionally distinct microdomains; receptor upregulation, shedding, and internalization; recruitment of signalling proteins; and changes to phospholipid content. Characterizing these changes has important implications for understanding the procoagulant activity of platelets.

Challenges that hamper the identification of new signalling proteins recruited to the PM upon platelet activation include (1) low relative abundance and (2) the transient, low-affinity nature of the interactions. The inner surface of the PM is lined with an abundance of cytoskeletal proteins, including actin, tubulin, and filamin that can easily mask the presence of low levels of signalling proteins. In addition, low-affinity interactions can be easily disrupted even with mild solublization conditions. Recently, Gibbins and co-workers have devised two different approaches for enriching PMPs recruited to the PM of collagen-related peptide (CRP)- and thrombin-stimulated platelets (CRP is a GPVI-specific agonist) (54, 55). CRP-stimulated platelets were lysed by repeated freezing–thawing in a physiological buffer and gentle sonication, without the use of detergents and high-salt buffers that would likely result in the loss of PMPs. All platelet membranes were subsequently pelleted using high-speed centrifugation and PMPs eluted from membranes with 100 mM sodium carbonate. A gentle elution step was used to avoid eluting IMPs and tightly associated cytoskeletal proteins that would have increased sample complexity. Phosphatase and kinase inhibitors (sodium vanadate and staurosporine, respectively) were included during the lysis step to prevent changes in phosphorylation that could disrupt the delicate balance of membrane-associated proteins. Sample complexity was further reduced by a 2D system, involving liquid-phase IEF in the first dimension and SDS-PAGE in the second dimension. A total of 105 proteins were identified by FT-ICR mass spectrometry in 26 bands that change in intensity in CRP-stimulated platelets, of which 44% were signalling proteins. Other proteins identified included cytoskeletal structural or regulatory proteins (19%), secreted proteins (10%), and proteins involved in metabolism (26%). Only 1% of the proteins identified were transmembrane proteins (P-selectin), demonstrating the relative specificity of the elution step. Surprisingly, the collagen chaperone protein Hsp47 (colligin) was detected in the PMP fraction of CRP-stimulated platelets and was shown to be exposed on the surface of activated platelets (54). Functional evidence demonstrated that surface bound Hsp47 facilitates collagen-mediated platelet aggregation. These findings highlight the importance of further characterizing PM associated PMPs in resting and activated platelets. Although this strategy provides an excellent starting point, a more sensitive approach is necessary as numerous GPVI signalling proteins known to be recruited to the PM (e.g. Syk, PLCγ2, and SLP-76) failed to be identified. Perhaps the combinatorial ligand library affinity chromatography-based approach recently described by Guerrier et al. for concentrating and identifying low-abundance soluble platelet proteins can be applied to identifying low-abundance PMPs recruited to the PM of activated platelets (56).

A different strategy was employed by Tucker et al. in a comparative analysis of the membrane proteomes of resting and thrombin-stimulated platelets (55). Platelet PMs were enriched by biotin-avidin affinity chromatography following freezing–thawing and sonication in 0.1% v/v Triton X-100. PMPs were eluted from biotinylated PMs bound to NeutrAvidin conjugated beads with 4% w/v DTT and 8% w/v CHAPs. Sample complexity was further reduced by solution-phase IEF followed by SDS-PAGE. Image analysis of silver stained gels revealed 65 bands of increased intensity in the thrombin-stimulated surface fraction compared to the resting surface fraction, from which 88 different proteins were identified (55). Twenty-five percent of the proteins identified were IMPs, 31% were known signalling proteins, and 8% were cytoskeleton proteins. One of the novel platelet signalling proteins identified was the adaptor HIP-55. Further analysis of HIP-55 in platelets demonstrated increased association with Syk and the β3 integrin subunit upon collagen and thrombin stimulation, suggesting that HIP-55 may be involved in regulating $\alpha_{IIb}\beta_3$ signalling. Platelets from HIP-55-deficient mice exhibited reduced fibrinogen binding in response to thrombin, which supports a role of HIP-55 in regulating integrin signalling (55).

2.2. Signalling Pathways and Systems Biology

One of the most active fields in platelet research is the study of signalling cascades. Ligand-mediated receptor occupancy/clustering initiates a signalling cascade that leads to downstream effects, which in the case of platelets are mostly related to activation, shape change, adhesion and aggregation. A detailed proteome analysis of signalling cascades in human platelets could potentially lead to a better understanding of the mechanisms that underlie platelet activation and function. One of the main ways in which signalling cascades are regulated is by reversible PTMs, with phosphorylation playing a major role. That is why the identification of phosphorylated proteins and mapping precise phosphorylation sites is so relevant in platelet proteomics research, as mentioned previously.

In recent years, gel-based proteomics was immediately applied to the study of platelet signalling, comparing the proteome of basal and activated platelets (57). Most of the studies were based on 2-DE for protein separation. However, in recent studies immunoprecipitations were combined with SDS-PAGE to improve the coverage of selective groups of low abundance proteins and those more difficult to analyze by 2-DE (e.g. very hydrophobic proteins, such as membrane proteins).

Among the main intracellular signalling cascades related to platelet activation/aggregation, the thrombin-receptor signalling pathway has been the most widely investigated by proteomics so far. In those studies, 2-DE was the primary method of choice in most cases. Thus, in 1998, Immler et al. identified by LC-MS/MS a group of spots on 2D gels that were differentially regulated in thrombin-activated platelets. Those features were related to different isoforms and phosphorylation states of myosin light chain (58). Two years later, Gevaert and collaborators studied cytoskeletal preparations of basal and thrombin-stimulated platelets by a combination of 2-DE and MALDI-MS. They reported the identification of 27 proteins, most of which were F-actin binding proteins, which translocate to the cytoskeleton upon thrombin stimulation (59).

In 2003, Marcus and collaborators combined different separation and analytical methods to obtain a more comprehensive proteome analysis of thrombin-activated platelets (60). In one approach, proteins were radiolabeled with ^{32}P and separated by 2-DE using different pI ranges. Phosphorylated proteins were detected by autoradiography. As part of the same study, Marcus and colleagues also separated the proteins by 2-DE following inmunoprecipitation with the 4G10 antiphosphotyrosine antibody. Differentially regulated proteins were identified by MALDI-TOF and nano-LC-ESI-MS/MS. Overall, the authors identified 55 proteins in the pI 4–7 and 6–11 ranges, most of which corresponded to different isoforms of pleckstrin and cytoskeletal proteins.

One of the approaches that has been recently used with success to analyze signalling cascades in platelets involves the separation of the whole proteome, without any pre-fractionation, by high-resolution 2-DE to detect differentially regulated features that can be analyzed by MS. García and colleagues have used this approach to investigate intracellular signalling cascades in platelets stimulated with thrombin-receptor activating peptide (TRAP), which is able to activate the main thrombin receptor PAR-1 (48). Working under non-aggregating conditions, basal and TRAP activated platelet proteins were extracted and separated by narrow range pI 4–7 and 6–11 high resolution 2D gels (18×18 cm). Gels were stained with a fluorescent dye prior to differential image analysis. That process led to the identification of 41 differentially regulated protein features, which were found to derive from 31 different genes, most of which corresponded to signalling proteins. Several of the proteins identified had not previously been reported in platelets, including the adapter downstream of tyrosine kinase 2 (Dok-2). Further studies revealed that the change in mobility of Dok-2 was due to tyrosine phosphorylation. García and collaborators also provided the first demonstration of phosphorylation of the regulator of G-protein signalling, RGS18, and mapped one of the phosphorylation sites by MS/MS (48). This report set the stage for future studies and illustrates the potential of 2-DE-based proteomics to study platelet signalling.

Following the success of the above study, García and colleagues carried out a similar platelet signalling proteomics study, this time by analyzing the proteome of CRP-activated platelets compared to the proteome of unstimulated platelets (49). Glycoprotein VI (GPVI) is the major activation inducing receptor for collagen in platelets. CRP is a GPVI-specific agonist that has been extensively used to elucidate the GPVI signalling cascade. Since tyrosine kinases and phosphatases play a fundamental role in the GPVI signalling cascade, the study of the tyrosine phosphoproteome was one of the main objectives of the study. García et al. used two main approaches for protein separation to address the question above: 2-DE (pI 4–7 and 6–11) of whole cell lysates, similar to the approach previously used by the same group to study PAR-1 signalling cascade; and phosphotyrosine immunoprecipitations (with agarose-conjugated 4G10 monoclonal antibody) followed by 1D-PAGE. In both cases, proteins were identified by LC-MS/MS. By using this global approach, 96 proteins were found to undergo PTMs in response to CRP in human platelets, including 11 novel platelet proteins, such as the adapters Dok-1, SPIN90, and osteoclast-stimulating factor 1 (OSF1). Interestingly, the 2-DE-based approach yielded the detection of 111 differentially regulated protein features, corresponding to 72 different ORFs. Overall, there was a shift of many proteins towards a more acidic region of the proteome following CRP-stimulation, which was consistent with protein phosphorylation. A high proportion of the proteins identified through the 2-DE differential analysis were signalling (36%) and cytoskeletal (14%) proteins, consistent with the critical role played by these two groups of proteins in mediating platelet function and their regulation by phosphorylation. Some signalling proteins identified were adapters, including Dok-1, Dok-2, Gads, Grb2, and SKAP-HOM. However, the 2-DE approach failed to identify several proteins well known to participate in the GPVI signalling cascade. This could be a consequence of a number of factors, including a low level of expression of many tyrosine phosphorylated signalling proteins, their co-localization in the gels with other more highly expressed proteins, changes in spot volume below the selected threshold of twofold, or inability of certain proteins to be properly resolved by 2-DE due to solubility problems during isoelectric focusing (IEF).

In order to overcome the above limitations of the 2-DE-based study of the GPVI signalling cascade, García and colleagues decided to use a complementary approach. Central to this modified approach was the introduction of an affinity step prior to electrophoresis (49). Indeed, as mentioned above, the GPVI signalling cascade was investigated in detail by using a combination of phosphotyrosine immunoprecipitation and 1D-PAGE followed by MS analysis. An advantage of this approach is the availability of reasonably good antiphosphotyrosine antibodies, such as the monoclonal antibody 4G10. Agarose-conjugated 4G10 was used to reduce masking by the IgG heavy chain. In this study, phosphotyrosine proteins immunoprecipitated from basal and CRP-stimulated platelets were separated on 4–12% Bis-Tris SDS-PAGE gels and stained with the specific phosphoprotein gel stain Pro-Q Diamond (Invitrogen), which detects all sites of protein phosphorylation. Equivalent gels were run in parallel and stained for total protein with a fluorescent dye similar to SYPRO Ruby. Specific bands corresponding to tyrosine-phosphorylated proteins were excised from the gels and the proteins in-gel trypsin digested and identified by LC-MS/MS. This approach identified 30 different tyrosine phosphorylated proteins in response to CRP. Most of the known GPVI signalling proteins were identified, including three novel proteins, namely, β-Pix, the SH3 adapter protein SPIN90, and the type I transmembrane protein G6f. Interestingly, G6f was found to be specifically phosphorylated on Tyr-281 in response to platelet activation with CRP, providing a docking site for the adapter Grb2 (49). All of the above data was validated by traditional biochemical approaches (e.g. immunoprecipitations and western blotting), which provided valuable information on the novel proteins identified and led to more detailed mechanistic studies.

Complementary to the above study, Schulz and colleagues analyzed the GPVI signalling cascade by 2D-DIGE and MS (61). In 2D-DIGE, proteins of two samples are labelled prior to IEF with spectrally distinct fluorescent cyanine minimal dyes (CyDyes). The samples are mixed together and run on the same 2D gel. In this way, proteins from different samples migrate under identical conditions and the gel-to-gel variations are eliminated. After electrophoresis the gel is scanned with a fluorescent imager at different wavelengths. Quantification is possible because there is a linear relationship between a labelled protein and the signal measured with the imager. This is because, on excitation of the dye by monochromatic light, the dye emits light in proportion to the amount of labelled compound in the sample. Three spectrally distinct dyes are available for DIGE: Cy2, Cy3, and Cy5. They can be excited with a blue, red, and green laser, respectively. Schulz and colleagues compared the proteome of basal platelets and platelets stimulated with an activating monoclonal antibody specific for GPVI. They identified eight differentially abundant proteins associated with various functions. One of these proteins, aldose reductase (AR), had an enzymatic activity significantly increased upon GPVI activation; interestingly, inhibition of AR resulted in reduced platelet aggregation.

In 2009, a group of researchers, based at Birmingham (UK), Oxford (UK), and Santiago de Compostela (Spain), carried out a detailed proteomic analysis of integrin $\alpha_{IIb}\beta_3$ outside-in signalling in human platelets (50). They took advantage of the success of their previous analyses of the TRAP-1 and GPVI signalling cascades, mentioned above. Since the integrin $\alpha_{IIb}\beta_3$ signalling pathway is based on tyrosine phosphorylation events – as happens with the GPVI cascade, the chosen experimental approach was also based on phosphotyrosine immunoprecipitations. Washed platelets (1.5 mL at 5×10^8 mL^{-1}) were plated on either a BSA- or fibrinogen-coated surface for 45 min at 37°C. Proteins were immunoprecipitated with an antiphosphotyrosine antibody (4G10, agarose-conjugated) and resolved on 4–12% NuPAGE Bis-Tris gradient gels (Invitrogen).

Following gel staining with a fluorescent dye equivalent to SYPRO Ruby, bands of interest were excised, proteins trypsin digested, and analyzed by MS. The approach led to the identification of 27 proteins, 17 of which were not previously known to be part of a tyrosine phosphorylation-based signalling cascade downstream of $\alpha_{IIb}\beta_3$. The group of proteins identified included the novel immunoreceptors G6f, and G6b-B, and two members of the Dok family of adapters, Dok-1 and Dok-3, which underwent increased tyrosine phosphorylation following platelet spreading on fibrinogen. Recently, the Dok-1 PTB domain has been shown to bind to the cytoplasmic tail of the integrin β3 subunit. However, in contrast to the talin PTB domain, Dok-1 negatively regulates integrin activation (62). Oxley and colleagues demonstrated that the inhibitory effect mediated by Dok-1 on $\alpha_{IIb}\beta_3$ signalling occurs through the phosphorylation of the β3 integrin Tyr747. Phosphorylation of this site favours formation of a Dok-1-integrin complex, over a talin–integrin complex (63). These findings demonstrate how phosphorylation can act as a "molecular switch" that modulates integrin activation, and illustrates the importance of Dok proteins in integrin signalling. García and colleagues, therefore, focused on Dok proteins for further mechanistic and functional studies. Indeed, they showed that tyrosine phosphorylation of Dok-1 and Dok-3 was primarily Src kinase-independent downstream of the integrin. Moreover, both proteins inducibly interacted with Grb2 and SHIP-1 in fibrinogen-spread platelets (50). Based on previous reports from B and T cells, the authors hypothesized that Dok-1 and Dok-3 participate in a multi-molecular signalling complex, together with SHIP-1 and Grb2, which may negatively regulate $\alpha_{IIb}\beta_3$ outside-in signalling.

All the above studies allowed the creation of databases with proteins relevant for platelet activation. In that context, a systems biology approach would help to predict phosphorylation events and how all those proteins interact with each other to exert their function in a particular signalling cascade. Dittrich and colleagues took this challenge and generated a functional interaction map of platelet phosphorylations and kinases after assembly of a comprehensive proteome and transcriptome database of human platelets (64). The complete data on the platelet proteome, interactome, and phosphorylation state used in this study is available online at the "PlateletWeb–Knowledgebase" (http://plateletweb.bioapps.biozentrum.uni-wuerzburg.de), where all interacting proteins are hyperlinked allowing an easy navigation through the platelet interactome network.

2.3. Platelet Clinical Proteomics

It has been only very recently that platelet proteomics started being applied to more clinical orientated studies. One of the fields where platelet clinical proteomics has proven to be of great utility is transfusion medicine. Platelets stored for transfusion produce pro-thrombotic and pro-inflammatory mediators implicated in adverse transfusion reactions. During the last 3 years a combination of two-dimensional gel electrophoresis and mass spectrometry-based proteomic approaches have been used to profile alterations in platelet proteins during storage (65, 66). These studies have shed light on the molecular mechanisms regulating the deterioration of blood platelets during storage. The proteomic data are now being translated into platelet biochemistry to connect the results to platelet function. In the future, proteomics in blood banking will aim to make use of protein markers identified for platelet storage lesion development to monitor proteome changes when alterations such as the use of additive solutions or pathogen reduction strategies are put in place to improve platelet quality for patients (66).

Besides the transfusion medicine reports highlighted above, only a handful of other articles have focused on platelet clinical proteomics studies. Those include the study of different protein expression in normal and dysfunctional platelets from uremic patients (67), the diagnosis of a family with Quebec Platelet Disorder (68), and the study of the α-granule proteome in grey platelet syndrome patients – already highlighted in the secretome section (34).

Platelet proteomics has been recently applied to cardiovascular disease. Besides the signalling studies mentioned above, our group at the Universidade de Santiago is applying 2-DE-based proteomics to search for platelet biomarkers in acute coronary syndromes (ACS). In an initial study, the proteome of platelets from patients with non-ST elevation acute coronary syndrome (NSTE-ACS) was compared with that from matched controls with ischemic chronic cardiomyopathy. Proteins were separated in large 2D gels, by using 24-cm pI 4–7 IPG strips for the first dimension and large 10% SDS-PAGE gel for the second. Following protein staining with SyproRuby, and differential image analysis, 40 differentially regulated features were detected. From those, 40 proteins were identified by MALDI-MS/MS corresponding to 22 different genes. Major groups of proteins identified corresponded to cytoskeletal, signalling and proteins either secreted or involved in vesicles or secretory trafficking pathway. The study highlights proteins involved in $\alpha_{IIb}\beta_3$ and GPVI signalling as differentially regulated in NSTE-ACS (69). Interestingly, the number of differences decreased with time as demonstrated in a patients' follow-up study (69). Another recent study from a different group also used 2-DE to compare the proteome of platelets from aspirin-resistant and aspirin-sensitive stable coronary ischemic patients (70). The conclusion is that platelets from both groups of patients differ in the expression levels of proteins associated with mechanisms such as energetic metabolism, cytoskeleton, oxidative stress and cell survival, which may be associated with their different ability to respond to aspirin. The above studies highlight the potential of platelet proteomics to identify possible biomarkers in cardiovascular disease.

3. Overview and Future Directions

Over the last decade, proteomics technology has been successfully applied to platelet research, contributing to the emerging field of platelet proteomics. Studies have ranged from general platelet proteome mapping and characterizing the proteomes of subcellular compartments, to more specific studies investigating secretomes and post-translational modifications as well as the proteome of signalling cascades. Those studies led to the identification of a considerable amount of novel platelet proteins, including numerous signalling proteins and novel receptors, many of which have been further studied at functional level in the search for drug targets for platelet-based diseases. Although the development of the field has been impressive, there are many possibilities for future progress.

1. Quantitative proteomics: the development of DIGE has allowed for an improvement in the reproducibility of traditional 2-DE analyses, allowing quantitation of differences in protein levels when comparing two or more groups. In addition, the continuous development of MudPIT platforms, based in the unbeatable analytical tools offered by MS, will continue helping to bypass analytical problems inherent to gel-based separation methods, especially 2-DE. The latter will be especially useful for the investigation of other PTMs besides N-linked glycosylation and phosphorylation, such as O-linked glycosylation, lipidation (palmitoylation, myristoylation, and prenylation), ubiquitination, and SUMOylation, all of which play a relevant role in platelet biology.

2. Membrane compartments: The study of specific membrane compartments can also take advantage from the above developments. For instance, there is a need for a better depiction of glycolipid-enriched membrane domains (GEMs; also known as *rafts*). GEMs are highly dynamic platelet membrane structures involved in critical signalling mechanisms linked to platelet activation processes. A better characterization of these domains would help to gain insights into the molecular mechanisms associated with platelet activation.

3. Systems biology: All the advances in platelet proteomics will come in parallel with the recent development of systems biology tools, which permit to integrate the vast amount of information generated by proteomics in a more comprehensive way. The depiction of interactome maps and protein networks will

permit a better interpretation of the proteomic data gathered in a particular study (e.g. signalling studies).

4. Clinical studies: Platelet clinical proteomics is still in the early days and although it has a lot of potential, it must face real challenges, primarily related to sample limitation. Weak experimental designs, particularly in a field where technical challenges remain in the production of high-quality data, can make it difficult or impossible to determine if differences reported between two or more samples are likely to reflect variation in a biological system or are solely analytically derived. In line with the above, most proteomics journals recommend running several biological replicates when studying clinical samples; however, in some cases there is a difficulty in obtaining enough platelet protein to run technical replicates, whereas in other cases the problem is to obtain a sufficient number of biological replicates (71). A good interaction between basic researchers, clinicians, and statisticians is key for a successful platelet clinical proteomics study. The starting point must be a detailed characterization of the clinical parameters of all patients entering the study. Patients and controls must be matched in the best possible way to make sure the differentially regulated proteins identified are not related to differences in parameters such as sex, age or treatments. Factors such as platelet preparation and platelet purity, protein extraction, and protein separation and identification approaches should be also very well controlled. It should be kept in mind that establishing an experimental design is a compromise between availability of biological material, the technical difficulties of the approach and the reliability of the expected results. We are at the beginning of a new era for platelet proteomics and its application to the study of platelet-related diseases. This new era will hopefully increase our knowledge of platelet disorders and improve their diagnosis.

In conclusion, platelet proteomics is still a young field that has contributed enormously to the development of platelet biology in recent years. Previous success should serve as motivation for scientists to push the field forward to address the new challenges ahead.

Acknowledgements

Y.A.S. is a British Heart Foundation Intermediate Research Fellow (FS/08/034/25085). A.G. is a Ramón y Cajal Research Fellow (Spanish Ministry of Science and Innovation, MICINN) and acknowledges grants from the MICINN (grant No. SAF2010-22151), Consellería de Economía e Industria (Xunta de Galicia, Spain), and Fundación Mutua Madrileña (Spain).

References

1. García A, Watson SP, Dwek RA, Zitzmann N. (2005) Applying proteomics technology to platelet research. *Mass Spectrom Rev* **24**, 918–930.

2. Marcus K, Immler D, Sternberger J, Meyer HE. (2000) Identification of platelet proteins separated by two-dimensional gel electrophoresis and analyzed by matrix assisted laser desorption/ionization-time of flight-mass spectrometry and detection of tyrosine-phosphorylated proteins. *Electrophoresis* **21**, 2622–2636.

3. O'Neill EE, Brock CJ, von Kriegsheim AF, Pearce AC, Dwek RA, Watson SP, Hebestreit HF. (2002) Towards complete analysis of the platelet proteome. *Proteomics* **2**, 288–305.

4. García A, Prabhakar S, Brock CJ, Pearce AC, Dwek RA, Watson SP, Hebestreit HF, Zitzmann N. (2004) Extensive analysis of the human platelet proteome by two-dimensional gel electrophoresis and mass spectrometry. *Proteomics* **4**, 656–668.

5. Weyrich AS, Schwertz H, Kraiss LW, Zimmerman GA. (2009) Protein synthesis by platelets: historical and new perspectives. *J Thromb Haemost* **7**, 1759–1766.

6. Yu Y, Leng T, Yun D, Liu N, Yao J, Dai Y, *et al.* (2010) Global analysis of the rat and human platelet proteome - the molecular blueprint for illustrating multi-functional platelets and cross-species function evolution. *Proteomics* **10**, 2444–57.

7. Gevaert K, Goethals M, Martens L, Van Damme J, Staes A, Thomas GR, Vandekerckhove J. (2003) Exploring proteomes and analyzing protein processing by mass spectrometric identification of sorted N-terminal peptides. *Nat Biotechnol* **21**, 566–569.

8. Gevaert K, Ghesquiere B, Staes A, Martens L, Van Damme J, Thomas GR, Vandekerckhove J. (2004) Reversible labeling of cysteine-containing peptides allows their specific chromatographic isolation for non-gel proteome studies. *Proteomics* **4**, 897–908.

9. Martens L, Van Damme P, Van Damme J, Staes A, Timmerman E, Ghesquière B, *et al.* (2005) The human platelet proteome mapped by peptide-centric proteomics: a functional protein profile. *Proteomics* **5**, 3193–204.

10. Almen, M.S., Nordstrom, K.J., Fredriksson, R., and Schioth, H.B. (2009). Mapping the human membrane proteome: a majority of the human membrane proteins can be classified according to function and evolutionary origin. *BMC Biol* **7**, 50.

11. Tan, S., Tan, H.T., and Chung, M.C. (2008). Membrane proteins and membrane proteomics. *Proteomics* **8**, 3924–3932.

12. Hopkins, A.L., and Groom, C.R. (2002). The druggable genome. *Nat Rev Drug Discov* **1**, 727–730.

13. Speers, A.E., and Wu, C.C. (2007). Proteomics of integral membrane proteins--theory and application. *Chemical reviews* **107**, 3687–3714.

14. Lu, B., McClatchy, D.B., Kim, J.Y., and Yates, J.R., 3 rd (2008). Strategies for shotgun identification of integral membrane proteins by tandem mass spectrometry. *Proteomics* **8**, 3947–3955.

15. Lu, B., Xu, T., Park, S.K., and Yates, J.R., 3 rd (2009). Shotgun protein identification and quantification by mass spectrometry. *Methods Mol Biol* **564**, 261–288.

16. Lewandrowski, U., Wortelkamp, S., Lohrig, K., Zahedi, R.P., Wolters, D.A., Walter, U., and Sickmann, A. (2009). Platelet membrane proteomics: a novel repository for functional research. *Blood* **114**, e10-19.

17. Moebius, J., Zahedi, R.P., Lewandrowski, U., Berger, C., Walter, U., and Sickmann, A. (2005). The Human Platelet Membrane Proteome Reveals Several New Potential Membrane Proteins. *Mol Cell Proteomics* **4**, 1754–1761.

18. Senis, Y.A., Tomlinson, M.G., Garcia, A., Dumon, S., Heath, V.L., Herbert, J., Cobbold, S.P., Spalton, J.C., Ayman, S., Antrobus, R., *et al.* (2007). A Comprehensive Proteomics and Genomics Analysis Reveals Novel Transmembrane Proteins in Human Platelets and Mouse Megakaryocytes Including G6b-B, a Novel Immunoreceptor Tyrosine-based Inhibitory Motif Protein. *Mol Cell Proteomics* **6**, 548–564.

19. Lewandrowski, U., Moebius, J., Walter, U., and Sickmann, A. (2006). Elucidation of N-glycosylation sites on human platelet proteins: a glycoproteomic approach. *Mol Cell Proteomics* **5**, 226–233.

20. Borges, L.G., Seifert, R.A., Grant, F.J., Hart, C.E., Disteche, C.M., Edelhoff, S., Solca, F.F., Lieberman, M.A., Lindner, V., Fischer, E.H., *et al.* (1996). Cloning and characterization of rat density-enhanced phosphatase-1, a protein tyrosine phosphatase expressed by vascular cells. *Circ Res* **79**, 570–580.

21. de la Fuente-Garcia, M.A., Nicolas, J.M., Freed, J.H., Palou, E., Thomas, A.P., Vilella, R., Vives, J., and Gaya, A. (1998). CD148 is a membrane protein tyrosine phosphatase present

in all hematopoietic lineages and is involved in signal transduction on lymphocytes. *Blood* **91**, 2800–2809.

22. Mori, J., Pearce, A.C., Spalton, J.C., Grygielska, B., Eble, J.A., Tomlinson, M.G., Senis, Y.A., and Watson, S.P. (2008). G6b-B inhibits constitutive and agonist-induced signaling by glycoprotein VI and CLEC-2. *J Biol Chem* **283**, 35419–35427.

23. Newland, S.A., Macaulay, I.C., Floto, A.R., de Vet, E.C., Ouwehand, W.H., Watkins, N.A., Lyons, P.A., and Campbell, D.R. (2007). The novel inhibitory receptor G6B is expressed on the surface of platelets and attenuates platelet function in vitro. *Blood* **109**, 4806–4809.

24. Senis, Y.A., Tomlinson, M.G., Ellison, S., Mazharian, A., Lim, J., Zhao, Y., Kornerup, K.N., Auger, J.M., Thomas, S.G., Dhanjal, T., *et al.* (2009). The tyrosine phosphatase CD148 is an essential positive regulator of platelet activation and thrombosis. *Blood* **113**, 4942–4954.

25. Amisten, S., Braun, O.O., Bengtsson, A., and Erlinge, D. (2008). Gene expression profiling for the identification of G-protein coupled receptors in human platelets. Thromb Res *122*, 47–57.

26. Italiano, J.E., Jr., and Battinelli, E.M. (2009). Selective sorting of alpha-granule proteins. *J Thromb Haemost* 7 **Suppl 1**, 173–176.

27. Nurden, A.T. (2005). Qualitative disorders of platelets and megakaryocytes. *J Thromb Haemost* 3, 1773–1782.

28. Coppinger, J.A., Cagney, G., Toomey, S., Kislinger, T., Belton, O., McRedmond, J.P., Cahill, D.J., Emili, A., Fitzgerald, D.J., and Maguire, P.B. (2004). Characterization of the proteins released from activated platelets leads to localization of novel platelet proteins in human atherosclerotic lesions. *Blood* **103**, 2096–2104.

29. Hernandez-Ruiz, L., Valverde, F., Jimenez-Nunez, M.D., Ocana, E., Saez-Benito, A., Rodriguez-Martorell, J., Bohorquez, J.C., Serrano, A., and Ruiz, F.A. (2007). Organellar proteomics of human platelet dense granules reveals that 14-3-3zeta is a granule protein related to atherosclerosis. *J Proteome Res* 6, 4449–4457.

30. Piersma, S.R., Broxterman, H.J., Kapci, M., de Haas, R.R., Hoekman, K., Verheul, H.M., and Jimenez, C.R. (2009). Proteomics of the TRAP-induced platelet releasate. *J Proteomics* 72, 91–109.

31. Coppinger, J.A., O'Connor, R., Wynne, K., Flanagan, M., Sullivan, M., Maguire, P.B., Fitzgerald, D.J., and Cagney, G. (2007). Moderation of the platelet releasate response by aspirin. *Blood* **109**, 4786–4792.

32. Della Corte, A., Maugeri, N., Pampuch, A., Cerletti, C., de Gaetano, G., and Rotilio, D. (2008). Application of 2-dimensional difference gel electrophoresis (2D-DIGE) to the study of thrombin-activated human platelet secretome. *Platelets* **19**, 43–50.

33. Maynard, D.M., Heijnen, H.F., Horne, M.K., White, J.G., and Gahl, W.A. (2007). Proteomic analysis of platelet alpha-granules using mass spectrometry. *J Thromb Haemost* **5**, 1945–1955.

34. Maynard DM, Heijnen HFG, Gahl WA, Gunay-Aygun M. (2010) The alpha granule proteome: novel proteins in normal and ghost granules in Gray Platelet Syndrome. *J Thromb Haemost* **8**, 1786–1796.

35. Feng, S., Christodoulides, N., Resendiz, J.C., Berndt, M.C., and Kroll, M.H. (2000). Cytoplasmic domains of GpIbalpha and GpIbbeta regulate 14-3-3zeta binding to GpIb/IX/V. *Blood* **95**, 551–557.

36. Pula G, Perera S, Prokopi M et al. (2008) Proteomic analysis of secretory proteins and vesicles in vascular research. *Proteomics Clin Appl* **2**, 882–891.

37. Mallat Z, Benamer H, Hugel B, et al. (2000) Elevated levels of shed membrane microparticles with procoagulant potential in the peripheral circulating blood of patients with acute coronary syndromes. *Circulation* **101**, 841–843.

38. Garcia BA, Smalley DM, Cho H, Shabanowitz J, Ley K, Hunt DF. (2005) The platelet microparticle proteome. *J Proteome Res* **4**, 1516–1521.

39. Benameur T, Andriantsitohaina R, Martinez MC (2009) Therapeutic potential of plasma membrane-derived microparticles. *Pharmacol Rep* **61**, 49–57.

40. Jin M, Drwal G, Bourgeois T, Saltz J, Haifeng MW (2005) Distinct proteome features of plasma microparticles. *Proteomics* **5**, 1940–1952.

41. Smalley DM, Root KE, Cho H, Ross MM, Ley K. (2007) Proteomic discovery of 21 proteins expressed in human plasma-derived but not platelet-derived microparticles. *Thromb Haemost* **97**, 67–80

42. Fox KAA (2004) Management of acute coronary syndromes: an update. *Heart* **90**, 698–706.

43. Nomura S, Ozaki Y, Ikeda Y. (2008) Function and role of microparticles in various clinical settings. *Thromb Res* **123**, 8–23.

44. Nomura S, Shouzu A, Taomoto K, Togane Y, Goto S, Ozaki Y, Uchiyama S, Ikeda Y. (2009) Assessment of an ELISA kit for platelet-derived microparticles by joint research at many

institutes in Japan. *J Atheroscler Thromb* **16**, 878–887.

45. Perez-Pujol S, Marker PH, Key NS. (2007) Platelet microparticles are heterogeneous and highly dependent on the activation mechanism: studies using a new digital flow cytometer. *Cytometry A.* **71**, 38–45.

46. Piersma SR, Broxterman HJ, Kapci M *et al.* (2009) Proteomics of the TRAP-induced platelet releasate. *J Proteomics* **72**, 91–109.

47. Reinders J, Sickmann A. (2005) State-of-the-art in phosphoproteomics. *Proteomics* **5**, 4052–4061.

48. García A, Prabhakar S, Hughan S, Anderson TW, Brock CJ, Pearce AC, Dwek RA, Watson SP, Hebestreit HF, Zitzmann N. (2004) Differential proteome analysis of TRAP-activated platelets: involvement of DOK-2 and phosphorylation of RGS proteins. *Blood* **103**, 2088–2095.

49. García A, Senis YA, Antrobus R, Hughes CE, Dwek RA, Watson SP, Zitzmann N. (2006) A global proteomics approach identifies novel phosphorylated signaling proteins in GPVI-activated platelets: involvement of G6f, a novel platelet Grb2-binding membrane adapter. *Proteomics* **6**, 5332–5343.

50. Senis YA, Antrobus R, Severin S, Parguiña AF, Rosa I, Zitzmann N, Watson SP, García A. (2009) Proteomic analysis of integrin alphaIIb-beta3 outside-in signaling reveals Src-kinase-independent phosphorylation of Dok-1 and Dok-3 leading to SHIP-1 interactions. *J Thromb Haemost* **7**, 1718–1726.

51. Zahedi RP, Lewandrowski U, Wiesner J, Wortelkamp S, Moebius J, Schutz C, et al. Phosphoproteome of resting human platelets. (2008) *J Proteome Res* **7**, 526–534.

52. Qureshi AH, Chaoji V, Maiguel D, Faridi MH, Barth CJ, Salem SM, *et al.* (2009) Proteomic and phospho-proteomic profile of human platelets in basal, resting state: insights into integrin signaling. *PLoS One* **4(10)**, e7627.

53. Lewandrowski, U., Zahedi, R.P., Moebius, J., Walter, U., and Sickmann, A. (2007). Enhanced N-glycosylation site analysis of sialoglycopeptides by strong cation exchange prefractionation applied to platelet plasma membranes. *Mol Cell Proteomics* **6**, 1933–1941.

54. Kaiser, W.J., Holbrook, L.M., Tucker, K.L., Stanley, R.G., and Gibbins, J.M. (2009). A functional proteomic method for the enrichment of peripheral membrane proteins reveals the collagen binding protein Hsp47 is exposed on the surface of activated human platelets. *J Proteome Res* **8**, 2903–2914.

55. Tucker, K.L., Kaiser, W.J., Bergeron, A.L., Hu, H., Dong, J.F., Tan, T.H., and Gibbins, J.M. (2009). Proteomic analysis of resting and thrombin-stimulated platelets reveals the translocation and functional relevance of HIP-55 in platelets. *Proteomics* **9**, 4340–4354.

56. Guerrier, L., Claverol, S., Fortis, F., Rinalducci, S., Timperio, A.M., Antonioli, P., Jandrot-Perrus, M., Boschetti, E., and Righetti, P.G. (2007). Exploring the platelet proteome via combinatorial, hexapeptide ligand libraries. *J Proteome Res* **6**, 4290–4303.

57. García A. (2006) Proteome analysis of signaling cascades in human platelets. *Blood Cells Mol Dis* **36**, 152–156.

58. Immler D, Gremm D, Kirsch D, Spengler B, Presek P, Meyer HE. (1998) Identification of phosphorylated proteins from thrombin-activated human platelets isolated by two-dimensional gel electrophoresis by electrospray ionization-tandem mass spectrometry (ESI-MS/MS) and liquid chromatography-electrospray ionization-mass spectrometry (LC-ESI-MS). *Electrophoresis* **19**, 1015–1023.

59. Gevaert K, Eggermont L, Demol H, Vandekerckhove J. (2000) A fast a convenient MALDI-MS based proteomic approach: identification of components scaffolded by actin cytoskeleton of activated human thrombocytes. *J Biotechnol* **78**, 259–6269.

60. Marcus K, Moebius J, Meyer HE. (2003) Differential analysis of phosphorylated proteins in resting and thrombin-stimulated human platelets. *Anal Bioanal Chem* **376**, 973–993.

61. Schulz C, Leuschen NV, Fröhlich T, Lorenz M, Pfeiler S, Gleissner CA, *et al.* (2010) Identification of novel downstream targets of platelet glycoprotein VI activation by differential proteome analysis: implications for thrombus formation. *Blood* **115**, 4102–4110.

62. Wegener KL, Partridge AW, Han J, Pickford AR, Liddington RC, Ginsberg MH, Campbell ID. (2007) Structural basis of integrin activation by talin. *Cell* **128**, 171–182.

63. Oxley CL, Anthis NJ, Lowe ED, Vakonakis I, Campbell ID, Wegener KL. (2008) An integrin phosphorylation switch: the effect of beta3 integrin tail phosphorylation on Dok1 and talin binding. *J Biol Chem* **283**, 5420–5426.

64. Dittrich M, Birschmann I, Mietner S, Sickmann A, Walter U, Dandekar T. (2008) Platelet protein interactions: map, signaling components, and phosphorylation groundstate. *Arterioscler Thromb Vasc Biol* **30**, 843–850.

65. Thiele T, Steil L, Gebhard S, Scharf C, Hammer E, Brigulla M, Lubenow N, Clemetson KJ,

Völker U, Greinacher A. (2007) Profiling of alterations in platelet proteins during storage of platelet concentrates. *Transfusion* **47**, 1221–1233.

66. Schubert P, Devine DV. (2010) Proteomics meets blood banking: identification of protein targets for the improvement of platelet quality. *J Proteomics* **73**, 436–444.

67. Marques M, Sacristán D, Mateos-Cáceres PJ, Herrero J, Arribas MJ, González-Armengol JJ, Villegas A, Macaya C, Barrientos A, López-Farré AJ. (2010) Different protein expression in normal and dysfunctional platelets from uremic patients. *J Nephrol.* **23**, 90–101.

68. Maurer-Spurej E, Kahr WH, Carter CJ, Pittendreigh C, Cameron M, Cyr TD. (2008) The value of proteomics for the diagnosis of a platelet-related bleeding disorder. *Platelets* **19**, 342–351.

69. Parguiña AF, Grigorian-Shamajian L, Agra RM, Teijeira-Fernández E, Rosa I, Alonso J, *et al.* (2010) Proteins involved in platelet signalling are differentially regulated in acute coronary syndrome: a proteomics study. *PLoS ONE* **5**(10), e13404.

70. Mateos-Cáceres PJ, Macaya C, Azcona L, Modrego J, Mahillo E, Bernardo E, Fernandez-Ortiz A, López-Farré AJ. (2010) Different expression of proteins in platelets from aspirin-resistant and aspirin-sensitive patients. *Thromb Haemost* **103**, 160–170.

71. García A (2010) Clinical proteomics in platelet research: challenges ahead. *J Thromb Haemost* **8**, 1784–1785

INDEX

Jonathan M. Gibbins and Martyn P. Mahaut-Smith (eds.), *Platelets and Megakaryocytes:*
Volume 3, Additional Protocols and Perspectives, Methods in Molecular Biology, vol. 788,
DOI 10.1007/978-1-61779-307-3, © Springer Science+Business Media, LLC 2012

Printed by Publishers' Graphics LLC USA
MO20120306-095
2012